T0298325

The Science of Carbon Sequestration and Capture

The Science of Carbon Sequestration and Capture examines the current scientific underpinnings of carbon capture and storage (CCS) and provides readers with sufficient background on the basics of geology, natural science, and chemical and environmental engineering so that they can understand the current state and art of the CCS field. Moreover, this book provides a wide-ranging discussion presented in the author's comprehensible conversational style describing the impact of CCS on climate, health, society in general, and the energy landscape. This book is directed at undergraduate and graduate students, professionals, scientists, and the general reading public who would like to gain a broad multidisciplinary view of one of the greatest challenges of our generation.

Features:

- Aims to fill the gap of missing information in published texts dealing with the carbon sequestration and capture revolution currently underway
- Provides an understanding of current science buttressing carbon capture and sequestration practices
- Explains the complexities of carbon sequestration and capture systems in basic and understandable terms

The Science of Carbon Sequestration and Capture

Frank R. Spellman

CRC Press is an imprint of the
Taylor & Francis Group, an **informa** business

Designed cover image: Shutterstock

First edition published 2024
by CRC Press
2385 NW Executive Center Drive, Suite 320, Boca Raton FL 33431

and by CRC Press
4 Park Square, Milton Park, Abingdon, Oxon, OX14 4RN

CRC Press is an imprint of Taylor & Francis Group, LLC

ISBN: 978-1-032-55899-8 (hbk)
ISBN: 978-1-032-55901-8 (pbk)
ISBN: 978-1-003-43283-8 (ebk)

DOI: 10.1201/9781003432838

Typeset in Times
by codeMantra

Contents

Preface xvi
Acronyms xvii
Abbreviations and Units xvii
Geologic Time Scale xix
Precambrian xx
Conversion Factors and SI Units xxi
Table A of SI Prefixes xxii
Conversion Factors xxiii
About the Author xxxvi

PART I Carbon Fundamentals

Chapter 1 Cave of the Dog 3

The Cave 3
A Historical Perspective 3
Carbon Dioxide: Physical Properties 6
Carbon Dioxide: Uses 6
If You Wish to Converse with Me … 10
Key Terms and Definitions 10
Absorber 10
Absorption (aka Scrubbing) 10
Absorption Units 11
AC 11
Acid Rain 11
Adaptation 11
Adaptive Management 11
Adsorption 11
Agglomerating Character 11
Agriculture 12
Albedo 12
Air Pollution Abatement Equipment 12
Air Temperature Adjustment 12
Alternating Current (AC) 12
Alternative Fuel 12
Ambient 12
Ampere (Amp) 13
Anaerobic Decomposition 13
Anion 13
Annual Removals 13
Anode 13

Anthropogenic 13
API Gravity 13
Appropriate Use 14
Bare Rock Succession 14
Benzene (C_6H_6) 14
Biobased Product 14
Biochemical Conversion 14
Biodiesel 14
Bioenergy 14
Biofuels 14
Biogas 15
Biogenic 15
Biogenic Emissions 15
Biological Diversity or Biodiversity 15
Biological Integrity 15
Biomass 15
Biomass Gas (Biogas) 15
Biomass Waste 15
Biomaterials 16
Biopower 16
Biorefinery 16
Biota 16
British Thermal Unit (Btu) 16
Btu Conversion Factor 16
Btu per Cubic Foot 16
Bulk Density 17
Bunker Fuels 17
Burnup 17
Butane (C_4H_{10}) 17
Butylene (C_4H_8) 17
Calcium Sulfate 17
Calorie 17
Carbon 18
Carbon Accounting 18
Carbon Budget 18
Carbon Cycle 18
Carbon Dioxide (CO_2) 18
Carbon Dioxide Equivalent (CDE) 18
Carbon Intensity 18
Carbon Monoxide (CO) 19
Carbon Output Rate 19
Carbon Sequestration 19
Carbon Sinks 19
Carnot Cycle 19
Cathode 19
Cation 19

Cellulose ... 19
Cellulosic Ethanol ... 19
Chlorofluorocarbon (CFC) ... 20
Climate ... 20
Climate Change ... 20
Climate Effects ... 20
Cloud Condensation Nuclei ... 20
Coarse Materials ... 20
Co-Firing ... 20
Cogeneration ... 20
Cogeneration System ... 21
Compressed Natural Gas (CNG) ... 21
Concentrating Solar Power or Solar Thermal Power System ... 21
Concentrator ... 21
Conservation ... 21
Conservation Program ... 21
Conservation Status ... 21
Conventional Oil and Natural Gas Production ... 21
Criteria Pollutant ... 22
Critical Habitat ... 22
Crust ... 22
DC ... 22
Deforestation ... 22
Degasification System ... 22
Degradable Organic Carbon ... 22
Demand Indicator ... 22
Dependable Capacity ... 22
Depleted Resources ... 23
Depletion Factor ... 23
Desulfurization ... 23
Diesel Fuel ... 23
Diversity Exchange ... 23
E-10 ... 23
E-85 ... 23
E-95 ... 23
Ecological System ... 23
Ecosystem ... 23
Efficiency ... 24
Electric Energy ... 24
Electric Utility ... 24
Emissions ... 24
Emissions Coefficient ... 24
Emissions Factor ... 24
Endothermic ... 24
Energy ... 24
Energy Efficiency ... 25

Energy Loss....25
Energy Source....25
Enthalpy....25
Environment....25
Environmental Health (Abiotic Aspects)....25
Environmental Impact Statement....25
Environmental Restoration....26
Environmental Restrictions....26
Ethane (C_2H_6)....26
Ethanol (Also Known as Ethyl Alcohol or Grain Alcohol, CH_3-CH_2OH)....26
Ether....26
Ethylene (C_2H_4)....26
Evapotranspiration (ET)....27
Exothermic....27
Fahrenheit....27
Fermentation....27
Fischer-Tropsch Fuels....27
Flux Material....27
Flyway....27
Forest Land....27
Fossil Fuels....28
Fractionation....28
Fuel Cell....28
Fuel Cycle....28
Fuel Ratio....28
Fumarole....28
Gallon....28
Gas....28
Gasification....29
Gasohol....29
Generation (Electricity)....29
Geologic Hazards....29
Geographic Information Systems (GIS)....29
Geologic Sequestration....29
Gigawatt (GW)....30
Gigawatt-Electric (GWe)....30
Gigawatt-hour (GWh)....30
Global Positioning System (GPS)....30
Global Warming....30
Greenhouse Effect....30
Greenhouse Gas....30
Growing Stock....30
Habitat....30
Habitat Conservation....31
Habitat Fragmentation....31

Hardwoods......31
Heat Rate......31
Heating Value......31
Hydraulic Fracturing......31
Hydrocarbon......31
Idle Cropland......31
Industrial Wood......32
Ion Exchange......32
Isobutane (C_4H_{10})......32
Isobutylene (C_4H_8)......32
Isohexane (C_6H_{14})......32
Isomerization......32
Joule......32
kBtu......32
Kilowatt (kW)......33
Kilowatt-hour (kWh)......33
Kinetic Energy......33
Low-Speed Shaft......33
Mantle......33
Megawatt (WM)......33
Methane......33
Methanogens......33
Methanol......33
Mineral......33
Mole......34
Naphtha......34
Natural Gas......34
Natural Processes......34
Natural Sinks......34
Niche......34
Peat......34
Permeability......34
Petrochemical Feedstocks......35
Photosynthesis......35
Planetary Albedo......35
Prescribed Fire......35
Propane (C_3H_8)......35
Propylene (C_3H_6)......35
Pyrolysis......35
Quadrillion Btu (Quad)......35
Radiant Energy......35
Reclamation......36
Recycling......36
Renewable Energy......36
Renewable Energy Resources......36
Riparian......36

- Sequestration ... 36
- Terrestrial Sequestration ... 36
- Watt (Electric) ... 36
- Watt (Thermal) ... 37
Not Exactly the Final Word on Carbon Dioxide ... 37
References ... 37

Chapter 2 Carbon Cycle ... 38

Let's Talk About Carbon: The King of Elements ... 38
Carbon Compounds, Alloys, and Bonds ... 39
Biogeochemical Cycles ... 41
411 on the Carbon Cycle ... 41
- The Slow Carbon Cycle ... 42
- The Fast Carbon Cycle ... 43
- It Must Go Somewhere ... 50
- Atmosphere Reservoir ... 50
- Marine Reservoir ... 51
- Land Reservoir ... 51
The Bottom Line ... 52
References ... 52

Chapter 3 Carbon Sources ... 54

Periodic Table ... 54
Biology of Carbon ... 55
- Primary Producers ... 56
- Other Tropes ... 56
The Bottom Line ... 56
Reference ... 56

Chapter 4 Carbon Sinks ... 57

Part and Parcel of the Carbon Cycle ... 57
Examples of Carbon Sinks ... 58
- Oceans ... 58
- Plants ... 59
- Soil ... 61
- Peat Bogs ... 62
- Fens ... 62
The Bottom Line ... 63
References ... 63

Chapter 5 The Color of Carbon ... 66

Introduction ... 66
The Carbon Rainbow ... 66

Red Carbon 67
The Blue Carbon Sink 68
White Knight versus Climate Change 70
The Bottom Line 71
References 71

Chapter 6 Adsorption vs Absorption 72

Introduction 72
Adsorption and Absorption and Ion Exchange 72
Reference 73

PART 2 Fundamentals of Carbon Capture and Sequestration

Chapter 7 Carbon Capture and Sequestration 77

Introduction 77
The 411 on Carbon Capture and Sequestration 77
Terrestrial Carbon Sequestration 78
Geologic Carbon Sequestration 82
Potential Impact of Terrestrial Sequestration 83
Potential Impact of Geologic Sequestration 84
The Bottom Line 84
References 84

PART 3 The Green Knight

Chapter 8 Taller than the Trees 89

Largest Carbon Sink 89
Carbon Storage 89
Reference 91

Chapter 9 Wood-Based Biomass: Heat Energy and Weight 92

Introduction 92
Fuel/Heat Value of Forests 92
Lower/Higher Heating Values 92
Effect of Fuel Moisture on Wood Heat Content 93
Moisture Content (MC) Wet and Dry Weight Basis Calculations 94
Forestry Volume Unit to Biomass Weight Considerations 95
Estimation of Biomass Weights from Forestry Volume Data 95

Biomass Expansion Factors (BEF) 96
Stand Level Biomass Estimation 96
Biomass Equations 97
References and Recommended Reading 101

Chapter 10 Forest Tree Carbon 103

Introduction 103
Forest Biomass Sampling 104
Terms and Concepts 105
Simple Random Sampling 106
Simple Random Sampling Methods 107
Stratified Random Sampling 115
Estimation of Number of Sampling Units 119
Optimum Allocation with Varying Sampling Costs 120
Sample Size in Stratified Random Sampling 121
Regression Estimation 124
Double Sampling 131
Sampling Protocols and Vegetation Attributes 134
Vegetation Attributes 134
Matrix of Monitoring Techniques/Vegetation Attributes and Monitoring Methods 139
Matrix of Monitoring Techniques and Vegetation Attributes 139
Monitoring Methods 140
The Bottom Line 147
Cited References and Recommended Reading 147

Chapter 11 Soil Carbon 151

Introduction 151
Soil Basics 151
Soil: What is it? 152
Definitions of Key Terms 154
About Soil 159
Functions of Soil 159
Back to Soil Basics 165
Soil: Physical Properties 167
Soil Separates 168
Soil Formation 170
Soil Characterization 172
Diagnostic Horizons and Temperature and Moisture Regimes 173
Soil Taxonomy 174
Soil Orders 175
Soil Suborders 176
Soil Great Groups and Subgroups 176
Soil Families and Series 176

Soil Mechanics/Physics 177
Soil Mechanics 177
Soil Particle Characteristics 179
Soil Stress 180
Soil Compressibility 181
Soil Compaction 181
Soil Failure 181
Soil Types 181
Soil Physics 182
Water and Soil 183
Water: What is it? 183
Soil Physical Properties 183
The Water Cycle (Hydrologic Cycle) 184
Soil Water 185
Soil Chemistry 186
What is Chemistry? 186
Elements and Compounds 187
Classification of Elements 187
Physical and Chemical Changes 188
Structure of the Atom 189
Periodic Classification of the Elements 189
Molecules and Ions 190
Chemical Bonding 190
Chemical Formulas and Equations 191
Molecular Weights, Formulas, and the Mole 192
Physical and Chemical Properties of Matter 192
States of Matter 193
The Gas Laws 193
Liquids and Solutions 194
Thermal Properties 195
Specific Heat 195
ACID + BASES and SALTS 196
pH Scale 196
Organic Chemistry 197
Organic Compounds 198
Hydrocarbons 199
Aliphatic Hydrocarbons 199
Aromatic Hydrocarbons 200
Environmental Chemistry 200
The Bottom Line 200
References 201

Chapter 12 Soil Organic Matter and Carbon 202

Introduction 202
Key Terms and Definitions 202

Soil Organic Carbon (SOC) 203
Soil C Accounting 206
Mineral Soil Organic Carbon 207
Assessing Soil Carbon Vulnerability 209
The Bottom Line 211
References and Recommended Reading 211

Chapter 13 Forest Soil Carbon Disturbance 214

Introduction 214
Vulnerability versus Disturbance 215
The Bottom Line 221
References and Recommended Reading 221

Chapter 14 Forest Carbon Status 223

Introduction 223
Carbon Status 223
A Vital Army 224
Flux or Flow of Carbon Dioxide 224
Climate Change and Forest SOC 226
The Bottom Line 227
References 227

Chapter 15 Forest SOC Disturbance: Fire Carbon is Measured and Reported at Different Scales and Units 230

Introduction 230
Forest Carbon Dynamics 230
Fire Effects on SOC 231
Note: The 411 on Meta-Analysis 233
Management and SOC 234
The Bottom Line 235
References and Recommended Reading 235

Chapter 16 Forest Disturbance: Harvesting and Thinning 238

Introduction 238
Forest Harvesting 238
Thinning Forest Stands 239
The Bottom Line—The Forest Paradox 240
References 240

Chapter 17 Forest Disturbance: Ungulate Herbivory 243

Introduction 243
TBCF and Others 245
The Basics 245

Livestock Grazing 246
The Bottom Line 247
References 248

Chapter 18 Forest Disturbance: Nutrient Additions 249

Forest Fertilization 249
Agroforestry 249
Fixing Nitrogen 250
It's All about Trees 252
The Bottom Line 253
References 253

Chapter 19 Forest Disturbance: Tree Mortality 255

Introduction 255
The Paradox: Dead or Alive 256
When Dead is Not Really Dead 257
The Carbon Fractions 257
Net Primary Production and Gross Primary Production 258
Atmospheric Chemicals and Productivity 260
Vegetation Species Composition (Genotype Effects) on Soil Productivity 260
Site Disturbance 260
The Bottom Line 261
References 262

Chapter 20 Forest Disturbance: Invasive Species 265

Introduction 265
U.S. Presidential Executive Order 1311—Invasive Species 265
Invasive Species: The Problem 266
Plant-Fungal Symbiosis 267
The Bottom Line 267
References 267

Index 269

Preface

This book, *The Science of Carbon Capture and Storage*, is the eleventh volume in the acclaimed series that includes *The Science of Ocean Pollution* (in production), *The Science of Lithium* (Li), *The Science of Electric Vehicles (EVs): Concepts and Applications, The Science of Rare Earth Elements: Concepts and Applications, The Science of Water, The Science of Air, The Science of Environmental Pollution, The Science of Renewable Energy, The Science of Waste*, and *The Science of Wind Power*, all of which bring this highly successful series fully into the twenty-first century. This book continues the series mantra based on good science and not-feel-good science. It also continues to be presented in the author's trademark conversational style—making sure communication is certain—not a failure. My aim, my goal, is to be comprehensive in coverage and comprehensible in what I deliver.

Following the successful format of the other editions in this series, the aim of this no-holds-barred book is to provide an understanding of the current science underpinning carbon capture and storage (CCS) and to provide students and interested readers with sufficient background on the basics of geology, natural science, and chemical and environmental engineering so that they can understand the current state and art of the CCS field. Moreover, this book provides a wide-ranging discussion, again, presented in the author's comprehensible conversational style describing the impact of CCS on climate, health, society in general, and the energy landscape. This book is directed at undergraduate and graduate students, professionals, scientists, and the general reading public who would like to gain a broad multidisciplinary view of one of the greatest challenges of our generation.

Frank Spellman, Norfolk, VA

Acronyms

ACT—America	Atmospheric Carbon and Transfer—American Program
AFOLU	agriculture, forestry, and other land use
AIM	atmospheric inverse modeling
ALT	active layer thickness
CAA	Clean Air Act
CCS	carbon capture and storage
CDU	carbon dioxide utilization
CFC	chlorofluorocarbons
DIS	dissolved inorganic carbon
DOC	dissolved organic carbon
DOE	U.S. Department of Energy
DOI	U.S. Department of the Interior
DOM	dissolved organic matter
EIA	U.S. Energy Information Administration
FAO	Food and Agriculture Organization
GHG	greenhouse gas
GWP	global warming potential
IPPC	Intergovernmental Panel on Climate Change
NASA	National Aeronautics and Space Administration
OMB	Office of Management and Budget
PIC	particulate inorganic carbon
POC	particulate organic carbon
POM	particulate organic matter
RE	renewable energy
TOC/td>	total organic carbon
USDA	U.S. Department of Agriculture
USFWS	U.S. Fish and Wildlife Service
VOC	volatile organic compounds

ABBREVIATIONS AND UNITS

C	carbon
$CaCO_3$	calcium carbonate
CH_4	methane
CO	carbon monoxide
CO_2	carbon dioxide
CO_2e	carbon dioxide equivalent
CO_3^{2-}	carbon monoxide
COS	carbon sulfide
°C	degrees Celsius
G	gram
Gt	gigaton

GW	gigawatt
ha	hectare
$\mathbf{H_2CO_3}$	carbonic acid
$\mathbf{HCO_3}$	bicarbonate ion
J	Joule
kg	kilogram
µatm	microatmosphere
µ	micrometer
µ	micromole
mol	mole
megawatt	Mw
N	nitrogen
$\mathbf{NH_3}$	ammonia
$\mathbf{NO_x}$	nitrogen oxides
OH	hydroxyl radical
ppb	parts per billion—think of ppb as being equivalent to one drop of water into an Olympic-size swimming pool.
ppm	parts per million (*Part per million*—is an alternative (but numerically equivalent) unit used in chemistry is milligrams per liter (mg/L). As an analogy think of a ppm as being equivalent to a full shot glass in an Olympic-size swimming pool.)
Tg	teragram
W	watts

DID YOU KNOW?

Most people are familiar with CO_2, but CO_2e is a foreign term to most people. CO_2e is a carbon dioxide equivalent which is a standard measurement of greenhouse gas emissions.

Geologic Time Scale

Erathem or Era	System, Sub-system or Period, Sub-period	Series or Epoch
	Quaternary 1.8 million years ago to the Present	**Holocene** 11,477 years ago (+/− 85 years) to the Present—Greek "holos" (entire) and "ceno" (new)
Cenozoic 65 million years ago to present "Age of Recent Life"		**Pleistocene** 1.8 million to approx. 11,477 (+/− 85 years) years ago—The Great Ice Age—Greek Words "pleistos" (most) and "ceno" (new).
		Pliocene 5.3 to 1.8 million years ago—Greek "pleion" (more) and "ceno" (new).
	Tertiary 65.5 to 1.8 million years ago	**Miocene** 23.0 to 5.3 million years ago—Greek "meion" (less) and "ceno" (new).
		Oligocene 33.9 to 23.0 million years ago—Greek "oligos" (little, few) And "ceno" (new).
		Eocene **55.8 to 33.9 million years ago—Greek "eos" (dawn) and "ceno" (new).**
		Paleocene 65.5 to 58.8 million years ago—Greek "palaois" (old) and "ceno" (new)
	Cretaceous 145.5 to 65.5 million years ago "The Age of Dinosaurs"	Late or Upper Early or Lower
	Jurassic 199.6 to 145.5 million years ago	
Mesozoic 251.0 to 65.5 million years ago—Greek means "middle life"		Late or Upper Middle Early or Lower
	Triassic 251.0 in 199.6 million years ago	Late or Upper Middle Early or Lower
	Permian 299.0 to 251.0 million years ago	Lopingian Guadalupian Cisuralian

(Continued)

(***Continued***)

	Pennsylvanian 318.1 to 299.0 million years ago "The Coal Age"	Late or Upper Middle Early or Lower
Paleozoic 542.0 to 251.0 million years ago, "Age of Ancient Life"	**Mississippian** 359.2 to 318.1 million years ago	Late or Upper Middle Early or Lower
	Devonian 416.0 to 359.2 million years ago	Late or Upper Middle Early or Lower
	Silurian 443.7 to 416.0 million years ago	Pridoli Ludlow Wenlock Llandovery
	Ordovician 488.3 to 443.7 million years ago	Late or Upper Middle Early or Lower
	Cambrian 542.0 to 488.3 million years ago	Late or Upper Middle Early or Lower

PRECAMBRIAN

Approximately 4 billion years ago to 542.0 million years ago

Conversion Factors and SI Units

The units most used by environmental engineering professionals, ecologists, forestry professionals, and those actively involved with studying, tracking, and measuring carbon storage and sequestration are based on the complicated English System of Weights and Measures. However, bench work is usually based on the metric system or the International System of Units (SI) due to the convenient relationship between milliliters (mL), cubic centimeters (cm^3), and grams (g).

The SI is a modernized version of the metric system established by international agreement. The metric system of measurement was developed during the French Revolution and was first promoted in the U.S. in 1866. In 1902, proposed congressional legislation requiring the U.S. government to use the metric system exclusively was defeated by a single vote. Although we use both systems in this text, SI provides a logical and interconnected framework for all measurements in engineering, science, industry, and commerce. The metric system is much simpler to use than the existing English system since all its units of measurement are divisible by 10.

For the benefit of the general non-professional reader, it is important to provide a listing of the various units and conversion factors commonly used by environmental engineers, environmental engineering professionals, ecologists, forestry professionals, and those actively involved with studying, tracking, and measuring carbon storage and sequestration. Enabling comprehension of these units and conversion factors includes providing a description of the prefixes commonly used in the SI system. These prefixes are based on the power 10. For example, a "kilo" means 1000 g, and a "centimeter" means one-hundredth of 1 m. The 20 SI prefixes used to form decimal multiples and submultiples of SI units are given in Table A of SI Prefixes.

Note that the kilogram is the only SI unit with a prefix as part of its name and symbol. Because multiple prefixes may not be used, in the case of the kilogram, the prefix names of Table A of SI Prefixes are used with the unit name "gram" and the prefix symbols are used with the unit symbol "g." With this exception, any SI prefix may be used with any SI unit, including the degree Celsius and its symbol °C.

TABLE A OF SI PREFIXES

Factor	Name	Symbol
10^{24}	Yotta	Y
10^{21}	Zetta	Z
10^{18}	Exa	E
10^{15}	Peta	P
10^{12}	Tera	T
10^{9}	Giga	G
10^{6}	Mega	M
10^{3}	Kilo	k
10^{2}	Hecto	h
10^{1}	Deka	da
10^{-1}	Deci	d
10^{-2}	Centi	c
10^{-3}	Milli	m
10^{-6}	Micro	µ
10^{-9}	Nano	n
10^{-12}	Pico	p
10^{-15}	Femto	f
10^{-18}	Atto	a
10^{-21}	Zepto	z
10^{-24}	Yocto	y

Example 1.1

10^{-6} kg = 1 mg (one milligram), but not 10^{-6} kg = 1 µkg (one microkilogram)

Example 1.2

Consider the height of the Washington Monument. We may write $h_w = 169{,}000$ mm = 16,900 cm = 169 m = 0.169 km using the millimeter (SI prefix "milli," symbol "m"); centimeter (SI prefix "centi," symbol "c"); or kilometer (SI prefix "kilo," symbol "k").

DID YOU KNOW?

The Fibonacci sequence is the following sequence of numbers:

1, 1, 2, 3, 5, 8, 13, 21, 34, 55, 89, 144, …

Or, alternatively,

0.1,1, 2, 3, 5, 8, 13, 21, 34, 55, 89, 144, …

Two important points, the first obvious: each term from the third onward is *the sum of the previous two*. Another point to notice is that if you divide each number in the sequence by the next number, beginning with the first, an interesting thing appears to be happening:

1/1 = 1, 1/2 = 0.5, 2/3 = 0.66666 …, 3/5 = 0.6, 5/8 = 0.625, 8/13 = 0.61538 …, 13/21 = 0.61904 … (the first of these ratios appear to be converging to a number just a bit larger than 0.6).

CONVERSION FACTORS

Conversion factors are given below in alphabetical order (Table B) and in unit category listing order (Table C).

Example 1.3

Problem: Find degrees Celsius of water at 72°F.
Solution:

$$°C = (F - 32) \times 5/9 = (72 - 32) \times 5/9 = 22.2$$

TABLE B
Alphabetical Listing of Conversion *Factors*

Factors	Metric (SI) or English Conversions
1 atm (atmosphere) =	1.013 bars
	10.133 newtons/cm^2 (newtons/square centimeter)
	33.90 ft. of H_2O (feet of water)
	101.325 kp (kilopascals)
	1,013.25 mg (millibars)
	13.70 psia (pounds/square inch – absolute)
	760 torr
	760 mm Hg (millimeters of mercury)
1 bar =	0.987 atm (atmospheres)
	1×10^6 dynes/cm^2 (dynes/square centimeter)
	33.45 ft of H_2O (feet of water)
	1×10^5 pascals [nt/m^2] (newtons/square meter)
	750.06 torr
	750.06 mm Hg (millimeters of mercury)
1 Bq (becquerel) =	1 radioactive disintegration/second
	2.7×10^{-11} Ci (curie)
	2.7×10^{-8} mCi (millicurie)
1 BTU (British Thermal Unit) =	252 cal (calories)
	1,055.06 j (joules)
	10.41 liter-atmosphere
	0.293 watt-hours
1 cal (calories) =	3.97×10^{-3} BTUs (British Thermal Units)
	4.18 j (joules)
	0.0413 liter-atmospheres
	1.163×10^{-3} watt-hours
1 cm (centimeters) =	0.0328 ft (feet)
	0.394 in (inches)
	10,000 microns (micrometers)
	100,000,000 Å = 10^8 Å (Ångstroms)
1 cc (cubic centimeter) =	3.53×10^{-5} ft^3 (cubic feet)

(*Continued*)

TABLE B (*Continued*)
Alphabetical Listing of Conversion *Factors*

Factors	Metric (SI) or English Conversions
	0.061 in^3 (cubic inches)
	2.64×10^{-4} gal (gallons)
	52.18 ℓ (liters)
	52.18 ml (milliliters)
1 ft^3 (cubic foot) =	28.317 cc (cubic centimeters)
	1,728 in^3 (cubic inches)
	0.0283 m^3 (cubic meters)
	7.48 gal (gallons)
	28.32 ℓ (liters)
	29.92 qts (quarts)
1 in^3	16.39 cc (cubic centimeters)
	16.39 ml (milliliters)
	$5.79 \times 10^{-4} ft^3$ (cubic feet)
	$1.64 \times 10^{-5} m^3$ (cubic meters)
	4.33×10^{-3} gal (gallons)
	0.0164 ℓ (liters)
	0.55 fl oz (fluid ounces)
1 m^3 (cubic meter) =	1,000,000 cc = 10^6 cc (cubic centimeters)
	33.32 ft^3 (cubic feet)
	61,023 in^3 (cubic inches)
	264.17 gal (gallons)
	1,000 ℓ (liters)
1 yd^3 (cubic yard) =	201.97 gal (gallons)
	764.55 ℓ (liters)
1 Ci (curie) =	3.7×10^{10} radioactive disintegrations/second
	3.7×10^{10} Bq (becquerel)
	1,000 mCi (millicurie)
1 day =	24 hrs (hours)
	1,440 min (minutes)
	86,400 sec (seconds)
	0.143 weeks
	2.738×10^{-3} yrs (years)
1°C (expressed as an interval) =	1.8°F = [9/5] °F (degrees Fahrenheit)
	1.8°R (degrees Rankine)
	1.0 K (degrees Kelvin)
°C (degree Celsius) =	[(5/9)(°F – 32°)]
1°F (expressed as an interval) =	0.556°C = [5/9]°C (degrees Celsius)
	1.0°R (degrees Rankine)
	0.556 K (degrees Kelvin)
°F (degree Fahrenheit) =	[(9/5)(°C) + 32°]
1 dyne =	1×10^{-5} nt (newton)

(*Continued*)

TABLE B (*Continued*)
Alphabetical Listing of Conversion *Factors*

Factors	Metric (SI) or English Conversions
1 ev (electron volt) =	1.602×10^{-12} ergs
	1.602×10^{-19} j (joules)
1 erg =	1 dyne-centimeters
	1×10^{-7} j (joules)
	2.78×10^{-11} watt-hours
1 fps (feet/second) =	1.097 kmph (kilometers/hour)
	0.305 mps (meters/second)
	0.01136 mph (miles/hour)
1 ft (foot) =	30.48 cm (centimeters)
	12 in (inches)
	0.3048 m (meters)
	1.65×10^{-4} nt (nautical miles)
	1.89×10^{-4} mi (statute miles)
1 gal (gallon) =	3,785 cc (cubic centimeters)
	0.134 ft^3 (cubic feet)
	231 in^3 (cubic inches)
	3.785 ℓ (liters)
1 gm (gram)	0.001 kg (kilogram)
	1,000 mg (milligrams)
	1,000,000 ng = 10^6 ng (nanograms)
	2.205×10^{-3} lbs (pounds)
1 gm/cc (grams/cubic cent.) =	62.43 lbs/ft^3 (pounds/cubic foot)
	0.0361 lbs/in^3 (pounds/cubic inch)
	8.345 lbs/gal (pounds/gallon)
1 Gy (gray) =	1 j/kg (joules/kilogram)
	100 rad
	1 Sv (sievert) – [unless modified through division by an appropriate factor, such as Q and/or N]
1 hp (horsepower) =	745.7 j/sec (joules/sec)
1 hr (hour) =	0.0417 days
	60 min (minutes)
	3,600 sec (seconds)
	5.95×10^{-3} weeks
	1.14×10^{-4} yrs (years)
1 in (inch) =	2.54 cm (centimeters)
	1,000 mils
1 inch of water =	1.86 mm Hg (millimeters of mercury)
	249.09 pascals
	0.0361 psi (lbs/in^2)
1 j (joule) =	9.48×10^{-4} BTUs (British Thermal Units)
	0.239 cal (calories)

(*Continued*)

TABLE B (*Continued*)

Alphabetical Listing of Conversion *Factors*

Factors	Metric (SI) or English Conversions
	10,000,000 ergs = 1×10^7 ergs
	9.87×10^{-3} liter-atmospheres
1.0	nt-m (newton-meters)
1 kcal (kilocalories) =	3.97 BTUs (British Thermal Units)
	1,000 cal (calories)
	4,186.8 j (joules)
1 kg (kilogram) =	1,000 gms (grams)
	2,205 lbs (pounds)
1 km (kilometer) =	3,280 ft (feet)
	0.54 nt (nautical miles)
	0.6214 mi (statute miles)
1 kw (kilowatt) =	56.87 BTU/min (British Thermal Units)
	1.341 hp (horsepower)
	1,000 j/sec (kilocalories)
1 kw-hr (kilowatt-hour) =	3,412.14 BTU (British Thermal Units)
	3.6×10^6 j (joules)
	859.8 kcal (kilocalories)
1 ℓ (liter) =	1,000 cc (cubic centimeters)
	1 dm^3 (cubic decimeters)
	0.0353 ft^3 (cubic feet)
	61.02 in^3 (cubic inches)
	0.264 gal (gallons)
	1,000 ml (milliliters)
	1.057 qts (quarts)
1 m (meter) =	1×10^{10} Å (Ångstroms)
	100 cm (centimeters)
	3.28 ft (feet)
	39.37 in (inches)
	1×10^{-3} km (kilometers)
	1,000 mm (millimeters)
	1,000,000 $\mu = 1 \times 10^6 \mu$ (micrometers)
	1×10^9 nm (nanometers)
1 mps (meters/second) =	196.9 fpm (feet/minute)
	3.6 kmph (kilometers/hour)
	2.237 mph (miles/hour)
1 mph (mile/hour) =	88 fpm (feet/minute)
	1.61 kmph (kilometers/hour)
	0.447 mps (meters/second)
	1 kt (nautical mile) = 6,076.1 ft (feet)
	1.852 km (kilometers)
	1.15 mi (statute miles)

(*Continued*)

TABLE B (*Continued*)

Alphabetical Listing of Conversion *Factors*

Factors	Metric (SI) or English Conversions
	2,025.4 yds (yards)
1 mi (statute mile) =	5,280 ft (feet)
	1.609 km (kilometers)
	1,609.3 m (meters)
	0.869 nt (nautical miles)
	1,760 yds (yards)
1 miCi (millicurie) =	0.001 Ci (curie)
	3.7×10^{10} radioactive disintegrations/second
	3.7×10^{10} Bq (becquerel)
1 mm Hg (mm of mercury) =	1.316×10^{-3} atm (atmosphere)
	0.535 in H_2O (inches of water)
	1.33 mb (millibars)
	133.32 pascals
	1 torr
	0.0193 psia (pounds/square inch – absolute)
1 min (minute) =	6.94×10^{-4} days
	0.0167 hrs (hours)
	60 sec (seconds)
	9.92×10^{-5} weeks
	1.90×10^{-6} yrs (years)
1 N (newton) =	1×10^{5} dynes
1 N-m (newton-meter) =	1.00 j (joules)
	2.78×10^{-4} watt-hours
1 ppm (parts/million-volume) =	1.00 ml/m^3 (milliliters/cubic meter)
1 ppm [wt] (parts/million-weight) =	1.00 mg/kg (milligrams/kilograms)
1 pascal =	9.87×10^{-6} atm (atmospheres)
	4.015×10^{-3} in H_2O (inches of water)
	0.01 mb (millibars)
	7.5×10^{-3} mm Hg (milliliters of mercury)
1 lb (pound) =	453.59 g (grams)
	16 oz (ounces)
1 lbs/ft^3 (pounds/cubic foot) =	16.02 g/l (grams/liter)
1 lbs/ft^3 (pounds/cubic inch) =	27.68 gms/cc (grams/cubic centimeter)
	1,728 lbs/ft^3 (pounds/cubic feet)
1 psi (pounds/square inch)=	0.068 atm (atmospheres)
	27.67 in H_2O (inches or water)
	68.85 mb (millibars)
	51.71 mm Hg (millimeters of mercury)
	6,894.76 pascals
1 qt (quart) =	946.4 cc (cubic centimeters)
	57.75 in^3 (cubic inches)

(Continued)

TABLE B (*Continued*)
Alphabetical Listing of Conversion *Factors*

Factors		Metric (SI) or English Conversions
		0.946 ℓ (liters)
1 rad =		100 ergs/gm (ergs/gram)
		0.01 Gy (gray)
		1 rem [unless modified through division by an appropriate factor, such as Q and/or N]
1 rem		1 rad [unless modified through division by an appropriate factor, such as Q and/or N]
1 Sv (sievert) =		1 Gy (gray) [unless modified through division by an
		appropriate factor, such as Q and/or N]
1 cm^2 (square centimeter)	=	1.076×10^{-3} ft^2 (square feet)
		0.155 in^2 (square inches)
		1×10^{-4} m^2 (square meters)
1 ft^2 (square foot) =		2.296×10^{-5} acres
		9.296 cm^2 (square centimeters)
		144 in^2 (square inches)
		0.0929 m^2 (square meters)
1 m^2 (square meter) =		10.76 ft^2 (square feet)
		1,550 in^2 (square inches)
1 mi^2 (square mile) =		640 acres
		2.79×10^{7} ft^2 (square feet)
		2.59×10^{6} m^2 (square meters)
1 torr =		1.33 mb (millibars)
1 watt =		3.41 BTI/hr (British Thermal Units/hour)
		1.341×10^{-3} hp (horsepower)
		52.18 j/sec (joules/second)
1 watt-hour =		3.412 BTUs (British Thermal Unit)
		859.8 cal (calories)
		3,600 j (joules)
		35.53 liter-atmosphere
1 week =		7 days
		168 hrs (hours)
		10,080 min (minutes)
		6.048×10^{5} sec (seconds)
		0.0192 yrs (years)
1 yr (year) =		365.25 days
		8,766 hrs (hours)
		5.26×10^{5} min (minutes)
		3.16×10^{7} sec (seconds)
		52.18 weeks

TABLE C
Conversion Factors by Unit Category

	Units of Length
1 cm (centimeter) =	0.0328 ft (feet)
	0.394 in (inches)
	10,000 microns (micrometers)
	100,000,000 Å = 10^8 Å (Ångstroms)
1 ft (foot) =	30.48 cm (centimeters)
	12 in (inches)
	0.3048 m (meters)
	1.65 $\times 10^{-4}$ nt (nautical miles)
	1.89 $\times 10^{-4}$ mi (statute miles)
1 in (inch) =	2.54 cm (centimeters)
	1,000 mils
1 km (kilometer) =	3,280.8 ft (feet)
	0.54 nt (nautical miles)
	0.6214 mi (statute miles)
1 m (meter) =	1 $\times 10^{10}$ Å (Ångstroms)
	100 cm (centimeters)
	3.28 ft (feet)
	39.37 in (inches)
	1 $\times 10^{-3}$ km (kilometers)
	1,000 mm (millimeters)
	1,000,000 μ = 1 $\times 10^6$ μ (micrometers)
	1 $\times 10^9$ nm (nanometers)
1 kt (nautical mile) =	6,076.1 ft (feet)
	1.852 km (kilometers)
	1.15 km (statute miles)
	2.025.4 yds (yards)
1 mi (statute mile) =	5,280 ft (feet)
	1.609 km (kilometers)
	1.690.3 m (meters)
	0.869 nt (nautical miles)
	1,760 yds (yards)
	Units of Area
1 cm^2 (square centimeter) =	1.076 $\times 10^{-3}$ ft^2 (square feet)
	0.155 in^2 (square inches)
	1 $\times 10^{-4}$ m^2 (square meters)
1 ft^2 (square foot) =	2.296 $\times 10^{-5}$ acres
	929.03 cm^2 (square centimeters)
	144 in^2 (square inches)

(Continued)

TABLE C (*Continued*)
Conversion Factors by Unit Category

	0.0929 m^2 (square meters)
1 m^2 (square meter) =	10.76 ft^2 (square feet)
	1,550 in^2 (square inches)
1 mi^2 (square mile) =	640 acres
	2.79×10^7 ft^2 (square feet)
	2.59×10^6 m^2 (square meters)
Units of Volume	
1 cc (cubic centimeter) =	3.53×10^{-5} ft^3 (cubic feet)
	0.061 in^3 (cubic inches)
	2.64×10^{-4} gal (gallons)
	0.001 ℓ (liters)
	1.00 ml (milliliters)
1 ft^3 (cubic foot) =	28,317 cc (cubic centimeters)
	1,728 in^3 (cubic inches)
	0.0283 m^3 (cubic meters)
	7.48 gal (gallons)
	28.32 ℓ (liters)
	29.92 qts (quarts)
1 in^3 (cubic inch) =	16.39 cc (cubic centimeters)
	16.39 ml (milliliters)
	5.79×10^{-4} ft^3 (cubic feet)
	1.64×10^{-5} m^3 (cubic meters)
	4.33×10^{-3} gal (gallons)
	0.0164 ℓ (liters)
	0.55 fl oz (fluid ounces)
1 m^3 (cubic meter) =	1,000,000 cc = 10^6 cc (cubic centimeters)
	35.31 ft^3 (cubic feet)
	61,023 in^3 (cubic inches)
	264.17 gal (gallons)
	1,000 ℓ (liters)
1 yd^3 (cubic yards) =	201.97 gal (gallons)
	764.55 ℓ (liters)
1 gal (gallon) =	3,785 cc (cubic centimeters)
	0.134 ft^3 (cubic feet)
	231 in^3 (cubic inches)
	3.785 ℓ (liters)
1 ℓ (liter) =	1,000 cc (cubic centimeters)
	1 dm^3 (cubic decimeters)
	0.0353 ft^3 (cubic feet)
	61.02 in^3 (cubic inches)
	0.264 gal (gallons)
	1,000 (milliliters)

(*Continued*)

TABLE C (*Continued*)
Conversion Factors by Unit Category

	1.057 qts (quarts)
1 qt (quart) =	946.4 cc (cubic centimeters)
	57.75 in^3 (cubic inches)
	0.946 ℓ (liters)
	Units of Mass
1 gm (grams) =	0.001 kg (kilograms)
	1,000 mg (milligrams)
	1,000,000 mg = 10^6 ng (nanograms)
	2.205 $\times 10^{-3}$ lbs (pounds)
1 kg (kilogram) =	1,000 gms (grams)
	2.205 lbs (pounds)
1 lbs (pound) =	453.59 gms (grams)
	16 oz (ounces)
	Units of Time
1 day =	24 hrs (hours)
	1440 min (minutes)
	86,400 sec (seconds)
	0.143 weeks
	2.738 $\times 10^{-3}$ yrs (years)
1 hr (hours) =	0.0417 days
	60 min (minutes)
	3,600 sec (seconds)
	5.95 $\times 10^{-3}$ yrs (years)
1 hr (hour) =	0.0417 days
	60 min (minutes)
	3,600 sec (seconds)
	5.95 $\times 10^{-3}$ weeks
	1.14 $\times 10^{-4}$ yrs (years)
1 min (minutes) =	6.94 $\times 10^{-4}$ days
	0.0167 hrs (hours)
	60 sec (seconds)
	9.92 $\times 10^{-5}$ weeks
	1.90 $\times 10^{-6}$ yrs (years)
1 week =	7 days
	168 hrs (hours)
	10,080 min (minutes)
	6.048 $\times 10^{5}$ sec (seconds)
	0.0192 yrs (years)
1 yr (year) =	365.25 days
	8,766 hrs (hours)
	5.26 $\times 10^{5}$ min (minutes)

(*Continued*)

TABLE C (*Continued*)
Conversion Factors by Unit Category

	3.16×10^7 sec (seconds)
	52.18 weeks
	Units of the Measure of Temperature
°C (degrees Celsius) =	[(5/9)(°F - 32°)]
1°C (expressed as an interval) =	1.8°F = [9/5]°F (degrees Fahrenheit)
	1.8°R (degrees Rankine)
1.0	K (degrees Kelvin)
°F (degree Fahrenheit) =	[(9/5)(°C) + 32°]
1°F (expressed as an interval) =	0.556°C = [5/9]°C (degrees Celsius)
	1.0°R (degrees Rankine)
	0.556 K (degrees Kelvin)
	Units of Force
1 dyne =	1×10^{-5} nt (newtons)
1 nt (newton) =	1×10^5 dynes
	Units of Work or Energy
1 BTU (British Thermal Unit) =	252 cal (calories)
	1,055.06 j (joules)
	10.41 liter-atmospheres
	0.293 watt-hours
1 cal (calories) =	3.97×10^{-3} BTUs (British Thermal Units)
	4.18 j (joules)
	0.0413 liter-atmospheres
	1.163×10^{-3} watt-hours
1 ev (electron volt) =	1.602×10^{-12} ergs
	1.602×10^{-19} j (joules)
1 erg =	1 dyne-centimeter
	1×10^{-7} j (joules)
	2.78×10^{-11} watt-hours
1 j (joule) =	9.48×10^{-4} BTUs (British Thermal Units)
	0.239 cal (calories)
	10,000,000 ergs = 1×10^7 ergs
	9.87×10^{-3} liter-atmospheres
	1.00 nt-m (newton-meters)
1 kcal (kilocalorie) =	3.97 BTUs (British Thermal Units)
	1,000 cal (calories)
	4,186.8 j (joules)
1 kw-hr (kilowatt-hour) =	3,412.14 BTU (British Thermal Units)
	3.6×10^6 j (joules)
	859.8 kcal (kilocalories)
1 Nt-m (newton-meter) =	1.00 j (joules)

(*Continued*)

TABLE C (*Continued*)
Conversion Factors by Unit Category

	2.78×10^{-4} watt-hours
1 watt-hour =	3.412 BTUs (British Thermal Units)
	859.8 cal (calories)
	3,600 j (joules)
	35.53 liter-atmospheres
	Units of Power
1 hp (horsepower) =	745.7 j/sec (joules/sec)
1 kw (kilowatt) =	56.87 BTU/min (British Thermal Units/minute)
	1.341 hp (horsepower)
	1,000 j/sec (joules/sec)
1 watt =	3.41 BTU/hr (British Thermal Units/hour)
	1.341×10^{-3} hp (horsepower)
	1.00 j/sec (joules/second)
	Units of Pressure
1 atm (atmosphere) =	1.013 bars
	10.133 newtons/cm^2 (newtons/square centimeters)
	33.90 ft. of H_2O (feet of water)
	101.325 kp (kilopascals)
	14.70 psia (pounds/square inch – absolute)
	760 torr
	760 mm Hg (millimeters of mercury)
1 bar =	0.987 atm (atmospheres)
	1×10^6 dynes/cm^2 (dynes/square centimeter)
	33.45 ft of H_2O (feet of water)
	1×10^5 pascals [nt/m^2] (newtons/square meter)
	750.06 torr
	750.06 mm Hg (millimeters of mercury)
1 inch of water =	1.86 mm Hg (millimeters of mercury)
	249.09 pascals
	0.0361 psi (lbs/in^2)
1 mm Hg (millimeter of merc.) =	1.316×10^{-3} atm (atmospheres)
	0.535 in H_2O (inches of water)
	1.33 mb (millibars)
	133.32 pascals
	1 torr
	0.0193 psia (pounds/square inch – absolute)
	1 pascal = 9.87×10^{-6} atm (atmospheres)
	4.015×10^{-3} in H_2O (inches of water)
	0.01 mb (millibars)
	7.5×10^{-3} mm Hg (millimeters of mercury)
1 psi (pounds/square inch) =	0.068 atm (atmospheres)

(*Continued*)

TABLE C (*Continued*)
Conversion Factors by Unit Category

	27.67 in H_2O (inches of water)
	68.85 mb (millibars)
	51.71 mm Hg (millimeters of mercury)
	6,894.76 pascals
1 torr =	1.33 mb (millibars)
	Units of Velocity or Speed
1 fps (feet/second) =	1.097 kmph (kilometers/hour)
	0.305 mps (meters/second)
	0.01136 mph (miles/hours)
1 mps (meters/second) =	196.9 fpm (feet/minute)
	3.6 kmph (kilometers/hour)
	2.237 mph (miles/hour)
1 mph (mile/hour) =	88 fpm (feet/minute)
	1.61 kmph (kilometers/hour)
	0.447 mps (meters/second)
	Units of Density
1 gm/cc (grams/cubic cent.) =	62.43 lbs/ft^3 (pounds/cubic foot)
	0.0361 lbs/in^3 (pounds/cubic inch)
	8.345 lbs/gal (pounds/gallon)
1 lbs/ft^3 (pounds/cubic foot) =	16.02 gms/ℓ (grams/liter)
1 lbs/in^2 (pounds/cubic inch) =	27.68 gms/cc (grams/cubic centimeter)
	1.728 lbs/ft^3 (pounds/cubic foot)
	Units of Concentration
1 ppm (parts/million-volume) =	1.00 ml/m^3 (milliliters/cubic meter)
1 ppm (wt) =	1.00 mg/kg (milligrams/kilograms)
	Radiation and Dose Related Units
1 Bq (becquerel) =	1 radioactive disintegration/second
	2.7×10^{-11} Ci (curie)
	2.7×10^{-8} (millicurie)
1 Ci (curie) =	3.7×10^{10} radioactive disintegration/second
	3.7×10^{10} Bq (becquerel)
	1,000 mCi (millicurie)
1 Gy (gray) =	1 j/kg (joule/kilogram)
	100 rad
	1 Sv (sievert) – [unless modified through division by an appropriate factor, such as Q and/or N]
1 mCi (millicurie) =	0.001 Ci (curie)
	3.7×10^{10} radioactive disintegrations/second
	3.7×10^{10} Bq (becquerel)
1 rad =	100 ergs/gm (ergs/gm)

(*Continued*)

TABLE C (*Continued*)
Conversion Factors by Unit Category

	0.01 Gy (gray)
	1 rem – [unless modified through division by an appropriate factor, such as Q and/or N]
1 rem =	1 rad – [unless modified through division by an appropriate factor, such as Q and/or N]
1 Sv (sievert) =	1 Gy (gray) – [unless modified through division by an appropriate factor, such as Q and/or N]

DID YOU KNOW?

Units and dimensions are not the same concepts. Dimensions are concepts like time, mass, length, weight, etc. Units are specific cases of dimensions, like hour, gram, meter, lb, etc. You can *multiply* and *divide* quantities with different units: 4 ft × 8 lb = 32 ft-lb, but you can *add* and *subtract* terms only if they have the same units: 5 lb + 8 kg = **NO WAY!!!**

About the Author

Frank R. Spellman, PhD, CSP, CHMM is a retired U.S. Naval Officer with 26 years active duty and also a retired full-time adjunct assistant professor of environmental health at Old Dominion University, Norfolk, Virginia, and the author of more than 161 books covering topics ranging from a 15-volume homeland security series, several safety, industrial hygiene, stormwater management, air pollution, and security manuals and also including concentrated animal feeding operations (CAFOs) to all areas of environmental science and occupational health and regulatory compliance. Many of his texts are readily available online at Amazon.com and Barnes and Noble.com, and several have been adopted for classroom use at major universities throughout the United States, Canada, Europe, and Russia; two have been translated into Chinese, Japanese, Arabic, and Spanish for overseas markets. Dr. Spellman has been cited in more than 850 publications. He serves as a professional expert witness for three law groups and as an incident/accident investigator and security expert for the U.S. Department of Justice and a northern Virginia law firm. In addition, he consults on homeland security vulnerability assessments for critical infrastructures including water/wastewater facilities nationwide and conducts pre-Occupational Safety and Health Administration (OSHA)/Environmental Protection Agency EPA audits throughout the country. Dr. Spellman receives frequent requests to co-author with well-recognized experts in several scientific fields; for example, he is a contributing author of the prestigious text *The Engineering Handbook,* 2nd ed. (CRC Press). Dr. Spellman lectures on wastewater treatment, water treatment, and homeland security and lectures on safety topics throughout the country and teaches water/wastewater operator/regulatory short courses at Virginia Tech (Blacksburg, Virginia). In 2011–2012, he traced and documented the ancient water distribution system at Machu Picchu, Peru, and surveyed several drinking water resources in Amazonia-Coco, Ecuador. Dr. Spellman also studied and surveyed two separate potable water supplies in the Galapagos Islands; he also studied and researched Darwin's finches while in the Galapagos. He holds a BA in public administration, a BS in business management, an MBA, and an MS and PhD in environmental engineering.

Part I

Carbon Fundamentals
1 Kilogram Carbon (C) = 3.664 Kilograms Carbon Dioxide (CO2)

1 Cave of the Dog

THE CAVE

This is not your typical caveperson's cave. No. No way. Oh, it is a cave, for sure. It is known as the Cave of the Dogs (Italian: Grotto del Cane) near Naples, Italy.

So, the question is why is the Cave of the Dogs not a typical caveperson's cave even though it dates to the time of cave dwellers? Well, to show the difference let's make a 15,000-year comparison and contrast to the present time (2023). Let's first look at a historical perspective.

A Historical Perspective[1]

According to today's calendar, it is June 26, 15,543 BC. The place, a large natural cave set deep under a solid outcropping that formed a significant mountain meadow before the last glacial icesheet gouged, gorged, ground, and pulverized it down to its present size and shape.

The colossal sheet of ice is in retreat. When it was at its full width, depth, and length, it extended several hundred miles beyond the cave site to a V-shaped valley, where glacial melt fed a youthful, raging river that ran through the valley's bottom land.

A small, steady stream of melt-water courses in a straight line past one side of the cave, down toward that valley, where it will join and feed that river.

On the other side is a sloping field of young grass, brush, and flowers—flowers everywhere. Up close, we see the stark remnants of the terminal moraine that formed this abrupt slope with its fresh cover of grass and blossoms. A closer look reveals a dark heap at the base of the slope—a heap of trash: skin, sinew, bone, decaying corpses, burnt remnants of past hunts, and feasts. We know extremely well the refuse, filth, and discards that people leave behind as their foulest signature; somebody, many bodies live close by. Where? Of course—in the cave.

Let's look inside.

We wander up to the huge hole in the rock that forms the mouth of the cave. We tread carefully—we do not want to disturb (startle, frighten—anger) the occupants. Remember, we're talking about cavepersons here. No language in common. No culture in common. Could we have anything in common with such … with such primitive people?

But all is quiet—and with no overt threat present, our curiosity overcomes our caution, and we walk into the opening chamber.

And something reaches out and grabs us—not cave dweller/woman, but a stench—a horrible stench—a stench too horrible to describe. With our fingers clamped tightly

[1] Adapted from Spellman, F.R. (2018). *The Science of Environmental Pollution*, 3rd ed. Boca Raton, FL: CRC Press.

DOI: 10.1201/9781003432838-2

to our noses, crushing to the bone (cartilage), we move on, and we are too interested in this strange environment, despite the reek, to retreat.

We can see well in this chamber because daylight pours in through the entrance. We take a few steps and stop to look around, religiously not breathing through our noses.

The walls of the cave are covered with black soot. A pit near the cave wall to the right, under attack by millions of flies and other insects provides much of the stench—the latrine. A heap of detritus like the dump outside provides the rest of the reek.

That this cave is abandoned dawns on us. We have no doubt as to why. The largest byproduct manufactured by humankind has taken over—the cave is a garbage dump—humankind's garbage (we are assuming that cave dwellers of this period were what we call human today).

Perhaps, back in other chambers, deeper in the cave, exist cave paintings and remnants of ceremony. But we don't have the tools to explore them with us today, and we retreat, grateful for a breath of fresh air.

Outside, a few hundred feet from the smelly cave, and within sight of the garbage heap, we stop to contemplate what we've seen. Seventeen thousand years from now, archaeologists will find this cave, and explore it thoroughly, learning information that will give us insights into the world of the people whose former home we have just visited. But the remains the archaeologists will find will be altered by 17,000 years of history. The picture they see will be incomplete, scattered by the natural interferences' life causes, giving mystery to the short and brutal lives of our ancestors.

But here, right now, we see similarities. We foul our environment in the same ways, and in more. But the cave dweller had a huge advantage over modern people, in that respect. When his living quarters became too foul for comfort, he could pick up whatever he considered of value that he and his clan could carry and move on. A fresh site was always just around the next bend in the river. The pollution he created was completely (eventually) naturally biodegradable—in a few years, this same foul cave could house humans again.

Although we have our similarities with those far-off ancestors, one stark difference is plain: modern man cannot destroy and pollute his environment with impunity. We can no longer simply pick up the stakes and move on. What we do to our environment has ramifications on a scale that we cannot ignore—or avoid.

So, let's get back to the here and now and the contrast between ancient caveperson's cave and the Cave of the Dogs.

Okay, here's deal—the Cave of the Dogs is about 33 ft (10 m) deep and does not have cave drawings within and has relatively clean walls—it is not the kind of place where you build campfires. The Cave of the Dogs does not have remnant relics of past occupants. The Cave of the Dogs does not have inquisitive archaeologists digging for memories of the distant past. The Cave of Dogs does not have waste piles buried in the floor; thus, there is not a noxious odor present, prevailing, encapsulating, and capturing one's senses.

No. The Cave of the Dogs is different—different from other caves—vastly different.

Okay, in what way or ways is the Cave of the Dogs different?

The Cave of the Dogs is different and uninhabitable because within the cave is a fumarole (aka a volcanic hydrothermal vent) that releases various gases but mostly carbon dioxide—again, of volcanic origin.

Carbon dioxide, what is it and is it important and why would it drive the troglodytes (aka cave dwellers) away from the Cave of the Dogs?

Well, here's the deal. Carbon dioxide (CO_2) was first found by Joseph Blake, a chemist and physician, in the 1750s. It is one of the gases that is found in abundance in the atmosphere (Young, 1878).

So, exactly what is carbon dioxide (CO_2)? It is a colorless, odorless gas (although it is felt by some people to have a slight pungent odor and biting taste), slightly soluble in water and denser than air (one and half times heavier than air), and a slightly acidic gas. Carbon dioxide gas in standard temperature and pressure conditions is relatively non-reactive and non-toxic. It will not burn, and it will not support combustion or life.

CO_2 is normally present in atmospheric air at about 0.035% by volume and cycles through the biosphere (carbon cycle) as shown in Figure 1.1. Carbon dioxide, along with water vapor, is primarily responsible for the absorption of infrared energy re-emitted by the Earth and, in turn, some of this energy is reradiated back to the Earth's surface. It is also a normal product of human and animal metabolism. Exhaled breath has up to 5.6% carbon dioxide. In addition, the burning of carbon-laden fossil fuels releases carbon dioxide into the atmosphere. Much of this carbon dioxide is absorbed by ocean water; some of it is taken up by vegetation through photosynthesis in the

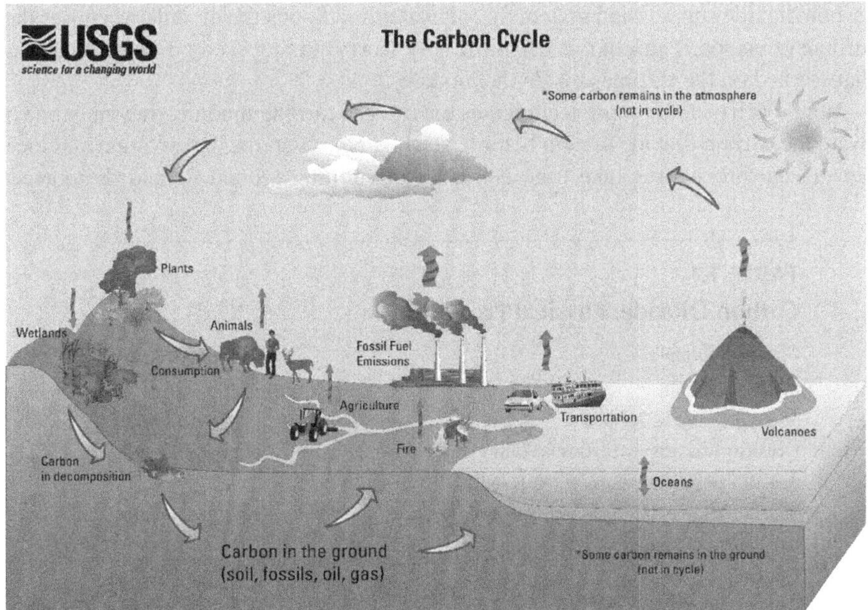

FIGURE 1.1 Public domain diagram of carbon cycle. USGS/USDOE/NOAA (2023). Accessed 3/15/23 @ https://www.gov/media/images/carbon-cycle.

carbon cycle (see Figure 1.1), and some stays in the atmosphere. Today (in 2023), it is estimated that the concentration of carbon dioxide in the atmosphere is approximately 410 parts per million (ppm) and is rising at a rate of approximately 20 ppm and maybe more every decade. The increasing rate of combustion of coal and oil has been primarily responsible for this occurrence, which (as we will see later in this text) may eventually have an impact on the global climate.

CARBON DIOXIDE: PHYSICAL PROPERTIES

The physical properties of carbon dioxide are noted in Table 1.1.

CARBON DIOXIDE: USES

Solid carbon dioxide is used quite extensively to refrigerate perishable foods while in transit. It is also used as a cooling agent in many industrial processes, such as grinding, rubber work, cold-treating metals, and vacuum cold traps. Gaseous carbon dioxide is used to carbonate soft drinks, for pH control in water treatment, in chemical processing, as a food preservative, and in pneumatic devices.

From a personal perspective, I can point out one of many safety uses of carbon dioxide. The safety use I know quite well and came to depend on had to do with modular LM-2500 gas turbine used in the propulsion system in a Spruance Class Naval destroyer. However, as the senior electrical officer in one of these vessels I also had a constant worry about carbon dioxide systems within the modules.

Why worry? Well, each gas turbine module held not only the turbine but also a carbon dioxide suppression system that activates and floods the module whenever the turbine or enclosed auxiliaries catch fire. My worry was focused on the technicians that worked on the turbines inside the modules.

Now it is true that before technicians enter a gas turbine module, they must inactivate the carbon dioxide system before entering. However, the deactivation and lock out tag out procedures take time and a lot of running around to find the correct

TABLE 1.1
Carbon Dioxide: Physical Properties

Chemical formula	CO_2
Molecular weight	44.01
Vapor pressure @ 70°F	838 psig
Density of the gas @ 70°F and 1 atm	0.1144 lb/ft^3
Specific gravity of the gas @ 70°F and 1 atm (air = 1)	1.522
Specific volume of the gas @ 70°F and 1 atm	8.741 ft^3/lb
Critical temperature	−109.3°F
Critical pressure	1070.6 psia
Critical density	29.2 lb/ft^3
Latent heat of vaporization @ 32°F	100.8 Btu/lb
Latent heat of fusion @ −69.9°F	85.6 Btu/lb

person to authorize the deactivation of the carbon dioxide system. So, because it is a cumbersome process to deactivate the carbon dioxide system, technicians would often take short cuts, ignore the danger, and enter the module to do their inspections or maintenance activities with the carbon dioxide system still active—not good and not smart and very dangerous. And when the carbon dioxide extinguishing agent is dumped, it is customary practice for the people inside to go to their knees and try to escape. Depending on their location inside the module they might have a slim chance of getting out and surviving—the key words here are "slim chance." Another problem with escape is that some of the module doors lock automatically when the carbon dioxide dumps. The two-person rule is called for, while one person is inside the module and the other is standing watch outside next to the door to help escape if necessary.

Okay, you might find the information provided to this point informative, but you may also be asking: what does any of this have to do with the Cave of the Dogs?

As early as 78 AD, there was mention of animals dying from poisonous fumes near Pozzuoli. Several centuries later, the cave became a tourist attraction. Several famous people including Mark Twain, Montesquieu, Goethe, and others were given the grand tour of the cave. They were able to enter the cave because the carbon dioxide leaving the fumarole hovered just inches above the cave floor. For humans of normal height, it was possible to enter the cave so long as they did not sink to the floor level. This is where the animals came into play. Tourists would watch chickens as they died instantly, and dogs became unconscious and could sometimes be revived if removed outside into fresh air and dumped into ice chilly water. However, some of the dogs died.

Probably the best early description of the Cave of the Dogs is provided by Young in 1878; he examined the cave for several days in 1877 and reported:

> On carrying a lighted torch into the cave, its smoke gradually falls, till it reaches the layer of gas, upon which it settles; and on looking in, the surface of the gaseous layer is seen, resembling that of water, and appears covered with beautiful undulations. On holding the head below the level of gas, holding the breath, and keeping the eyes open, an intolerable prickling sensation is produced upon the eyes by the carbonic acid.
>
> A dog brought into the cave, as is the custom there, appears, as it were, to drink the gas, lapping. Then its eyes begin to dilate to an unnatural size, and its lapping becomes more spasmodic; beyond this it does not seem to suffer. While in the cave, also, the dog was able to stand, but when carried and set on its feet outside in the fresh air, it fell, and lay struggling as if in paroxysms of suffocation, but recovered in two to three minutes. I was told, however, that the animal gets into such a nervous state with the prospects of its frequent ordeals, that it must be killed in three months.
>
> *(Young, 1878)*

Mark Twain's account of the Cave of the Dogs, written in 1869, is an interesting description that is included in his *Innocents Aboard.* He tells the following:

> … the Grotto of the Dog claimed our chief attention, because we had heard and so much about it. Everybody has written about the Grotto del Cane and its poisonous vapors, from Pliny down to Smith, and every tourist has held a dog over its floor by its

> legs to test the capabilities of the place. The dog dies in a minute and a half—a chicken instantly. As a general thing, strangers who crawl in there to sleep do not get up until they are called. And then they don't either. The stranger that ventures to sleep there takes a permanent contract. I longed to see this grotto. I resolved to take the dog and hold him myself; suffocate him a little, and time him; suffocate him some more and then finish him. We reached the grotto at about three in the afternoon and went ahead at once to do the experiments. But now, an important difficulty presented itself. We had no dog.
>
> *(Mark Twain, 1869)*

So, you might be wondering about the status of the Cave of the Dogs at the present time. Well, the best and most recent account from Mathews (2016), in his own words,

> the area degraded terribly after WW11 and because an eyesore for shoddy overbuilding and illegal waste dumping. I drove by the baths hundreds of times over the years and never knew about the lake, never knew that I was 100 yards from the Grotto of the Dog …

Case Study 1.1 Carbon Dioxide at Mauna Loa

In a 2021 report by NOAA, the atmospheric carbon dioxide measure at its Mauna Loa Atmospheric Baseline Observatory peaked in May of 2021 at a monthly average of 419 parts per million (ppm), the highest level since correct measurements began more than 60 years ago. The Mauna Loa observatory is perched on a baren volcan in the middle of the Pacific Ocean and is the benchmark sampling location for carbon dioxide.

What makes Mauna Loa the best location to track carbon dioxide levels in the atmosphere?

Note that the observatory is ideally situated for sampling well-mixed air—undisturbed by the influence of local pollution sources or vegetation, producing measurements that stand for the average state of the atmosphere in the northern hemisphere.

Because of the observatory's ideal location, Scripps' scientists started on-site measurements of carbon dioxide at NOAA's weather station on Mauna Loa in 1958. NOAA began measurements in 1974, and the two research institutions have made complementary, independent observations ever since.

In May of 2021, NOAA's measurements at the mountaintop observatory averaged 419.13 ppm. Scientists at Scripps calculated a monthly average of 418.92 ppm. The average in May 2020 was 417 ppm.

A senior scientist with NOAA's Global Monitoring Laboratory noted that CO_2 is by far the most abundant human-caused greenhouse gas and persists in the atmosphere and oceans for thousands of years after it is emitted.

Senior scientists at NOAA's Global Monitoring Laboratory note that roughly 40 billion tons of carbon dioxide are being added to the atmosphere per year, and that is the carbon that we dig up out of the Earth, burn, and release into the atmosphere as CO_2—year after year. Many scientists today feel that if we want to avoid catastrophic climate change, the highest priority must be to reduce carbon dioxide emissions to zero as soon as possible.

Observations and tracking have shown that the highest monthly mean carbon dioxide value of the year occurs in May. This makes sense, makes sense, when you consider that in the northern hemisphere just before plants start to remove large amounts of carbon dioxide from the atmosphere during the growing season. In the northern hemisphere during fall, winter, and early spring, plants and soils give off carbon dioxide, causing levels to rise through May. This seasonal rise and next fall in carbon dioxide levels every year was first seen by David Keeling; this dynamic is now known as the Keeling Curve (see Figure 1.2). Keeling was the first to recognize that despite the seasonal fluctuation, carbon dioxide levels were rising every year. Time and observation have shown that every single year since the start of the measurements, carbon dioxide was higher than the preceding year.

About world-wide measurement of carbon dioxide in the atmosphere, sampling stations around the globe, including those taken at Mauna Loa, are incorporated into NOAA's Global Greenhouse Gas Reference network, a foundational research dataset for international climate scientists and a benchmark for policymakers try to head off the impacts of climate change.

At the present time, the atmospheric burden of carbon dioxide is now comparable to where it was during the Pliocene Climatic Optimum, between 4.1 and 4.5 million year ago, when carbon dioxide levels were close to, or above 400 ppm. During that time, sea level was about 78 ft higher than today (Nature, 2019), and

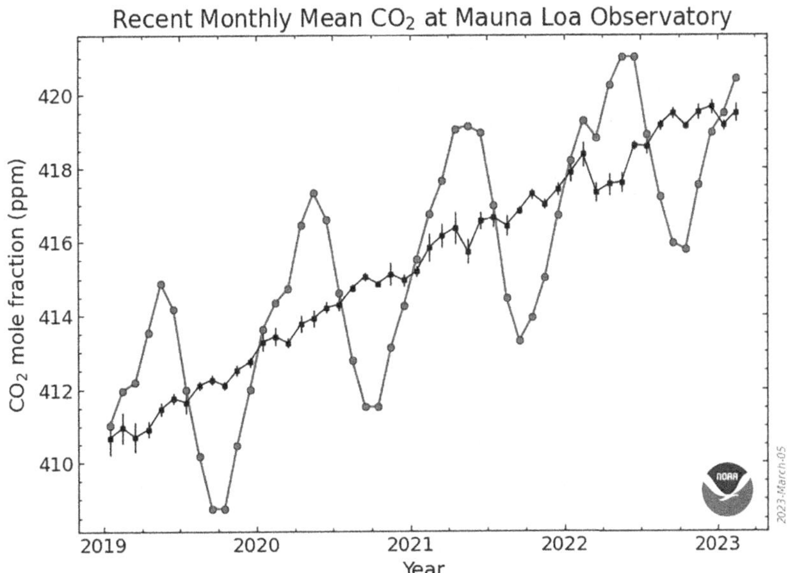

FIGURE 1.2 This graph depicts the upward trajectory of carbon dioxide in the atmosphere as measured at the Mauna Loa Atmospheric Baseline Observatory by NOAA and the Scripps Institution of Oceanography. The annual fluctuation is known as the Keeling Curve. From NOAA Global Monitoring Laboratory. Public domain image Accessed 3/21/23 @ https://gmo.noaa.gov/ccgg/trends/.

the average temperature was 7°F higher than that in pre-industrial times, and studies indicate (Brigham-Grette et al., 2013) large forests occupied areas of the Arctic that are now tundra.

IF YOU WISH TO CONVERSE WITH ME …

An incredibly wise person, Voltaire to be precise, once stated that "if you wish to converse with me, define your terms." It is widespread practice in a book like this one to include a glossary, a vocabulary or clavis at the end of the book made up of an alphabetical list of terms in a particular domain of knowledge with definitions for the terms—terms defined.

Well, based on years of teaching technical courses in the military and civilian venues, along with college lower- and upper-level classes, I have seen that students rarely refer to a glossary at the end of their textbook—in print of digital. Typically, the glossary terms that are newly introduced, uncommon, or specialized are listed.

So, the question becomes: "Why not introduce these terms and their definitions in the beginning of the text? If anyone is to understand subject matter presented, should they not know the terms and their meaning?"

My answer: yes. What I do is I give each student an exact copy of my lecture. And in this copy each new term is listed and defined—the goal is to completely understand the subject matter presented. I also listed symbols used in environmental studies related to my classes.

Well, you might say that if I hand out a copy of each lecture and post it online, won't this reduce the attendance in person in my classes? Never happened. I asked one of my female students (typically my students in my classes were 95% female) why it was that almost every class was always full of attendees even though they could have gotten paper or digital copies of each lecture without attending classes?

Her answer (abbreviated) was that everyone looked forward to their opportunity to say what they thought and thought what they said in my classes—without any repercussions. They could vent pent up feelings, mostly about the environment and what humans are doing to it but also on other topics of interest.

Anyway, the key terms and their definitions are presented in the following for one reason only: I wish to converse with you and you with me.

KEY TERMS AND DEFINITIONS

Absorber

This is part of the absorption process equipment in which the gas mixture that needs to be separated is brought into contact with the solvent. In a photovoltaic device, the material readily absorbs photons to generate charge carriers (free electrons or holes).

Absorption (aka Scrubbing)

(1) Movement of a chemical into a plant, animal, or soil. (2) Any process by which one substance penetrates the interior of another substance. In chemical spill cleanup, this process applies to the uptake of chemical by capillaries within certain sorbent materials.

Absorption Units

Devices or units designed to transfer the contaminant from a gas phase to a liquid phase.

AC

Alternating current.

Acid Rain

Also called acid precipitation or acid deposition, acid rain is precipitation having harmful amounts of nitric and sulfuric acids formed primarily by sulfur dioxide and nitrogen oxides released into the atmosphere when fossil fuels are burned. It can be wet precipitation (rain, snow, or fog) or dry precipitation (absorbed gaseous and particulate matter, aerosol particles or dust). Acid rain has a pH below 5.6. Normal rain has a pH of about 5.6, which is slightly acidic. The term pH is a measure of acidity or alkalinity and ranges from 0 to 14. A pH measurement of 7 is regarded as neutral. Measurements below 7 show increased acidity, while those above indicate increased alkalinity.

Adaptation

Adjustment to environmental conditions.

Adaptive Management

Focuses on learning and adapting, through partnerships of managers, scientists, and other stakeholders who learn together how to create and support sustainable ecosystems. Adaptive management helps science managers support flexibility in their decisions, knowing that uncertainties exist and provides managers that latitude to change direction will improve understanding of ecological systems to achieve management objectives is about taking action to improve progress toward desired outcomes.

Adsorption

Removing carbon dioxide from a mixture using the difference in adsorption of the different components in a solid.

Agglomerating Character

Agglomeration describes the caking properties of coal. Agglomerating character is figured out by examination and testing of the residue when a small, powdered sample is heated to 950°C under specific conditions. If the sample is "agglomerating," the residue will be coherent, show swelling or cell structures, and be capable of supporting a 500 g weight without pulverizing.

Agriculture

An energy-consuming subsector of the industrial sector that consists of all facilities and equipment engaged in growing crops and raising animals.

Albedo

The ratio of light reflected by a surface to the light falling on it.

Air Pollution Abatement Equipment

Equipment used to reduce or cut airborne pollutants, including particulate matter (dust, smoke, fly ash, dirt, etc.), sulfur oxides, nitrogen oxides (NOx), carbon monoxide, hydrocarbons, odors, and other pollutants.

Air Temperature Adjustment

Change in air temperature associated with change in tree canopy cover (°C per 1% change in tree canopy cover).

Alternating Current (AC)

An electric current that reverses its direction at regularly recurring intervals, usually 50 or 60 times per second.

Alternative Fuel

Alternative fuels, for transportation applications, include the following:

- methanol
- denatured ethanol and other alcohols
- fuel mixtures having 85% or more by volume of methanol, denatured ethanol, and other alcohols with gasoline or other fuels
- natural gas
- liquefied petroleum gas (propane)
- hydrogen
- coal-derived liquid fuels
- fuels (other than alcohol) derived from biological materials (biofuels such as soy diesel fuel)
- electricity (including electricity from solar energy)

Ambient

Natural condition of the environment at any given time.

Ampere (Amp)

A unit of electrical current; it can be thought of like the rate of water flowing through a pipe (liters per minute).

Anaerobic Decomposition

Decomposition in the absence of oxygen, as in the anaerobic digester or lagoon, which produces carbon dioxide and methane.

Anion

A negatively charged ion; an ion that is attracted to the anode.

Annual Removals

The net volume of growing stock trees removed from the inventory during a specified year by harvesting, cultural operations such as timber land improvement, or land clearing.

Anode

The electrode at which oxidation (a loss of electrons) takes place. For fuel cells and other galvanic cells, the anode is the negative terminal; for electrolytic cells, the anode is the positive terminal.

Anthropogenic

Made or generated by a human or caused by human activity. The term is used in the context of global climate change to refer to gaseous emissions that are the rules of human activities, as well as other potentially climate-altering activities, such as deforestation.

API Gravity

American Petroleum Institute measure of specific gravity of crude oil or condensate in degrees. An arbitrary scale expressing the gravity of density of liquid petroleum products. The measuring scale is calibrated in terms of degrees API; it is calculated as follows:

$$\text{Degrees API} = \left(141.5/\text{sp. Gr. } 60°\text{F}/60°\text{F}\right) - 131.5$$

Appropriate Use

A proposed or existing use on a refuge that meets at least one of the following conditions: (1) the use is a wildlife-dependent one, and (2) the use contributes to fulfilling the refuge purpose(s).

Bare Rock Succession

An ecological succession process whereby rock or parent materials are slowly degraded to soil by a series of bio-ecological processes.

Benzene (C_6H_6)

An aromatic hydrocarbon present in small proportion in some crude oils and made commercially from petroleum by the catalytic reforming of naphthenes in petroleum naphtha. Also made from coal in the manufacture of coke. Used as a solvent in the manufacture of detergents, synthetic fibers, petrochemicals, and as a part of high-octane gasoline.

Biobased Product

The term "biobased product," as defined by Farm Security and Rural Investment Act (FSRIA), means a product determined by the US Secretary of Agriculture to be a commercial or industrial product (other than food or feed) that is composed, in whole or in significant part, of biological products or renewable domestic agriculture materials (including plant, animal, and marine materials) or forestry materials.

Biochemical Conversion

The use of fermentation or anaerobic digestion to produce fuels and chemicals from organic sources.

Biodiesel

Fuel derived from vegetable oil or animal fat. It is produced when vegetable oil or animal fat is chemically reacted with alcohol.

Bioenergy

Useful, renewable energy produced from organic matter, which may either be used directly as a fuel or processed into liquids and gases.

Biofuels

Liquid fuels and blending components produced from biomass (plant) feedstocks, used primarily for transportation.

Biogas

A combustible gas derived from decomposing biological waste. Biogas normally consists of 50%–60% methane.

Biogenic

Produced by biological processes of living organisms.

Biogenic Emissions

Emissions that are naturally occurring and are not significantly affected by human actions or activity.

Biological Diversity or Biodiversity

The variety of life and its processes includes the variety of living organisms, the genetic differences among them, and the communities and ecosystems in which they occur.

Biological Integrity

Biotic composition, structure, and functioning at genetic, organism, and community levels comparable with historic conditions, including the natural biological processes that shape genomes, organisms, and communities.

Biomass

Any organic non-fossil material of biological origin constituting a renewable energy source. It is produced from organic matter that is available on a renewable or recurring basis, including agricultural crops and trees, wood and wood residues, plants (including aquatic plants), grasses, animal manure, municipal residues, and other residue materials. Biomass is generally produced in a sustainable manner from water and carbon dioxide by photosynthesis. There are three main categories of biomass: primary, secondary, and tertiary.

Biomass Gas (Biogas)

A medium Btu gas having methane and carbon dioxide, resulting from the action of microorganisms on organic materials such as a landfill.

Biomass Waste

Organic non-fossil material of biological origin that is a byproduct or a discarded product. Biomass waste includes municipal solid waste from biogenic sources, landfill gas, sludge waste, agricultural crop byproducts, straw, and other biomass solids, liquids, and gases, but it excludes wood and wood-derived fuels (including black liquor), biofuel feedstock, biodiesel, and fuel ethanol. Biomass waste also includes energy crops grown specifically for energy production, which would not normally be waste.

Biomaterials

Products derived from organic (as opposed to petroleum-based) products.

Biopower

The use of biomass feedstock to produce electric power to heat through direct combustion of the feedstock, through gasification and then combustion of the resultant gas, or through other thermal conversion processes. Power is generated with engines, turbines, fuel cells, or other equipment.

Biorefinery

A facility that processes and converts biomass into value-added products. These products can range from biomaterials to fuels such as ethanol or important feedstocks to produce chemicals and other materials. Biorefineries can be based on several processing platforms using mechanical, thermal, chemical, and biochemical processes.

Biota

The plant and animal life of a region.

British Thermal Unit (Btu)

This is a basic measure of thermal (heat) energy. A Btu is defined as the amount of energy needed to increase the temperature of 1 pound of water by 1°F, at normal atmospheric pressure. 1 Btu = 1,055 J.

Btu Conversion Factor

A factor for converting energy data between one unit of measurement and the British thermal unit (Btu). Btu conversion factors are generally used to convert energy data from physical units of measure (such as barrels, cubic feet, or short tons) into the energy-equivalent measure of Btu.

Btu per Cubic Foot

The total heating value, expressed in Btu, produced by the combustion, at constant pressure, of the amount of the gas that would occupy a volume of 1 cubic foot at a temperature of 60°F if saturated with water vapor and under a pressure equivalent to the of 30 inches of mercury at 32°F and under standard gravitational force (980.665 cm/s^2)

with air of the same temperature and pressure as the gas, when the products of combustion are cooled to the initial temperature of gas and air when the water formed by combustion is condensed to the liquid state. (Sometimes called gross heating value or total heating value.)

Bulk Density

Weight per unit of volume, usually specified in pounds per cubic foot.

Bunker Fuels

Fuel supplied to ships and planes, both domestic and foreign, consists primarily of residual and distillate fuel oil for ships and kerosene-based jet fuel for aircraft. The term "international bunker fuels" is used to denote the consumption of fuel for international transport activities. *Note*: For the purposes of greenhouse gas emissions inventories, data on emissions form combustion of international bunker fuels are subtracted from national emissions totals. Historically, bunker fuels have meant only ship fuel.

Burnup

The amount of thermal energy generated per unit mass of fuel, expressed as Gigawatt-Days Thermal per Metric Ton of Initial Heavy Metal (GWDT/MTIHM), rounded to the nearest gigawatt day.

Butane (C_4H_{10})

A straight-chain or branch-chain hydrocarbon extracted from natural gas or refinery gas streams, which is gaseous at standard temperature and pressure. It includes isobutene and normal butane and is appointed in ASTM Specification D1835 and gas Processors Association specifications for commercial butane.

Butylene (C_4H_8)

An olefinic hydrocarbon recovered from refinery or petrochemical processes, which is gaseous at standard temperature and pressure. Butylene is used in the production of gasoline and various petrochemical products.

Calcium Sulfate

A white crystalline salt, insoluble in water. Used in Keene's cement, in pigments, as a paper filler, and as a drying agent.

Calorie

The amount of heat needed to raise the temperature of 1 g of water 1°C.

Carbon

A non-metallic element (symbol C) was found in all organic substances and in some inorganic substances such as coal and natural gas. The atomic weight of C is 12, and that of CO_2 is 44. To convert emissions reported as mass or weight of C to mass of CO_2, to mass or weight of C, multiply CO_2 by 12/44.

Carbon Accounting

In general, it refers to processes undertaken to *measure* amounts of carbon dioxide equivalents emitted by an entity. It is used (i.e., among other things) by nation states, corporations, and individuals.

Carbon Budget

The balance of the exchanges (incomes and losses) of carbon between carbon sinks (e.g., atmosphere and biosphere) in the carbon cycle.

Carbon Cycle

All carbon sinks and exchanges of carbon from one sink to another by various chemical, physical, geological, and biological processes.

Carbon Dioxide (CO_2)

A product of combustion; the most common greenhouse gas.

Carbon Dioxide Equivalent (CDE)

The amount of carbon dioxide by weight emitted into the atmosphere would produce the same estimated radiative force as a given weight of another radiatively active gas. Carbon dioxide equivalents are computed by multiplying the weight of the gas being measured (for example, methane) by its estimated global warming potential (which is 21 for methane). "Carbon equivalent units" are defined as carbon dioxide equivalents multiplied by the carbon content of carbon dioxide (i.e., 12/44).

Carbon Intensity

The amount of carbon by weight emitted per unit of energy consumed. A common measure of carbon intensity is the weight of carbon per British thermal unit (Btu) of energy. When there is only one fossil fuel under consideration, the carbon intensity and the emission coefficient are identical. When there are several fuels, carbon intensity is based on their joint emissions coefficients weighted by their energy consumption levels.

Carbon Monoxide (CO)

A colorless, odorless gas produced by incomplete combustion. Carbon monoxide is poisonous inhaled.

Carbon Output Rate

The amount of carbon by weight per kilowatt-hour of electricity produced.

Carbon Sequestration

Process through which carbon dioxide is removed from the atmosphere, for example in forests through the process of photosynthesis. During this process, carbon dioxide is taken up through plants' leaves and incorporated into the plants' wood biomass.

Carbon Sinks

Carbon reservoirs and conditions that take in and store more carbon (carbon sequestration) than they release. Carbon sinks can serve to partially offset greenhouse gas emissions. Forests and oceans are common carbon sinks.

Carnot Cycle

An ideal heat engine (conceived by Sadi Carnot) in which the sequence of operations forming the working cycle consists of isothermal expansion, adiabatic expansion, isothermal compression, and adiabatic compression back to its first state.

Cathode

The electrode at which reduction (a gain of electrons) occurs. For fuel cells and other galvanic cells, the cathode is the positive terminal; for electrolytic bells (where electrolysis occurs), the cathode is the negative terminal.

Cation

A positively charged ion.

Cellulose

The main carbohydrate in living plants. Cellulose forms the skeletal structure of the plant cell wall.

Cellulosic Ethanol

Ethanol is derived from cellulosic and him-cellulosic parts of biomass.

Chlorofluorocarbon (CFC)

Any of various compounds consisting of carbon, hydrogen, chlorine, and fluorine used as refrigerants. CFCs are now found to be harmful to the Earth's atmosphere.

Climate

The average weather (usually taken over a 30-year period) for a particular region and time. Climate is different from weather, but rather, it is the average pattern of weather for a particular region. Weather describes the short-term state of the atmosphere. Climatic elements include precipitation, temperature, humidity, sunshine, wind velocity, and phenomena such as fog, frost, and hailstorms, and other measures of the weather.

Climate Change

The term "climate change" is sometimes used to refer to all forms of climatic inconsistency, but because the Earth's climate is never static, the term is more effectively used to imply a significant change from one climatic condition to another. In some cases, climate change has been used synonymously with the term, global warming; scientists, however, tend to use the term in the wider sense to also include natural changes in the climate.

Climate Effects

Impact on residential space heating and cooling (kg CO_2/tree/year) from trees found greater than approximately 15 m (50 ft) from a building (Far trees) due to associated reductions in wind speeds and summer air temperatures.

Cloud Condensation Nuclei

Aerosol particles provide a platform for the condensation of water vapor, resulting in clouds with higher droplet concentrations and increased albedo.

Coarse Materials

Wood residues suitable for chipping, such as slabs, edgings, and trimmings.

Co-Firing

Practice of introducing biomass into the boilers of coal-fired power plants.

Cogeneration

The production of electrical energy and another form of useful energy (such as heat of steam) through the sequential use of energy.

Cogeneration System

A system using a common energy source to produce both electricity and steam for other uses, resulting in increased fuel efficiency.

Compressed Natural Gas (CNG)

Mixtures of hydrocarbon gases and vapors, consisting principally of methane in gaseous form that has been compressed.

Concentrating Solar Power or Solar Thermal Power System

A solar energy conversion system characterized by the optical concentration of solar rays through an arrangement of mirrors to generate an elevated temperature working fluid. Concentrating solar power (but not solar thermal power) may also refer to a system that focuses solar rays on a photovoltaic cell to increase conversion efficiency.

Concentrator

A reflective or refractive device that focuses incident insolation onto an area smaller than the reflective or refractive surface, resulting in increased insolation at the point of focus.

Conservation

Managing natural resources (includes preservation, restoration, and enhancement) to prevent loss or waste.

Conservation Program

A program in which a utility company gives home weatherization services free or at reduced cost or provides free or low-cost devices for saving energy, such as energy efficient light bulbs, flow restrictors, weather stripping, and water heater insulation.

Conservation Status

Assessment of the status of ecological processes and of the viability of species or populations in an ecoregion.

Conventional Oil and Natural Gas Production

Crude oil and natural gas are produced by a well drilled into a geologic formation in which the reservoir and fluid characteristics allow the oil and natural gas to readily flow to the wellbore.

Criteria Pollutant

A pollutant determined to be hazardous to human health and regulated under EPA's National Ambient Air Quality Standards. The 1970 amendments to the Clean Air Act required EPA to describe the health and welfare impacts of a pollutant as the "criteria" for inclusion in the regulatory regime.

Critical Habitat

According to US Federal law, the ecosystems upon which endangered and threatened species depend; specific geographic areas, whether occupied by a listed species or not, that are essential for its conservation and that have been formally appointed by rule published in the Federal register.

Crust

Earth's outer layer of rock. Also called the lithosphere.

DC

Direct current.

Deforestation

The net removal of trees from forested land.

Degasification System

The methods employed for removing methane from a coal seam that could not otherwise be removed by standard ventilation fans and thus would pose a substantial hazard to coal miners. These systems may be used prior to mining or during mining activities.

Degradable Organic Carbon

The part of organic carbon present in such solid wastes as paper, food waste, and yard waste that is susceptible to biochemical decomposition.

Demand Indicator

A measure of the number of energy-consuming units, or the amount of service or output, for which energy inputs are required.

Dependable Capacity

The load-carrying ability of a station or system under adverse conditions for a specified period.

Depleted Resources

Resources that have been mined include coal recovered, coal lost in mining, and coal reclassified as sub-economic because of mining.

Depletion Factor

Annual percentage of the depletion of the thermal resource.

Desulfurization

The removal of sulfur, as from molten metals, petroleum oil, or flue gases.

Diesel Fuel

A fuel composed of distillates obtained in petroleum-refining operation of blends of such distillates with residual oil used in motor vehicles. The boiling point and specific gravity are higher for diesel fuels than for gasoline.

Diversity Exchange

An exchange of ability or energy, or both, between systems whose peak loads occur at various times.

E-10

A mixture of 10% ethanol and 90% gasoline based on volume.

E-85

A mixture of 85% ethanol and 15% gasoline based on volume.

E-95

A fuel having a mixture of 95% ethanol and 5% gasoline.

Ecological System

Dynamic assemblages of communities that occur together on the landscape at some spatial scale of resolution are tied together by similar ecological processes, and form a cohesive, distinguishable unit on the ground. Examples are spruce-fir forest, Great Lakes dune and swale complex, and Mojave Desert riparian shrublands.

Ecosystem

A natural community of organisms interacting with its physical environment, regarded as a unit.

Efficiency

The ratio of the useful energy output of a machine or other energy converting plant to the energy input.

Electric Energy

The ability of an electric current to produce work, heat, light, or other forms of energy. It is measured in kilowatt-hours.

Electric Utility

A corporation, person, agency, authority, or other legal entity or instrumentality aligned with distribution facilities for delivery of electric energy for use primarily by the public. Included are investor-owned electric utilities, municipal and State utilities, Federal electric utilities, and rural electric cooperatives. A few entities that are tariff based and corporately aligned with companies that own distribution facilities are also included.

Emissions

Anthropogenic releases gases into the atmosphere. In the context of global climate change, they consist of radiatively important greenhouse gases (e.g., the release of carbon dioxide during fuel combustion).

Emissions Coefficient

A unique value for scaling emissions to activity data in terms of a standard rate of emissions per unit of activity (e.g., pounds of carbon dioxide emitted per Btu of fossil fuel consumed).

Emissions Factor

A rate of carbon dioxide output resulting from the consumption of electricity, natural gas, or any other fuel source.

Endothermic

A chemical reaction that absorbs or requires energy (usually in the form of heat).

Energy

The ability to do work. That is, the ability to do work as measured by the capability of doing work (potential energy) or the conversions of this capacity to motion (kinetic energy). Energy has several forms, some of which are easily convertible and can be changed to another form useful for work. Most of the world's convertible energy comes from fossil fuels that are burned to produce heat that is then used as a transfer medium to mechanical or other means to accomplish tasks. Electrical energy is usually measured in kilowatt-hours, while heat energy is usually measured in British thermal units (Btu).

Energy Efficiency

A ratio of service provided to energy input (e.g., lumens to watts in the case of light bulbs). Services supplied can include building-sector end uses such as lighting, refrigeration, and heating: industrial processes or vehicle transportation. Unlike conservation, which involves some reduction of service, energy efficiency supplies energy reductions without sacrifice of service.

Energy Loss

Deleted because there is no need for a general term to encompass all forms of energy loss. Terms referring to losses specific to energy sources are defined separately.

Energy Source

Any substance or natural phenomenon that can be consumed or transformed to supply heat or power. Examples include petroleum, coal, natural gas, nuclear, biomass, electricity, wind, sunlight, geothermal, water movement, and hydrogen in fuel cells.

Enthalpy

A thermodynamic property of a substance, defined as the sum of its internal energy plus the pressure of the substance times its volume, divided by the mechanical equivalent of heat. The total heat content of air; the sum of enthalpies of dry air and water vapor, per unit weight of dry air; measured in Btu per pound (or calories per kilogram).

Environment

The sum of all biological, chemical, and physical factors to which organisms are exposed.

Environmental Health (Abiotic Aspects)

The composition, structure, and functioning of soil, water, air, and other abiotic features comparable with historic conditions, including the natural abiotic processes that shape the environment.

Environmental Impact Statement

A document created form a study of the expected environmental effects of a new development or installation.

Environmental Restoration

Although usually described as "cleanup," this function encompasses a wide range of activities, such as stabilizing contaminated soil; treating groundwater; decommissioning process buildings, nuclear reactors, chemical separations plants, and many other facilities; and exhuming sludge and buried drums of waste.

Environmental Restrictions

In reference to coal accessibility, land-use restrictions that constrain, postpone, or prohibit mining to protect environmental resources of an area; for example, surface- or ground water quality, air quality affected by mining, or plants or animals or their habitats.

Ethane (C_2H_6)

A straight-chain saturated (paraffinic) hydrocarbon extracted predominately from the natural gas stream, which is gaseous at standard temperature and pressure. It is a colorless gas that boils at a temperature of −127°F.

Ethanol (Also Known as Ethyl Alcohol or Grain Alcohol, CH_3-CH_2OH)

A clear, colorless flammable oxygenated hydrocarbon with a boiling point of 173.5°F in the anhydrous state. However, it readily forms a binary azeotrope with water, with a boiling point of 172.67°F at a composition of 9,557% by weight ethanol. It is used in the United States as a gasoline octane enhancer and oxygenate (largest 10% concentration). Ethanol can be used in higher concentrations (E85) in vehicles designed for its use. Ethanol is typically produced chemically from ethylene, or biologically from fermentation of various sugars from carbohydrates found in agricultural crops and cellulosic residues from crops or wood. The lower heating value, equal to 76,000 Btu per gallon, is assumed for estimates in this text.

Ether

A generic term applied to a group of organic chemical compounds composed of carbon, hydrogen, and oxygen, characterized by an oxygen atom attached to two carbon atoms (e.g., methyl tertiary butyl ether).

Ethylene (C_2H_4)

An olefinic hydrocarbon recovered from refinery or petrochemical processes, which is gaseous at standard temperature and pressure. Ethylene is used as petrochemical feedstock for many chemical applications and the production of consumer goods.

EVAPOTRANSPIRATION (ET)

The joint evaporation from the soil surface and transpiration from plants. Transpiration is the evaporation of water from internal surfaces of living plant organics and its later diffusion into the atmosphere. Evaporation is the physical process by which liquid water is converted to vapor.

EXOTHERMIC

A chemical reaction that gives off heat.

FAHRENHEIT

A temperature scale on which the boiling pint of water is at 212° above zero on the scale, and the freezing point is at 32° above zero at standard atmospheric pressure.

FERMENTATION

Conversion of carbon-containing compounds by microorganisms for production of fuels and chemicals such as alcohols, acids, or energy-rich gases.

FISCHER-TROPSCH FUELS

Liquid hydrocarbon fuels are produced by a process that combines carbon monoxide and hydrogen. The process is used to convert coal, natural gas, and low-value refiner products into a high-value diesel substitute.

FLUX MATERIAL

A substance used to promote fusion, e.g., of materials or minerals.

FLYWAY

Any one of several established migration routes of birds.

FOREST LAND

Land at least 10% stocked by forest trees of any size, including land that formerly had such tree cover and that will be naturally or artificially regenerated. Forest land includes transition zones, such as areas between heavy-forested and non-forested lands that are at least 10% stocked with forest trees and forest areas next to urban and built-up lands. Also included are pinyon-juniper and chaparral areas in the West and afforested areas. The minimum area for classification of forest land is 1 acre. Roadside, streamside, and shelterbelt strips of trees must have a crown width of at least 120 ft to qualify as forest land. Unimproved roads and trails, streams, and clearings in the forest areas are classified as forest if they are less than 120 ft wide.

Fossil Fuels

A general term for combustible geologic deposits of carbon in reduced (organic) form and of biological origin, including coal, oil, natural gas, oil shales, and tar sands. A major concern is that they emit carbon dioxide into the atmosphere when burnt, thus significantly contributing to the enhanced greenhouse effect.

Fractionation

The process by which saturated hydrocarbons are removed from natural gas and separated into distinct products, or "fractions," such as propane, butane, and ethane.

Fuel Cell

One or more cells can generate an electrical current by converting the chemical energy of a fuel directly into electrical energy. Fuel cells differ from conventional electrical cells in that the active materials such as fuel and oxygen are not contained within the cell but are supplied from outside.

Fuel Cycle

The complete set of sequential processes or stages involved in the utilization of fuel, including extraction, transformation, transportation, and combustion. Emissions generally occur at each stage of the fuel cycle.

Fuel Ratio

The ratio of fixed carbon to volatile matter in coal.

Fumarole

A vent or hole in the Earth's surface, usually in a volcanic region, from which steam, gaseous vapors, or hot gases release.

Gallon

A volumetric measure equal to 4 quarts (231 cubic inches) is used to measure fuel oil. One gallon equals 3,785 L; One barrel equals 42 gallons.

Gas

A non-solid, non-liquid combustible energy source that includes natural gas, coke-oven gas, blast-furnace gas, and refinery gas.

Gasification

A chemical or heat process to convert a solid fuel to a gaseous form.

Gasohol

A motor vehicle fuel that is a blend of 90% unleaded gasoline and 10% ethanol (by volume).

Generation (Electricity)

The process of producing electric energy from other forms of energy; also, the amount of electric energy produced, expressed in watt-hours (Wh).

Geologic Hazards

A geologic hazard is one of several types of adverse geologic conditions capable of causing damage or loss of property and life. These hazards include the following:

- avalanches
- earthquakes
- forest fires
- geomagnetic storms
- ice jams
- landslide
- mudslide
- rock falls
- torrents
- volcanic eruptions
- alluvial fans
- geyser deposits
- liquefaction
- sand dune migration
- thermal springs
- stream erosion

Geographic Information Systems (GIS)

A computerized system to compile, store, analyze, and display geographically referred information.

Geologic Sequestration

A type of engineered sequestration, where captured carbon dioxide is injected for permanent storage into underground geologic reservoirs, such as oil and natural gas fields, saline aquifers, or abandoned coal mines.

Gigawatt (GW)

One billion watts or one thousand megawatts.

Gigawatt-Electric (GWe)

One billion watts of electric capacity.

Gigawatt-hour (GWh)

One billion watt-hours.

Global Positioning System (GPS)

A navigation system uses satellite signals to fix the location of a radio receiver on or above the Earth's surface.

Global Warming

An increase in the near surface temperature of the Earth. Global warming has occurred in the distant past because of natural influence (it is a cyclical event that has occurred throughout Earth's history), but the term is most often used to refer to the warming predicted to occur because of increased emissions of greenhouse gases from commercial or industrial resources.

Greenhouse Effect

The effect produced as greenhouse gases allow incoming solar radiation to pass through the Earth's atmosphere but prevent most of the outgoing infrared radiation from the surface and lower atmosphere form escaping into outer space.

Greenhouse Gas

Gases that trap the heat of the sun in the Earth's atmosphere, producing the greenhouse effect. The two major greenhouse gases are water vapor and carbon dioxide. Other greenhouse gases include methane, ozone, chlorofluorocarbons, and nitrous oxide.

Growing Stock

A classification of timber inventory that includes live trees of commercial species meeting specified standards of quality or vigor. Cull tress are excluded. When associated with volume, it includes only trees 5.0 inches in dbh. and larger.

Habitat

The place or type of site where species and species assemblages are typically found and/or successfully reproduce.

Habitat Conservation

Protecting an animal or plant habitat to ensure that the use of that habitat by the animal or plant is not altered or reduced.

Habitat Fragmentation

The breaking up of a specific habitat into smaller, unconnected areas.

Hardwoods

Usually broad-leaved and deciduous trees.

Heat Rate

A measure of generating station thermal efficiency commonly stated as Btu per kilowatt-hour. Note: Heat rates can be expressed as either gross or net heat rates, depending on whether the electricity output is gross or net generation. Heat rates are typically expressed as net heat rates.

Heating Value

The maximum amount of energy that is available from burning a substance.

Hydraulic Fracturing

Fracturing of rock at depth with fluid pressure. Hydraulic fracturing at depth may be accomplished by pumping water into a well at extremely high pressures. Under natural conditions, vapor pressure may rise high enough to cause fracturing in a process known as hydrothermal brecciation.

Hydrocarbon

An organic chemical compound of hydrogen and carbon in the gaseous, liquid, or solid phase. The molecular structure of hydrocarbon compounds varies from the simplest (methane, a constituent of natural gas) to the heaviest and most complex.

Idle Cropland

Land in cover and soil improvement crops, and cropland on which no crops were planted. Some cropland is idle each year for various physical and economic reasons. Acreage diverted from corps to soil-conserving uses (if not eligible for and used as cropland pasture) under Federal farm programs is included in this component. Cropland enrolled in the Federal Conservation Reserve Program (CRP) is included in the idle cropland.

Industrial Wood

All commercial roundwood products except Fuelwood.

Ion Exchange

Reversible exchange of ions adsorbed on a mineral or synthetic polymer surface with ions in solution in contact with the surface. A chemical process used for recovery of uranium from solution by the interchange of ions between a solution and a solid, commonly a resin.

Isobutane (C_4H_{10})

A branch-chain saturated (paraffinic) hydrocarbon extracted from both natural gas and refinery gas streams, which is gaseous at standard temperature and pressure. It is a colorless gas that boils at a temperature of 11°F.

Isobutylene (C_4H_8)

A branch-chain olefinic hydrocarbon recovered from refinery or petrochemical process, which is gaseous at standard temperature and pressure. Isobutylene is used in the production of gasoline and various petrochemical products.

Isohexane (C_6H_{14})

A saturated branch-chain hydrocarbon. It is a colorless liquid that boils at a temperature of 156.2°F.

Isomerization

A refining process that alters the fundamental arrangement of atoms in the molecule without adding or removing anything from the original material. Used to convert normal butane into isobutane (C_4), an alkylation process feedstock, and normal pentane and hexane into isopentane (C_5) and isohexane (C_6), high-octane gasoline components.

Joule

This is the basic energy unit for the metric system, or in a later more comprehensive formulation, the International System of Units (SI). It is ultimately defined in terms of the meter, kilogram, and second.

kBtu

A unit of work or energy, measured as 1,000 British thermal units. One kBtu is equivalent to 0.293 kWh.

Kilowatt (kW)

One thousand watts of electricity.

Kilowatt-hour (kWh)

One thousand watt-hours.

Kinetic Energy

Energy available because of motion that varies directly in proportion to an object's mass and the square of its velocity.

Low-Speed Shaft

Connects the rotor to the gearbox.

Mantle

The Earth's inner layer of molten rock, lying beneath the Earth's crust and above the Earth's core of liquid iron and nickel.

Megawatt (WM)

One million watts of electricity.

Methane

A colorless, flammable, odorless hydrocarbon gas (CH_4) which is the major component of natural gas. It is also an important source of hydrogen in various industrial processes. Methane is a greenhouse gas.

Methanogens

Bacteria that synthesize methane required completely anaerobic conditions for growth.

Methanol

Also known as methyl alcohol or wood alcohol, having the chemical formula CH_3OH. Methanol is usually produced by chemical conversion at elevated temperature and pressure. Although usually produced from natural gas, methanol can be produced from gasified biomass.

Mineral

Any of the various naturally occurring organic substances, such as metals, salt, sand, stone sulfur, and water, usually obtained from the Earth. Note: For reporting on the Financial Reporting System, the term also includes organic non-renewable substances that are extracted from the Earth such as coal, crude oil, and natural gas.

Mole

The quantity of a compound or element that has a weight in grams numerically equal to its molecular weight. Also referred to as "gram molecule" or "gram molecular weight."

Naphtha

A generic term applied to a petroleum fraction with an approximate boiling range between 122°F and 400°F.

Natural Gas

A gaseous mixture of hydrocarbon compounds, the primary one being methane.

Natural Processes

A complex mix of interactions among animals, plants, and their environment that ensures maintenance of an ecosystem's full range of biodiversity. Examples include population and predator-prey dynamics, pollination and seed dispersal, nutrient cycling, migration, and dispersal.

Natural Sinks

In reference to greenhouse gases, it refers to any natural process in which these gases are absorbed from the atmosphere.

Niche

The specific part or smallest unit of a habitat occupied by an organism.

Peat

Peat consists of partially decomposed plant debris. It is considered an early stage in the development of coal. Peat is distinguished from lignite by the presence of free cellulose and a high moisture content (exceeding 70%). The heat content of air-dried peat (about 50% moisture) is about 9 million Btu per ton. Most US peat is used as a soil conditioner.

Permeability

The ability of a rock to transmit fluid through its pores or fractures when subjected to a difference in pressure. Typically, a measure in darcies or millidarcies.

Petrochemical Feedstocks

Chemical feedstocks delivered from petroleum principally for the manufacture of chemicals, synthetic rubber, and a variety of plastics.

Photosynthesis

The manufacture by plants of carbohydrates and oxygen for carbon dioxide and water in the presence of chlorophyll, with sunlight as the energy source. Carbon is sequestered, and oxygen and water vapor are released in the process.

Planetary Albedo

The fraction of incident solar radiation that is reflected by the Earth-atmosphere system and returned to space, mostly by back scatter from clouds in the atmosphere.

Prescribed Fire

The application of fire to wildland fuels, either by natural or intentional ignition, to achieve identified land-use objectives.

Propane (C_3H_8)

A straight-chain saturated (paraffinic) hydrocarbon extracted from natural gas or refinery gas steams, which is gaseous at standard temperature and pressure.

Propylene (C_3H_6)

An olefinic hydrocarbon recovered from refinery or petrochemical processes, which is gaseous at standard temperature and pressure. Propylene is an important petrochemical feedstock.

Pyrolysis

The thermal decomposition of biomass at elevated temperatures (great than 400°F, or 200°C) in the absence of air. The product of pyrolysis is a mixture of solids (char), liquids (oxygenated oils), and gases (methane, carbon monoxide, and carbon dioxide) with the proportions determined by operating temperature, pressure, oxygen content, and other conditions.

Quadrillion Btu (Quad)

Equivalent to 10th to the 15th power Btu.

Radiant Energy

Energy that transmits away from its source in all directions.

Reclamation

Process of restoring surface environment to acceptable pre-existing conditions. It includes surface contouring, equipment removal, well plugging, and revegetation.

Recycling

The process of converting materials that are no longer useful as designed or intended into a new product.

Renewable Energy

Energy that is produced using resources which regenerate quickly or are inexhaustible. Wind energy is considered inexhaustible, because, while it may blow intermittently, it will never stop.

Renewable Energy Resources

Energy resources that are naturally replenishing but flow limited. They are virtually inexhaustible in duration but limited in the amount of energy that is available per unit of time. Renewable energy resources include biomass, hydro, geothermal, solar, wind, ocean thermal, wave action, and tidal action.

Riparian

Referring to the interface between freshwater habitats and the terrestrial landscape.

Sequestration

In nature, the annual net rate that a tree removes carbon dioxide from the atmosphere through the processes of photosynthesis and respiration (kgCO_2/tree/year).

Terrestrial Sequestration

Basic sequestration of carbon in above- and below-ground biomass and soils.

Watt (Electric)

The electrical unit of power. The rate of energy transfer is equivalent to 1 A of electric current flowing under a pressure of 1 V at unity power factor.

Watt (Thermal)

A unit of power in the metric system, expressed in terms of energy per second, is equal to the work done at a rate of 1 J/s.

NOT EXACTLY THE FINAL WORD ON CARBON DIOXIDE

In 1949, G.S. Callendar published an article titled, *Can Carbon Dioxide Influence Climate*? He points out that in its simplest terms, carbon dioxide is almost completely transparent to solar radiation, but it is particularly opaque to the heat which is radiated back to space from the Earth. Because of this, it acts as a "heat trap, allowing the temperature near the Earth's surface to rise above the level it would attain if there were no carbon dioxide in the air."

As the measurements from Mauna Loa show, despite decades of negotiation, the global community has been unable to meaningfully slow, let alone reverse, the annual increase in atmospheric carbon dioxide levels (NOAA, 2021).

The bottom line: this is not the final word on carbon dioxide in this book and anywhere else, for that matter.

REFERENCES

Brigham-Grette, J. et al. (2013). Pliocene warmth, polar amplification, and stepped pleistocene cooing recorded in NE Arctic Russia. *Science* 340:1421.

Callendar, G.S. (1949). Can carbon dioxide influence climate? *Weather* 4(10):310–314.

Mathews, J. (2016). Naples: Life, Death & Miracles: Agnano & and the Grotto of the Dog. Accessed 3/20/23 @ https://www.naplesidm.com/agnano.html.

Nature (2019). Constraints on Global Mean Sea Level during Pliocene Warmth. Accessed 3/21/23 @ https://www.nature.com/articles/s41580-019-1543-2.

NOAA (2021). Carbon Dioxide Peaks Near 420 Parts Per Million at Mauna Loa Observatory. Accessed 3/21/23 @ https://gml.noaa.gov/obop/mio/

Twain, M. (1869). Innocents Abroad. Accessed 3/20/23 @ https://www.gutenberg.org/files/3176/3176-h/3176-h.htm#ch30.

Young, T.G. (1878). The Gas of the Grotto del Cane. Accessed 2/23/23 @ https://babel.hathitrust.org/cgiptid=hvd.hx1drj&view=1up&seq=71&skin=2021.

2 Carbon Cycle

LET'S TALK ABOUT CARBON: THE KING OF ELEMENTS

It goes without any hyperbole in stating that carbon is one of the most essential element for all living organisms—it is the chemical backbone of line on Earth. As shown in Figure 2.1, carbon is an element with atomic number 6 and element symbol C.

Chemistry is said to be the filter that eliminates many students from pursuing degrees not only in chemistry but also in the sciences. Having taught chemistry and organic chemistry I can say without hesitation that chemistry is indeed a filter but if the students squeeze by basic college chemistry and they next take organic chemistry it is the absolute filter for those who can't comprehend the concepts—the synthesis. Like algebra, a lot of organic chemistry is in the abstract. Either you understand the concepts, or you do not.

The bottom line: Organic chemistry is a rigorous and demanding workload.

Anyway, about organic chemistry it can be accurately said that carbon is the basis for organic chemistry as it occurs in all living organisms—all living organisms. The simplest organic molecules consist of carbon chemically bonded to hydrogen. For example, the carbon-hydrogen bond (C–H bond) is a common chemical bond between carbon and hydrogen atoms found in many organic compounds. This bond shares its outer valence electrons with up to four hydrogens—a covalent, single bond. Some of the most common carbon-hydrogen bonds include C_2H_6 (Ethane), C_2H_4 (Ethylene), and C_2H_2 (Acetylene).

By the way, there are other organics besides carbon—oxygen, nitrogen, phosphorus, and sulfur that come to mind. But it is carbon that has the highest melting point of all the elements. It also has the highest sublimation point of the elements—meaning carbon transitions directly from the solid to the gaseous state, without passing through the liquid state; its sublimation point is around 3,800°C.

Carbon, a non-metal, is often called the "King of the Elements" because it forms more compounds than any other element—millions of compounds.

Carbon compounds have limitless uses and forms. It can be soft like graphite and extremely hard like diamonds. A few examples of different forms of carbon—graphite—are shown in Figure 2.2.

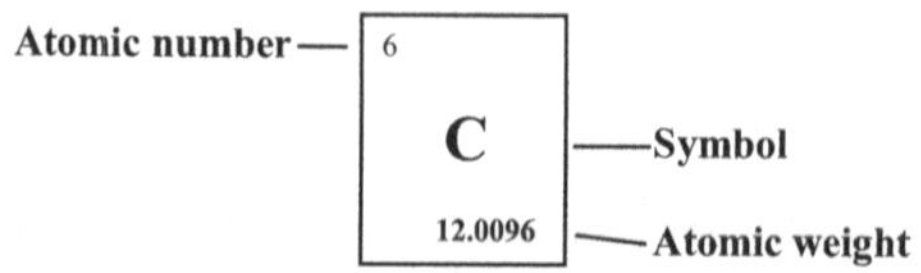

FIGURE 2.1 Symbol for the element carbon.

DOI: 10.1201/9781003432838-3

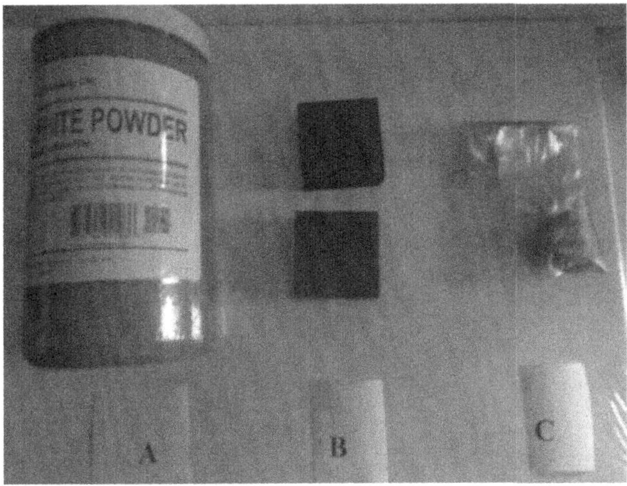

FIGURE 2.2 Shows examples of different forms of graphite including (a) raw graphite, a pure carbon specimen; (b) graphite ingot blocks of 99.9% pure graphite; and (c) pure graphite in powder form.

In addition to its various forms, carbon has almost limitless uses. Along with diamonds used for drilling and cutting, carbon in the form of graphite is used in pencils, a protector against rust and as a lubricant and as a filter to remove toxins, odors, and tastes, and to trap grease (e.g., vent filter system used above some kitchen stoves).

Note that there is nothing new about knowledge of carbon. Dating back to the era of cave persons, amorphous carbon (soot) was certainly apparent to cave dwellers when their caves became covered in soot from there fires.

On Earth, the amount of carbon is constant. It is transformed from one form to another via the carbon cycle, which is discussed in greater detail in the following section.

Note: In Chapter 1, you were introduced to the carbon and carbon dioxide cycle. In this chapter, we take a close look at the important biogeochemical cycle. In Figure 2.3 is shown a simplified carbon cycle which is commonly used to provide the fundamental operation of the cycle.

DID YOU KNOW?

The ability of carbon to form chains and rings is known as catenation; therefore, many carbon compounds contain rings or long chains or form polymers.

CARBON COMPOUNDS, ALLOYS, AND BONDS

The element carbon has friends; some of them are good friends. Carbon is present in several carbon chemical compounds. Note that inorganic carbon compounds also exist. Examples include the allotropes (different forms of a pure element) of carbon

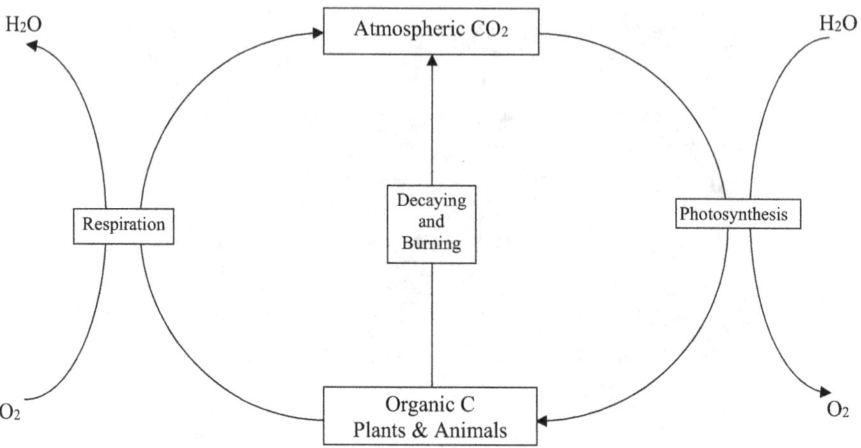

FIGURE 2.3 Simplified carbon cycle. This simple diagram also points out, except for volcanic outbreaks like the fumarole in the Cave of the Dogs, the sources of carbon including decay of organisms, exhalation by living beings, agriculture (plants and decay) of land, and combustion of fuels (burning).

(graphite, diamond, and buckminsterfullerene), carbon monoxide, carbon dioxide, carbides, and the following salts of inorganic anions: carbonates, cyanides, cyanates, and thiocyanates. Some examples of carbon compounds include the following:

- carbon dioxide (CO_2)
- glucose ($C_6H_{12}O_6$)
- methane (CH_4)
- carbon tetrachloride (CF_4)
- tetraethyl lead [$(CH_3CH_2)_4Pb$]

Carbon compounds may be classified as organic, organometallic, or inorganic:

- **Organic compounds:** these compounds always contain carbon and hydrogen. Proteins, lipids, carbohydrates, and nucleic acids are major classes of organic compounds. Although it is possible to synthesize them in the lab, usually, organic compounds exist in living organisms.
- **Organometallic compounds:** these compounds, such as ferrocene, tetraethyl lead, and Zeise's salt (aka potassium trichloro(ethylene) platinate (II) hydrate), contain at least one carbon-metal bond.
- **Inorganic carbon compounds:** these compounds contain carbon but not hydrogen. Inorganic compounds exist in minerals and gases. Examples include calcium carbonate ($CaCO_3$), carbon dioxide (CO_2), and carbon monoxide (CO).

Regarding alloys containing carbon, there are several of them including steel and cast iron. Partial carbon alloys, those smelted using coke, include zinc, aluminum, and chromium.

Note that carbon tends to form covalent chemical bonds with itself and other types of atoms. When carbon bonds to other carbon atoms, non-polar covalent bonds are formed. When carbon bonds to non-metals or metalloids, polar covalent bonds are formed.

DID YOU KNOW?

Many carbon compounds are combustible, but most carbon compounds have low reactivity at room temperature. However, even with their low reactivity at room temperature, they can react violently when heated, and this is plainly clear when you consider fuels that are stable until heated. Moreover, carbon compounds with nitrogen are often explosive. Unstable compounds of carbon are bonded between the atoms, are unstable, and release considerable energy when broken.

BIOGEOCHEMICAL CYCLES

Several chemicals are essential to life and follow predictable cycles through nature. In these natural cycles or *biogeochemical cycles*, the chemicals are converted from one form to another as they progress through the environment. The carbon cycle is a biogeochemical cycle—obviously, an especially important cycle.

Note: Smith (1983) categorizes biogeochemical cycles into two types: the *gaseous* and the *sedimentary*. Gaseous cycles include the carbon and nitrogen cycles. The main sink of nutrients in the gaseous cycle is the atmosphere and the ocean. Sedimentary cycles include the sulfur cycle. The main sink for sedimentary cycles is soil and rocks of the Earth's crust.

411 ON THE CARBON CYCLE

Carbon, which is an essential ingredient of all living things, is the basic building block of the large organic molecules necessary for life. Carbon is cycled into food chains from the atmosphere, as shown in Figures 2.1 and 2.3. From these figures, green plants obtain carbon dioxide (CO_2) from air and, photosynthesis, described by Asimov (1989) as the "most important chemical process on Earth," produces the food and oxygen that all organisms live by. Part of the carbon produced remains in living matter; the other part is released as CO_2 in cellular respiration. Miller (1988) points out that the carbon dioxide released by cellular respiration in all living organisms is returned to the atmosphere.

Some carbon is contained in buried dead and animal and plant materials. Many of these buried animal and plant materials were transformed into fossil fuels. Fossil fuels, coal, oil, and natural gas contain copious amounts of carbon. When fossil fuels are burned, stored carbon combines with oxygen in the air to form carbon dioxide, which enters the atmosphere. In the atmosphere, carbon dioxide acts as a beneficial heat screen as it does not allow the radiation of Earth's heat into space. This balance is important. The problem is that as more carbon dioxide from burning is released into the atmosphere, the balance can be and is being altered. Odum (1983) warns that the recent increase in consumption of fossil fuels "coupled with the decrease in "removal capacity" of the green belt is beginning to exceed the delicate balance." Massive increases of carbon dioxide into the atmosphere tend to increase the possibility of global warming. The consequences of global warming "would be catastrophic ... and the resulting climatic change would be irreversible" (Abrahamson, 1988).

The Slow Carbon Cycle

Holli Riebeek (2011) points out that carbon takes 100–200 million years to move between rocks, soil, ocean, and atmosphere in the slow carbon cycle. On average, 10–100 million metric tons of carbon move through the slow carbon cycle every year. Note that human emissions of carbon, in comparison with natural emissions, are about 10^{15} g. However, in the fast carbon cycle moves 10^{16}–10^{17} g of carbon per year.

Carbon moves from the atmosphere to the rocks (lithosphere) beginning with precipitation. What happens is the atmospheric carbon combines with water (rain) to form a weak acid—carbonic acid—that falls to the surface. When the carbonic acid precipitation hits the surface, chemical weathering takes place whereby rocks are dissolved and this action releases calcium, sodium, magnesium, or potassium ions. All water eventually finds its way to the oceans either by runoff in rivers or seeping groundwater or by evaporation that creates precipitation and much of it is deposited, many by rivers, into the oceans. Note that the rivers carry calcium ions into the ocean, where they react with carbonate dissolved in the water. As a point of interest, that reaction produces calcium carbonate, which is then deposition onto the ocean floor, where it becomes limestone. By the way, you may be familiar with calcium carbonate; it is that chalky white substance that dries on your facet if you live in a hard water area. In today's ocean, most of the calcium carbonate is made by calcifying (shell-building) organisms (corals) and coccolithophores and foraminifera (plankton). Upon their deaths, they sink to the ocean floor. Given the right amount of time, layers of shells and sediment are cemented together and turn to rock, storing the carbon in limestone and its derivatives.

DID YOU KNOW?

Limestone, or its metamorphic cousin, marble, is rock made primarily of calcium carbonate. These rock types are formed from the bodies of marine animals and plants, and their shells and skeletons can be preserved as fossils (Riebeek, 2011).

Note all carbon-containing rock is made this way. Some of the rock is made from the carbon in living things (organic carbon) that have been embedded in layers of mud. Over millions of years, the heat and pressure compress the mud, forming sedimentary shale rock.

In specific circumstances, when a plethora dead plant matter builds faster than it can break down during decay, layers of organic carbon become hydrocarbons such as oil, coal, or natural gas instead of forming sedimentary shale rock.

The mechanisms involved in the slow cycle return carbon to the atmosphere via volcanic activity. Note that land and ocean surfaces on Earth sit on several moving crustal plates. When these plates collide, and one sinks beneath the other, the rock it carries melts under the extreme heat and pressure. When the heated rock recombines into minerals such as silicate, carbon dioxide is released.

The cycle begins again anytime volcanoes erupt and vet the gas to the atmosphere and cover the land with fresh silicate rock. Volcanoes emit millions of tons of carbon dioxide per year. Sounds like a lot compared to the amount of carbon dioxide emitted by humans who emit 100–300 times more than volcanoes—by burning fossil fuels.

The waltz between ocean, land, and atmosphere is regulated by chemistry. The chemical weathering involved with carbon dioxide emission into the atmosphere because of volcanic activity makes temperatures rise, leading to more rain, which in turn leads to more dissolution of rock, and eventually creating more ions that will eventually deposit more carbon on the ocean floor. But again, this dance does not occur quickly.

There is one partner, however, that is a faster component of the slow carbon cycle: the ocean. Where air meets water (at the surface, of course), carbon dioxide gas dissolves in and ventilates out of the ocean in a steady give-and-take with the atmosphere. After arrival in the ocean, carbon dioxide gas reacts with water molecules to release hydrogen, making the ocean more acidic. The product created from this action is bicarbonate ions.

Before the Industrial Revolution, there was a balance between the vented carbon dioxide and the atmosphere with the carbon the ocean received during the weathering of rock. Here's the problem today: since carbon concentrations in the atmosphere have increased, the ocean now takes more carbon from the atmosphere that it releases. Over ages, the ocean will absorb up to 85% of the extra carbon people have put into the atmosphere by burning fossil fuel, however, the process is slow because it is tied to ocean water circulation—the movement of water from the ocean's surface to its depths.

Meanwhile, the rate at which the ocean takes carbon dioxide from the atmosphere is controlled by the winds, currents, and temperature. It is conceivable that changes in ocean currents and temperatures helped remove carbon from and restore carbon to the atmosphere over the few thousand years in which the Ice Age began and ended (Riebeek, 2011).

The Fast Carbon Cycle

We measure the time it takes for carbon to move through the fast carbon cycle in a lifespan. The fast carbon cycle is mainly the passage of carbon through life forms on Earth or the biosphere. Between 1,000 and 100,000 million metric tons (10^{15} and 10^{17}g) of carbon move through the fast carbon cycle every year (Riebeek, 2011).

What makes carbon important in its role in biology is its ability to form many bonds—up to four per atom—in what is felt to be an endless variety of complex organic molecules. Several organic molecules contain carbon atoms that have formed strong bonds to other carbon atoms, combining into long chains and rings.

Because of the substantial number of organic compounds, a proper systematic classification was needed. Thus, organic compounds can be broadly classified as acyclic (open chain) or cyclic (closed chain).

Okay, what is the difference between the open-chain and closed-chain organic compounds? Acyclic or open-chain compounds (aka aliphatic compounds) are also called straight-chain compounds such as ethane, isobutane, acetic acid, and acetaldehyde. Closed-chain organic compounds are characterized by moving in cycles or occurring at regular intervals and having chains of atoms arranged in a ring.

Let's get back to the fast carbon cycle.

The main components of the fast carbon cycle are plants and phytoplankton. The microscopic organisms in the ocean (phytoplankton) and plants take up carbon dioxide from the atmosphere by absorbing their cells. Both plants and animals use energy from the Sun in combining carbon dioxide and water to form sugar (CH_2O) and oxygen. The chemical reaction is as follows:

$$CO_2 + H_2O + \text{energy} = CH_2 + O_2$$

To move carbon from a plant and return it to the atmosphere, four things can happen, all of which involve the same chemical reaction. Plants break down sugar to get the energy they need to grow. Both animals and people eat the plants or plankton and break down the plant sugar to get energy. Plants and animals die, decay, or are eaten by bacteria at the end of the growing season. Also, plants can be consumed by fire. In each case, sugar is combined with oxygen to release water, carbon dioxide, and energy. This reaction is shown as follows:

$$CH_2O + O_2 = CO_2\ H_2O + \text{energy}$$

Note that in all the four process, carbon dioxide is released and usually ends up in the atmosphere. Also note that the fast carbon cycle is so tightly bonded to plant life that the growing seas can be seen by the way carbon dioxide fluctuates in the atmosphere. Atmospheric carbon dioxide concentrations clime in the Northern Hemisphere during winter. When you think about this, it makes sense because during winter, there are a few land plants growing and many of them are decaying; thus, there is a cyclical decline in carbon dioxide removal from the atmosphere. In contrast, during the spring, when plants begin growing again, concentration drops.

During the changing of seasons is when the ebb and flow of the fast carbon cycle are visible. Carbon is drawn out of the atmosphere when the Northern Hemisphere land masses turn green. This cycle peaks in August, with about 2 parts per million of carbon dioxide being drawn out of the atmosphere. As vegetation dies back in the fall and winter in the Northern Hemisphere decomposition and respiration return carbon dioxide to the atmosphere.

Left undisturbed and calm, the fast and slow carbon cycles maintain a relatively steady concentration of carbon in the atmosphere, land, plants, and ocean. However, whenever there is disturbance where anything changes the amount of carbon in one reservoir, the effect is felt through the others.

"Any time there is disturbance …?"

Yes. Consider, for example, that the carbon cycle has changed in response to climate change—whatever the disturbance (the causal factor) was. In the Earth's past, variations in Earth's orbit alter the amount of energy Earth receives from the Sun and leads to a cycle of ice ages and warm periods like our current climate. Speculation aside, what are the most probable causes of ice ages on Earth? According to the *Milankovitch hypothesis*, Ice Age occurrences are governed by a combination of factors: (1) the Earth's change of altitude in relation to the Sun (the way it tilts in a 41,000-year cycle and at the same time wobbles on its axis in a 22,000-year cycle), making the time of its closest approach to the Sun come at different seasons; and (2) the 92,000-year cycle of eccentricity in its orbit round the Sun, changing it from an elliptical to a near circular orbit, the most severe period of an Ice Age coinciding with the approach to circularity.

So is this hypothesis real, have meaning?

We have a lot of speculation about ice ages and their causes and their effects. This is the bottom line. We know that ice ages occurred—we know that they caused certain things to occur (e.g., formation of the Great Lakes), and although there is a lot we do not know, we recognize the possibility of recurrent ice ages. Lots of possibilities exist. Right now, no single theory is sound, and doubtless many factors are involved. Keep in mind that the possibility does exist that we are still in the Pleistocene Ice Age. It may reach another maximum in another 60,000 plus years or so.

One thing seems certain; the Earth's climate is changing. Our climate is changing based on observations, measurements, and historical comparisons. The evidence seems clear:

- changing temperature and precipitation factors (Vose et al., 2017; IPCC, 2013).
- increases in ocean temperatures, sea level, and acidity (EPA, 2023).
- melting of glaciers and sea ice (Hayhoe et al., 2018).
- changes in the frequency, intensity, and duration of extreme weather events (EPA, 2023).
- shifts in ecosystem characteristics, like the length of growing season, timing of blooming flowers, and migration of birds (EPA, 2023).

Additionally, we must consider that several factors including cooler temperatures and increased phytoplankton (algae) growth may have increased the amount of carbon the ocean took out of the atmosphere.

DID YOU KNOW?

Harmful algae blooms (HABs) are events involving the proliferation of toxicity of otherwise harmful phytoplankton. The events may occur naturally or may be the result of human activity (e.g., nutrient runoff). In the US, HABs frequently cause shellfish bed closures due to concerns over health risks associated with consumption of contaminated shellfish. A study conducted by researchers at the Woods Hole Oceanographic Institute found that HABs result in average annual commercial fishing losses of $18 million. Apart from this long-term average, HABs can result in acute losses to separate local fisheries (Spellman, 2020).

Another thing seems certain; when the Northern Hemisphere summers cooled, and the ice built up on land the carbon cycle slowed.

The increased temperatures and algae growth caused a drop in atmospheric carbon which caused additional cooling. Likewise, at the end of the last Ice Age, ~12,000–10,000 years ago, carbon dioxide in the atmosphere rose dramatically as the temperature warmed.

Scientists and researchers have been able to trace levels of carbon dioxide and Antarctic temperatures over the past 800,000 years. Note that from the geological record, we have discovered that temperature variations were touched off by variations in Earth's orbit, which increased global temperatures and released carbon dioxide into the atmosphere, which in turn warmed the Earth.

DID YOU KNOW?

Shifts in Earth's orbit are on-going and never-ending, in predictable cycles. In about 30,000 years, Earth's orbit will have changed enough to reduce sunlight in the Northern Hemisphere to the levels that led to the Ice Age (Reibeek, 2011).

In addition to Earth's on-going orbital changes, there are also changes occurring in the carbon cycle. The difference is that orbital changes are natural occurrences whereas changes to the carbon cycle are happening because of what we, the people, are doing. We disturb the carbon cycle by clearing land and burning fossil fuels (see Figure 2.4).

When pointing this out to one of my combined upper-level grad student college classes in a course called the Science of Environmental Pollution, one of my enterprising and supersmart students raised her hand and asked: "If fossil fuels are so bad for our environment and maybe to human beings in general then why did our creator whomever that might make fossil fuel part of the makeup of Earth. Also, since we have the oil, gas, and coal available to us for use, if we do not use any of the fossil

FIGURE 2.4 The burning of fossil fuels is the primary source of increased carbon dioxide in the atmosphere today. Photograph by F.R. Spellman.

fuels, what are we supposed to do with them? Do we simply ignore that? They exist and that their potential provides us with the energy we need now and in the future. Do we just let them rot?"

After asking those questions the only sound in that packed classroom was from those who were breathing and not and not holding their breath.

So, you are wondering how I replied to the student's questions about fossil fuels and the rest.

I said, "Good questions … and I want all of you in this course to think about her questions and formulate in your minds your answers. And, oh, by the way, your next assignment is to write what you think about her questions and your answers to them and put all that into a written classroom assignment of no more than 250 words … due two weeks from tonight." This was the only bailout, so to speak, for me because often I can't answer the unanswerable. This is not a "let the students eat cake" scenario; it is a safe bailout with parachute for me and makes the students open their minds. Is this not the purpose of education—to open one's mind and to make them think?

Case Study 2.1 Turn of the Gas Stoves!

At this time in March 2023, you may have heard the reports, the suggestions, the dire warnings—the hysteria—about gas stoves used by many in kitchens to cook food (see Figure 2.4). The Consumer Product Safety Commission is studying the safety hazards from indoor air pollution by gas stoves. The hysteria reported on social media and cable news programs is about the US Government possibly outlawing the use of gas stoves. While it is true that there are some health concerns related to the use of gas stoves, it is unlikely that any government agency is going to storm our homes and forcefully remove the stoves as I have heard a few of my former students suggest might be the case.

Anyway, let's get back to what we are presently doing, besides using our gas stoves to cook our food, which is changing the carbon cycle.

When we clear land, especially forests, we remove a dense growth of plants that had stored carbon in biomass—wood, stems, and leaves. Trees take carbon out of the atmosphere as they grow, and when we clear the forest we eliminate this vital function—we eliminate plants that would otherwise take carbon out of the atmosphere as they grow. Our tendency is to replace the dense growth with crops or pastures, which store less carbon. This tendency is also negative in that we expose soil that vents carbon from decayed plant matter into the atmosphere. These land use changes are currently emitting almost a billion tons of carbon into the atmosphere per year.

Note that without human meddling, the carbon in fossil fuels would leak slowly into the atmosphere through volcanic activity over millions of years in the slow carbon cycle. Carbon stored for millions of years is released into the atmosphere every year when we burn coal, oil, and natural gas. When we cause this acceleration of carbon release to the atmosphere, we move the carbon from the slow cycle to the fast cycle (see Figure 2.5). In 2014, 35 billion metric tons of carbon dioxide were released into the atmosphere due to the usage of fossil fuels (Conlen, 2023).

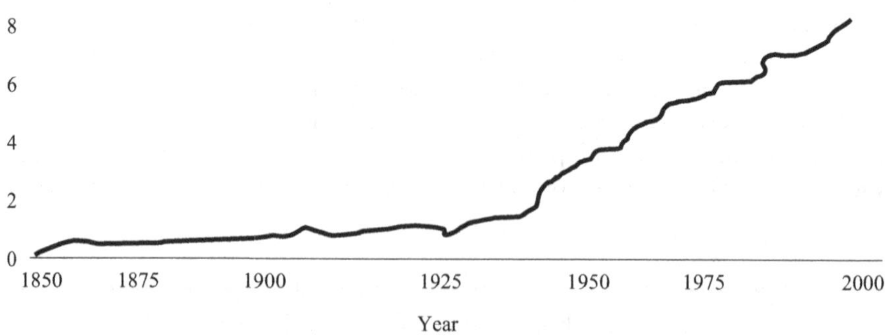

FIGURE 2.5 Emissions of carbon dioxide by humanity—primarily from the burning of fossil fuels, with a contribution from cement production—have been growing steadily since the onset of the Industrial Revolution. About half of these emissions are removed by the fast carbon cycle each year, and the rest of them remain in the atmosphere. Data for the graph based on USEPA, 2023 data.

DID YOU KNOW?

It was at the beginning of the Industrial Revolution, when humans first started during fossil fuels, carbon dioxide concentrations in the atmosphere have risen from about 280 to 387 ppm, almost a 40% increase. If you are wondering what this means, well, what it means is that for every million molecules in the atmosphere, 387 of them are now carbon dioxide—this is the highest concentration in two million years. From 1750 methane was at a concentration of 715–1,774 ppb in 2005, was the highest reading, at that time (Riebeek, 2011).

Case Study 2.2 Cement Production

It always came up; someone always raised the question. And to be honest about it, I remember being a bit unnerved waiting—for the question—until someone asked the question. Usually it came from one or two of the few students who had not completed their Junior-/Senior-level environmental chemistry courses (we offered two diverse types).

Anyway, the question: "Does cement really contribute to greenhouse gas?"

"Yes," I always replied, but also added that this mostly occurs in the production of cement and then provided the following explanation provided by USEPA (2019).

Cement plants operate elevated temperature kilns to produce clinkers, which are ground and blended with other materials to make cement, the bonding agent in cement (see Figure 2.6). Cement manufacturing is energy intensive and a major source of greenhouse gas (GHG) from the industrial sector (USEPA, 2019).

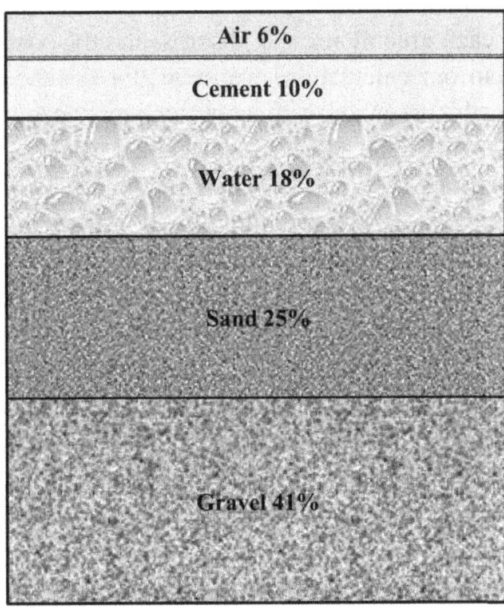

FIGURE 2.6 Typical cement concrete. Adapted from Smit (2014). *Introduction to Carbon Capture and Sequestration*. London: Imperial College Press.

It Must Go Somewhere

Additionally, liberated carbon must go somewhere. Well, in this regard we know that approximately 55% of the extra carbon is taken up by land plants and the ocean, but at the same time, 45% of the carbon stays in the atmosphere. Over time, the land and oceans will take up most of the extra carbon dioxide, but as much as 20% may remain in the atmosphere for many thousands of years.

The changes in the carbon cycle impact each reservoir—deep ocean, atmosphere, surface ocean, marine sediments, and the terrestrial biosphere. The excess carbon is problematic in two significant ways: excess carbon in the atmosphere warms the planet and helps plants on land grow more; also, excess carbon in the ocean makes the water more acidic, putting marine life in danger.

Let's take a closer look at the three main carbon reservoirs: atmosphere, ocean, and land.

Atmosphere Reservoir

Carbon dioxide in the atmosphere is the most important gas for controlling Earth's temperature. Carbon dioxide along with methane and halocarbons are greenhouse gases that absorb a wide energy including heat (infrared energy) emitted by Earth and then re-emitted. This re-emitted energy travels out in all directions, but a portion returns to Earth and heats the surface. Without greenhouse gases, Earth would be a frozen orb. With too many greenhouse gases, Earth would be the planet Venus where the greenhouse atmosphere keeps temperatures around 750°F (400°C)—ideally, a Goldilocks situation is preferred on Earth where greenhouse gases are not too thick or too thin, but exactly right.

We can calculate how much each gas contributes to warming the planet; it is based on wavelengths of energy each greenhouse gas absorbs, and the concentration of the gases in the atmosphere. From our calculations we know, for instance, that carbon dioxide causes about 20% of Earth's greenhouse effect; water vapor accounts for about 50%; and clouds account for 25%—methane and aerosols (small particles) cause the rest.

How about water vapor? Well, water vapor in the atmosphere and air are controlled by Earth's temperature. When temperatures are warmer more water evaporates from the oceans, air masses expand, and humidity increases. When water vapor is cooled, it condenses and falls out as rain, sleet, or snow.

Carbon dioxide, on the other hand, does not evaporate and remain as a gas at a wider range of atmospheric temperatures. It is the carbon dioxide molecule that provides the initial greenhouse heating needed to maintain water vapor concentrations. The Earth cools when carbon dioxide concentrations drop and some water vapor falls out of the atmosphere, and the greenhouse warming caused by water vapor drops. Similarly, when carbon dioxide concentrations rise, air temperatures too rise, and more water vapor evaporates into the atmosphere—this amplifies greenhouse heating.

Accordingly, while carbon dioxide contributes less to the overall greenhouse effect than water vapor, researchers have found that carbon dioxide is the gas that sets the temperature. It can be said that carbon dioxide is the controller that operates the greenhouse effect. Consider, for example, that carbon dioxide controls the amount of water vapor in the atmosphere and thus the size of the greenhouse effect. Moreover, the rising carbon dioxide concentrations are causing the Earth to heat up—this is in sync with greenhouse gases which have been increasing, and average global temperatures have risen ~0.8°C–0.9°C (1.45°F) since 1980.

This is just the beginning of global warming and will continue to increase depending on carbon dioxide concentration, over time. In the sense of time, this is a crucial factor to consider because greenhouse warming doesn't occur right away, and this is the case because of the great sink—oceans—soak up heat. What this means is that Earth's temperature will increase at least another 0.6°C (1°F) because the atmosphere already contains carbon dioxide. Note that the degree to which temperatures go up beyond that depends in part on how much more carbon humans release into the atmosphere in the future (Riebeek, 2011).

Marine Reservoir

Direct chemical exchange in our oceans works to diffuse about 30% of the carbon dioxide humans put into the atmosphere. When carbon dioxide is dissolved in the ocean, carbonic acid is created—this increases the acidity of the water. Since around 1750, the pH of the ocean's surface has dropped by about 0.1.

Ocean acidification has two impacts on marine organisms. First, carbonic acid reacts with carbonate ions in the water to form bicarbonate. The problem is for shell-building animals like coral that needs calcium carbonate for their shells. Animals need the calcium carbonate because it reduces the amount of energy they need to expend to build their shells—the shells end up being thinner and more fragile.

Second, more acidic water will dissolve the carbonate shells of marine organisms, making them pitted and weak. On the other hand, the more acidic water is, the better it dissolves calcium carbonate. So, over time, this reaction will allow the ocean to soak up excess carbon dioxide because the more acidic water will dissolve more rock, release more carbonate ions, and increase the ocean's capacity to absorb carbon dioxide.

DID YOU KNOW?

One of the products of the greenhouse effect is warmer oceans. When this occurs there is a decrease in phytoplankton (algae), which prefer cool, nutrient-rich waters. Another problem with a warmer climate is soil baking. When this occurs the rate at which organic matter decays increases and carbon in the form of methane and carbon dioxide seep into the atmosphere.

Land Reservoir

Plants on land, especially forest lands, have taken up approximately 25% of the carbon dioxide that humans have put into the atmosphere. Note that with increased amounts of atmosphere dioxide available to convert to plant matter in photosynthesis, plants can grow more—a lot more. This process is referred to as carbon fertilization. Various models predict that plants might grow anywhere from 12% to 76% more if atmospheric carbon dioxide is doubled. It is important to point out that this massive increase in plant growth with increased carbon dioxide in the atmosphere is limited by both water and fertile soil. Without these two vital ingredients, the increase in carbon dioxide has little to no effect.

DID YOU KNOW?

You know and maybe observed (often) that water by itself does not cut oil or grease. This is because carbon compounds are non-polar, meaning that they have low solubility.

THE BOTTOM LINE

Presently, carbon cycle operation (and function) is well understood—well, that is, to the best of current knowledge. The problem with our current knowledge is that it is current, and without any psychic-far-thinking, at all. The point is that the carbon cycle is changing. What will those changes look like? The bottom line: we do not know what we do not know about changes in the carbon cycle and the overall effect of the changes.

REFERENCES

Abrahamson, D.E., Ed. (1988). *The Challenge of Global Warming*. Washington, DC: Island Press.

Asimov, I. (1989). *How Did We Find Out about Photosynthesis?* New York: Walker & Company.

Conlen, M. (2023). How Much Carbon Dioxide Are We Emitting? Accessed 2/27/23 @ https://climate.nasa.gov/news-are-we-emitting.

Hayhoe, K., Wuebbles, D.J., Easterling, D.R., Fahey, D.W., Doherty, S., Kossin, J., Sweet, W., Vose, R., Wehner, M. (2018). Our changing climate. In: Readmillers, D.R., Avery, C.W., Easterling, D.R., Kunkel, K.E., Lewis, L.L.M., Maycock, T.K., Stewart, B.C. (eds) *Impacts, Risks, and Adaptation in the United States: Fourth National Climate Assessment*, volume II. U.S. Global Change Research Program, Washington, DC, pp. 91–94.

IPCC (2013). *Climate Change 2013: The Physical Science Basis*. Working Groups I, II and III Contribution to the Fifth Assessment Report of the Intergovernmental Panel on Climate Change. Stocker, T.F., Qin, D., Plattner, G.K., Tignor, M., Allen, S.K., Boschung, J., Naules, A., Xia, Y., Bex, V., Midgley, P.M. (eds). Cambridge University Press, Cambridge, UK and New York, pp. 187–189 and 201–208.

Jouzel, J. (2007). *EPICA Dome C Ice Core 800KYr Deuterium Data and Temperature Estimates*. IGBP PAGES/World Data Center for Paleoclimatology Data Contribution Series #2007-091. Boulder, CO: NOAA/NCDC.

Luthi, D., Le Floch, M., Bereiter, B., Blunier, T., Barnola, J.M., Siegenthaler, U., Rayand, D., Jouzel, J., Fischer, H., Kawamuar, K., Storcker, T.F. (2008). High-resolution carbon dioxide concentration recorded 650,000-800,000 years before present. *Nature* 411:379–382.

Miller, G.T. (1988). *Environmental Science: An Introduction*. Belmont, CA: Wadsworth Publishing Company.

Odum, E.P. (1983). *Basic Ecology*. Philadelphia, PA: Saunders College Publishing.

Riebeek, H. (2011). The Carbon Cycle. Accessed 3/31/23 @ https://www/earthobservatory/nasa/gov/features/CarbonCycle.

Sellers, P.J., Hall, F.G., Asrar, G., Strebel, D.E., Murphy, R.E. (1992). An Overview of the First International Satellite Land Surface Climatology Project (ISLSCP) Field Experiment (FIFE). *Journal of Geophysical Research* 97(D17):18345–18371.

Smith, R.M. (2003). *Elements of Ecology*, 7th ed. London, England: Pearson ed.

Spellman, F.R. (2020). *The Science of Water*, 4th ed. Boca Raton, FL: CRC Press.

USEPA (2019). U.S. Cement Industry Carbon Intensifies (2019). EPA 430-F-21-004. Washington, DC: United States Protection Agency.

USEPA (2023). Overview of Greenhouse Gas Emissions in 2018. Accessed 3/27/23 @ https://19January2021snapshot.epa.gov/ghgemissions/overview-greenhouse-gases_.html.

Vose, R.S., Easterling, D.R., Kunkel, K.E., LeGrande, A.M., Wehne, M.F. (2017). Temperature changes in the United States. In: Wuebbles, D.J., Fahey, D.W., Hibbard, K.A., Dokken, D.J., Stewart, B.C., Maycock, T.K. (eds) *Climate Science Special Report: Fourth National Climate Assessment*, Volume I. U.S. Global Research Program, Washington, DC, pp. 197–199. doi 10.7930/JON29V45.

3 Carbon Sources

PERIODIC TABLE

Before moving on to the sources of carbon we need to refer to the periodic table for insight into where carbon (C) resides in the scheme of chemical things, so to speak. As a quick review for those knowledgeable in chemistry and introductory for those who are not aware of the periodic table (see Figure 3.1), more than one hundred elements known today are organized in the periodic table. The horizontal rows of the period table are known as periods, which the vertical columns are known as groups (see Figure 3.1). Note that the elements in each group share similar characteristics. In Figure 3.1, if you count from left to right, you will see that carbon belongs in the fourteenth column (aka group 14). Notice that there are five elements in the group—carbon on top, and then silicon, germanium, tin, and lead.

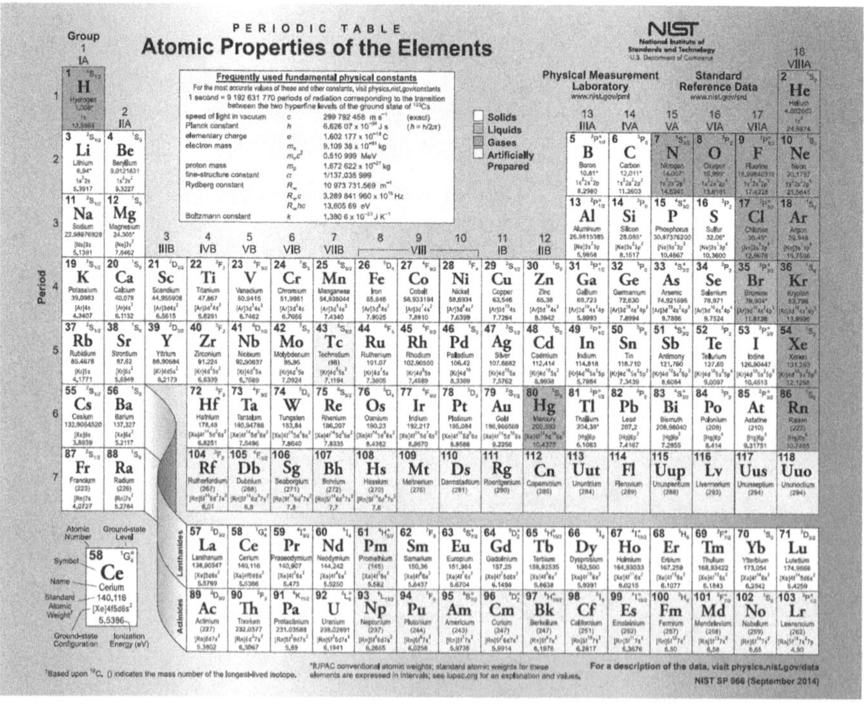

FIGURE 3.1 The periodic table. Lithium (Li) can be found on the far left-hand side of the periodic table in group 1 or 1 A in the alkali metals directly beneath hydrogen (H). From NIST periodic table. Accessed 10/17/2022 @ http://www.NIST.gov. (U.S. Department of Commerce).

DOI: 10.1201/9781003432838-4

It is important to point out that the elements of the carbon group all have several functions, and each element is unique. As mentioned earlier, carbon is the basis for all life on our planet. Keep in mind that carbon is not plainly important to us but is also necessary for all living creatures because they rely on carbon for structure, life processes, and energy. Regarding silicon (Si), it is the second most common ingredient in the Earth's crust after oxygen (O). Rock and sand are primarily made from silicon. Tin (Sn) and lead (Pb) are both everyday metals that have uses in many different industries. Although it is rare compared to the other carbon group elements, Germanium (Ge) is highly prized and used in electronics and computer systems.

DID YOU KNOW?

Compounds of carbon and nitrogen (especially amines) often have a distinct, unpleasant, fishy odor as liquids. Note, however, that usually the solids are odorless.

BIOLOGY OF CARBON

It is often thought that carbon is a "chemical thing." And that is true, of course. However, to properly define carbon and learn of its sources we also need to recognize that carbon is also a "biological thing." So, to properly define carbon and describe its sources we need to combine chemistry with biology to form biochemistry. With this combination of the two sciences we can say that carbon is an element with an atomic number of 6, and it is widely distributed forming organic compounds when combined with oxygen, hydrogen, and so forth.

Well, that is fine but where does carbon come from?

In biochemistry the molecules that an organism uses as its carbon source for generating biomass are referred to as "carbon sources." We pointed out earlier that it is possible for both organic and inorganic sources of carbon. Heterotrophs must derive their nutritional requirements from complex organic carbon substances as a source of both energy and carbon. On the other hand, autotrophs can use inorganic materials as both a source of carbon and an abiotic source of energy, such as, for instance, inorganic chemical energy or light (photoautotrophs) chemolithotrophs. The biology of all of this is part of the carbon cycle which begins with a carbon source that is inorganic (carbon dioxide) and progresses through the carbon fixation process (Tortora et al., 2019).

Note that anyplace carbon is produced, naturally or artificially, as well as any chemical molecules made of carbon, (e.g., carbon dioxide and methane), are sources of carbon. The point is that there is no solitary source of carbon; it comes from a variety of sources, such as fossil fuel combustion, animal respiration, plant deterioration, and wildfires. Also, when plastics are broken down, they are a growing source of carbon. These sources are located worldwide, but some are more productive than others.

Primary Producers

An organism is said to be a primary producer, an autotroph, if it uses carbon from simple components like making organic molecules using carbon dioxide. The motive force behind this activity is chemosynthesis or photosynthesis and inorganic chemical processes. Light and other abiotic sources convert energy into energy that may be used by other organisms by storing it in heterotrophs and other organic compounds.

Note: Keep in mind that defining an "organic" compound is not totally clear depending on who is giving the definition.

Note: About the food chain, **producers** construct organic substances through photosynthesis and chemosynthesis. **Consumers** and **decomposers** use organic matter as their food and convert it into abiotic components—that is, they dissipate energy fixed by producers through food chains.

Other Tropes

Heterotrophs are organisms—hetero (other) and trope (nutrition) that consume other plants or animals for food and energy. Two types of "other tropes" are organotrophs and lithotrophs. Organotrophs use reduced carbon molecules, including lipids, proteins, and carbohydrates from plants and animals. In contrast, lithoheterotrophs rely on inorganic materials like nitrite or sulfur.

Then there are microalgae, and there is a need for carbon sources which are the crucial components for their growth. Several sources of carbon such as acetate, methanol, carbon dioxide, glucose, or other organic substances are involved in the development of microalgae. Microalgae employ organic care sources as a source of carbon for photosynthesis to make chemical energy when grown in photoautotrophic conditions. Note, however, whether there is light or not some species of microalgae can use organic carbon as a carbon source.

THE BOTTOM LINE

There are several sources of carbon, and they release more carbon than they absorb. On the other hand, carbon sinks extract carbon dioxide from the atmosphere and absorb more carbon than they release. In the next chapter we discuss carbon sinks in detail.

REFERENCE

Tortora, G.J., Funke, B.R., Case, C.L. (2019). *Microbiology: An Introduction*. Boston: Pearson.

4 Carbon Sinks

PART AND PARCEL OF THE CARBON CYCLE

A sink is an activity or mechanism that removes a greenhouse gas or a precursor of greenhouse gas or aerosol from the atmosphere. Examples of natural carbon sinks include the following:

- Agricultural lands
- Grasslands
- Coral reefs
- Freshwater lakes and wetlands
- Peat bogs
- Seagrass beds, kelp forests, salt marshes and swamps (all coastal ecosystems)
- Artificial carbon sinks include landfills and carbon capture and storage processes (discussed later).

As part of the carbon cycle, carbon sinks, natural or otherwise, work to accumulate and store some carbon-containing chemical compounds for an indefinite period and thereby removing carbon dioxide (CO_2) from the atmosphere (Global Carbon Budget, 2021). These sinks form an important part of the carbon cycle.

From the global perspective, the two most important carbon sinks are vegetation and the ocean (National Geographic Society, 2020; Spellman, 2023). Another important carbon storage medium is soil. Most of the organic carbon compounds retained in the soil of agricultural areas have been depleted due to intensive farming. The truth be told approximately 20% of the world's grasslands have been converted to cultivated crops (Ramankutty et al., 2008; Ontl and Janowiak, 2017).

Okay what is the point here?

The point here is that soil disturbance, such as farming and cultivation, is a widespread practice in annual row crop production that leads to greatly accelerated losses of organic matter. For instance, consider the Midwest (USA), many soils have lost 30%–50% of carbon (25–40 metric tons of carbon per hectare) from convers to agriculture (Lal, 2002). Note that these losses occur primarily through the disturbance of soil. When soil disturbance occurs, it disrupts the soil's structure, which exposes organic carbon to decomposition from soil microbes and invertebrates and increases soil aeration and temperature, enhancing the activity of decomposers. Another factor that can't be overlooked when soil is disturbed is erosion, leading to deposition elsewhere on the landscape and additional losses of carbon dioxide to the atmosphere (Lal, 1995).

DOI: 10.1201/9781003432838-5

EXAMPLES OF CARBON SINKS

Oceans

Oceans are the main natural carbon sinks, absorbing approximately 50% of the carbon emitted into the atmosphere. The contributors to this extraction of carbon include plankton, fish, corals, algae, and other photosynthetic bacteria.

With the passage of time, organic carbon settles into the deep ocean—this process is referred to as the "biological pump." The ocean's biological pump is one of the three ocean pumps discussed in the next section.

DID YOU KNOW?

The total active pool of carbon at the Earth's surface for durations of less than 10,000 years is 40,000 gigatons C (Gt C, a gigaton is one billion tons, or the weight of approximately 6 million blue whales), and about 95% (~38,000 Gtr C) is stored in the ocean, mostly as dissolved inorganic carbon (Falkowski et al., 2000).

Ocean Pumps

There are three main pumps (or processes) involved in the carbon cycle that brings atmospheric carbon dioxide into the ocean interior and distributes it through the oceans. These three pumps are: (1) the biological pump, (2) the solubility pump, (3) and the carbonate pump.

Biological Pump

In general, this process moves carbon dioxide from the atmosphere to the ocean surface to deeper water where it is stored. Phytoplankton float on the surface of oceans and use photosynthesis to produce the energy they need to live. Like green land plants, phytoplankton suck up carbon dioxide from the atmosphere. Phytoplankton play a key role in the ocean's food chain, but if they aren't devoured, they die and start the second stage of the biological pump when they start striking through the ocean. Clumps of these dead organisms sink and generally avoid being consumed. Finally, when the clumps of dead phytoplankton reach the ocean floor they and the carbon dioxide they contain remain for thousands of years.

Solubility Pump

The solubility pump is a physico-chemical counterpart of the biological pump. This pump conveys significant amounts of carbon known as and in the form of dissolved inorganic carbon (DIC) from the ocean's surface to its interior. This pump's process is not biological, instead it involves physical and chemical processes only (Raven and Falkowski, 1999).

The driver of the solubility pump is two processes in the ocean:

- A strong inverse function of seawater temperature is the solubility of carbon dioxide
- The deep water at high latitudes where seawater is usually cooler and denser drives the thermohaline circulation.

Seawater in the ocean's interior (deep water) is formed under the same surface conditions that promote carbon dioxide solubility; it contains a higher concentration of dissolved inorganic carbon than might be expected from average surface concentrations. Accordingly, the carbon from the atmosphere is pumped into the ocean's interior when these two processes act together. Note that one consequence of this is that when deep water upwells in warmer, equatorial latitudes, it strongly outgasses carbon dioxide to the atmosphere because of the reduced solubility of the gas (Raven and Falkowski, 1999).

Carbonate Pump Sometimes referred to as the "hard tissue" component of the biological pump, the carbonate pump fixes inorganic bicarbonate and causes a net release of carbon dioxide (Hain et al., 2014). Coccolithophores, like some other surface marine organisms, produce hard structures out of calcium carbonate, which is a form of particulate inorganic carbon, by fixing bicarbonate (Rost and Reibessel 2004). This is DIC fixation that is an important part of the oceanic carbon cycle.

$$Ca^{2+} + 2\,HCO_{3^-} \rightarrow CaCO_3 + CO_2 + H_2O$$

Note that regarding organisms, once this carbon is fixed into hard tissue, the organisms either stay in the euphotic zone to be recycled as part of the regenerative nutrient cycle or once they die, continue to the second phase of the biological pump, and begin to sink to the ocean floor. The sinking particles often form aggregates as they sink, greatly increasing the sinking rate. These aggregates give particles a better chance of avoiding predation and decomposition in the water column and eventually reaching the ocean floor (de la Rocha and Passow, 2014).

DID YOU KNOW?

Biological carbon pump's budget calculations are based on the ratio between sedimentation (i.e., carbon export to the ocean floor) and remineralization (i.e., release of carbon to the atmosphere).

The Bottom Line: The biological pump is not the result of a single process but is the combination of several processes each of which influences biological pumping. Taken as a whole, the pump transfers about 11 gigatons of carbon every year into the ocean's interior. This removes carbon from contact with the atmosphere for thousands of years or longer. An ocean without a biological pump would result in atmospheric carbon dioxide levels about 414 parts per million (for short "ppm") higher than the present day.

Plants

Green plants are a natural carbon sink and play a huge role in controlling carbon (carbon dioxide). With plants, it's all about photosynthesis—the process by which they live. The process of photosynthesis occurs in the cells of microscopic organisms and within the leaves of a plant. The process involves the use of solar radiation energy, or sunlight, to change carbon dioxide and water into glucose (carbon sugar).

DID YOU KNOW?

It is accepted that approximately 50% of all carbon extraction from the atmosphere is done through the process of photosynthesis.

Note that along with plants, trees pull vast amounts of carbon dioxide out of the atmosphere during photosynthesis, incorporating some of that carbon into structures like wood.

Even when a tree is harvested and turned into many types of the construction and manufacturing of diverse forest products (e.g., building wood-framed houses, structures, and furniture), they continue to store a substantial amount of carbon for the duration of their useful life (Skog, 2008; Skog and Nicholson, 2000). Of course no two wood products are the same because there are differences in the type of wood product, its production, its use, and its disposal have substantial influences on the among and duration of carbon storage (see Figure 4.1).

Standard methods are available for estimating the carbon that is sequestered in harvested wood products (Smith et al., 2006), and life-cycle assessment approaches

HALF-LIFE OF CARBON

100 years	Single family homes (post 1980)
80 years	Single family homes (pre-1980)
70 years	Multifamily homes
30 years	Furniture
6 year	Free sheet of paper

Half-life of carbon

100 years--------Single Family Homes (post 1980)

80 years----------Single-family homes (pre-1980)

70 years----------Multifamily homes

30 years----------Furniture

6 year------------Free sheet of paper

FIGURE 4.1 Shows carbon stocks averaged across the United States. Adapted from Skog and Nicholson (2000) *The Impact of Climate Change on American's Forests*. Fort Collins, CO: Rocky Mountain Research Station, USDA Forest Service.

can be used for more in-depth analysis of carbon gains and emissions (Ingerson, 2011; Perez-Garcia et al., 2007; Bergman et al., 2014).

In 2015, the amount of carbon stored in harvested wood products in the United States was more than 2,600 million metric tons—note that this is equivalent to approximately 3% of the among of carbon stored in U.S. forest lands). Carbon stored in these wood products is further divided into two different pools: carbon that is stored in products currently in use and carbon that is stored in landfills (Skog and Nicholson, 2000). Carbon in wood products is currently stored in products in use at about the 60% level—this includes items with different decay rates such as paper, pallets, and the lumber used to construct buildings. Note that landfills store the remaining 40% of the wood product carbon. Also note that landfills periodically cover and oxygen is not able to enter and facilitate decay; as a result, the total amount of carbon released from wood products in landfills is substantially reduced (Skog and Nicholson, 2000). Because of anaerobic conditions the carbon that is released a greater proportion of it is in the form of methane (Skog and Nicholson, 2000).

Soil

If we were to take all the carbon contained in terrestrial vegetation and the atmosphere combined and compare their total carbon content to soil, soil wins the comparison, hands down, not even close. Having said this keep in mind that soil is a short to long-term storage medium (Swift, 2001; Batjes, 1996; 2016). Charcoal, plant litter, and other biomass accumulate in the soil. These soil residents are degraded by chemical and biological processes. Note that organic carbon polymers such as cellulose, hemicellulose, lignin, aliphatic compounds, waxes, and terpenoids are collectively retained as humus (Lorenza et al., 2007).

In the colder boreal forests of North America and the Taiga of Russia is where organic matter tends to accumulate in litter and soils. It's a bit different in sub-tropical and tropical climate regions due to their elevated temperatures and extensive leaching by rainfall resulting in the rapid oxidizing of humus and leaf litter. In areas where slash and burn or shifting cultivation agriculture are practiced are generally only fertile for 2–3 years before they are deserted.

Soil organic matter (SOM) is stored in grasslands and mainly in their extensive fibrous root mats. The cooler temperatures and semi-arid to arid climatic conditions in certain regions enable these soils to accumulate significant quantities of organic matter. This can vary based on precipitation, the length of the winter season, and the frequency of naturally occurring lightning-induced grassfires. It is true that these fires release carbon dioxide, but they improve the quality of the grasslands overall. They accomplish this by increasing the amount of carbon retained in the humic material. Moreover, they also deposit carbon directly into the soil in the form of biochar that does not significantly degrade back to carbon dioxide (Woolf et al., 2010).

Note that biochar, as defined by the International Biochar Initiative (2015), is the lightweight black residue, made of carbon and ashes, remaining after the pyrolysis of biomass. It is a stable solid that is rich in pyrogenic carbon (PyC) and can endure in soil for thousands of years (Lean, 2008).

DID YOU KNOW?

Pervasive in the environment, pyrogenic carbon (PyC) including soot, char, black carbon, and biochar is produced by the incomplete combustion of organic matter accompanying biomass burning and fossil fuel consumption. PyC is distributed throughout the atmosphere as well as soils, sediments, and water in both the marine and terrestrial environment (Bird et al., 2015).

Production of biochar may be a means to mitigate climate change due to its potential for sequestering carbon with little effort (Balal et al., 2016; The Royal Society, 2009; Woolf et al., 2010). Moreover, biochar, the high-carbon, fine-grained residue produced via pyrolysis, may increase the soil fertility of acidic soils and increase agricultural productivity.

Note that biomass burning and natural decomposition releases copious amounts of carbon dioxide plus methane to the atmosphere. The biochar production process also releases carbon dioxide (approximately 50% of the biomass); however, the remaining carbon content becomes indefinitely stable (Woolf et al., 2010). Biochar slows the growth in atmospheric greenhouse gas levels in the atmosphere simply because it remains in the ground for centuries. At the same time, biochar in the soil has the benefit of improving water quality, increasing soil fertility, raising agriculture productivity, and, most importantly, reducing pressure on old-growth forests—we must save our old-growth forests (Spellman, 2020).

Note: Like coal, biochar can sequester carbon in the soil for hundreds to thousands of years (Lehmann, 2013; Winsley, 2007).

Woolf et al. (2010) reported that the sustainable use of biochar could reduce the global net emissions of carbon dioxide, methane, and nitrous oxide by up to 1.8 billion tons of carbon dioxide equivalent per year (compared to about 50 billion tons emitted in 2021), without endangering food security, habitats, or soil conservation. Note, however, that a 2018 study doubted that enough biomass would be available to achieve significant carbon sequestration (EGTOP, 2018).

Peat Bogs

One of the major carbon sinks is often ignored, is little known, is a mystery, or is only plain not given a first or second or any thought at all. This often-ignored carbon sink—often called the "missing sink"—is the world's peat bogs. Organic matter in peat bogs undergoes slow anaerobic decomposition below the surface. This process is slow enough that in many cases the bog grows rapidly and fixes more carbon from the atmosphere than is released. Little by little, the peat grows deeper. Important to note, peat bogs hold about one-quarter of the carbon stored in land plants and soils (Chester, 2000).

Fens

Earlier it was stated that one of the major carbon sinks is often ignored, is little known, is a mystery, or is only plain not given a first or second or any thought at all and that is the peat bogs. Note, however, if you mention the term "fen" to the average person, he or she does not have a clue what a fen is. Additionally, there are those who might be familiar with the term fen and if asked they probably will reply that a fen is a pet bog or is a type of bog or is like one. Now, it gets dicey if you ask anyone for a definition of what a fen is? Could be a real mind twister for many—to correctly define it.

Okay, while it is true that some fens and bogs can appear quite similar, and may even contain similar plant and animal species, though by definition they are quite the opposite of one another. Well, the fact is that some confusion is understandable, particularly when given that there are several different wetland classification systems around the globe.

Fens are peat-forming wetlands that receive nutrients from sources other than precipitation (it does not take precipitation to form or maintain a fen); instead, they usually receive nutrients from upslope sources through drainage from surrounding mineral soils and form groundwater movement. Another difference of note between a bog and a fen is that a fen is less acidic and has higher nutrient levels. As a result, they are more suited and able to support a much more diverse plant and animal community. Typically, these systems are often covered by grass, sedges, rushes, and wildflowers. Note that some fens are characterized by parallel ridges of vegetation separated by less productive hollows. In these patterned fens the ridges run from perpendicular to the downslope direction of water movement. Over time, a fen can morph into a bog where peat builds up and separates the fen from its groundwater supply and thus reduces the nutrients available.

THE BOTTOM LINE

To date, the Earth's oceans, forests, and other terrestrial ecosystems (like peat bogs) have absorbed a substantial amount of the carbon emitted by human activities. The problem is that carbon emissions are likely to increase and land-clearing, ocean pollution, and peat bog draining are also likely to occur. Thus, the absorption of carbon by sinks like bogs, fens, and others may be radically reduced.

REFERENCES

Balal, Y., Liu, G., Wang, R., Abbas, Q., Imtiaz, M., Liu, R. (2016). Investigating the biochar effects on C-mineralization and sequestration of carbon in soil compared with conventional amendments using stable isotope approach. *Global Change Biology Bioenergy* 9(6):1085–1099.

Batjes, N.H. (1996). Total carbon and nitrogen in the soils of the world. *European Journal of Soil Science* 47(2):151–163.

Batjes, N.H. (2016). Harmonized soil property values for broad-scale modelling with estimates of global soil carbon stocks. *Geoderma* 269:61–68.

Bergman, R., Puettmann, M., Taylor, A., Skog, K.E. (2014). The carbon impacts of wood products. *Forest Products Journal* 64(7–8):220–231.

Bird, M.I., Wynn, J.G., Saiz, G., Wurster, C.M., McBeath, A. (2015). The pyrogenic carbon cycle. *Annual Review of Earth and Planetary Sciences* 43:273–298.

Chester, B. (2000). The Case of the Missing Sink. Accessed 4/13/23 @ https://reporter-archive.mcgill.Ca/32/15/routlet/index.html.

de la Rocha, C.L. and Passow, U. (2014). The Biological Pump. *Treatise on Geochemistry*, pp. 93–122. Accessed 02/3/23 @ https://www.elsevier.com/books/treatise-on.

EGTOP (2018). Final Report on Fertilizers. Accessed 4/12/23 @ https://ec.europa.eu/info/sites/efault/files/food-farmngn-fisheries/farming/documents/final-report-egtop-fetilizers-iii_en.pdf.

Falkowski, P., Scholes, R.J., Boyle, E., Canadell, J., Canfield, D., Eiser, J., Gruber, N., Hubbard, K., Hogberg, P. (2000). The global caron cycle: A test of our knowledge of earth as a system. *Science* 290(5490):291–296.

Global Carbon Budget (2021). Global Carbon Project. Assessed 4/9/23 @ https://www.globalcarbonproject.org/carbonbudget/21/files/GCP.

Hain, M.P., Sigman, D.M., Haug, G.H. (2014). The biological pump in the past. *Treatise on Geochemistry* 8:485–517.

Hamilton, J. (2009). The Sole Option is to Adapt, the Climate Author Says. Accessed 4/12/23 @ https://www.thestar.com/sciencetech/article/654444.

Ingerson, A. (2011). Carbon storage potential of harvested wood: Summary and polity implications. *Mitigation and Adaptation Strategies for Global Change* 16(3):307–323.

International Biochar Initiative (2015). Standardized Production Definition and Product Testing Guidelines for Biochar that is Used in Soil. Accessed 4/11/23 @ https://www.biochar-international.org/wp-content/uploads/2018/04/IBI_Biochar_Standards_V2.1_Final.pdf.

Kern, D.C., Ruivo de, L.P., Frazao, F.J.L. (2009). *Amazonian Dark Earths: Wim Sombroek's Vision*. Dordrecht, Netherlands: Springer.

Lal, R. (1995). Global soil erosion by water and carbon dynamics. In: Kimble, J.M., Levine, E. R., Stewart, B.A. (eds) *Soils and Global change*. Boca Raton, FL: CRC Press, pp. 131–142.

Lal, R. (2002). Soil carbon dynamics in cropland and rangeland. *Environmental Pollution* 116:353–362.

Lean, G. (2008). Ancient Skills Could Reverse Global Warming. Accessed 4/11/23 @ https://web.archive.org/web/20110913052413/https://www.Independent.co.uk/environment/climate-change/ancient-skills-could-reverse-global-warming-1055700.html.

Lehmann, J. (2013). Soil Biochemistry. Accessed 4/12/23 @ https://www.css.cornell.edu/faculty/lehmann/reserach/terra%20preta/terrapretamain.html.

Lorenza, K., Lala, R., Prestonb, C.M., Nieropc, K.G.J. (2007). Strengthening the soil organic carbon pool by increasing contributions form recalcitrant aliphatic bio(macro)molecules. *Geoderma* 142(1–2):1–10.

National Geographic Society (2020). Carbon Sources and Sinks. Accessed 4/9/23 @ https://www.nationalgeographic.org/encyclopedia/carbon-sources-and-sinks.

Ontl, T. and Janowiak, M. (2017). *Grassland and Carbon Management*. Washington, DC: U.S. Department of Agriculture, Forest Service. Climate Change Resource Center.

Perez-Garcia, J., Lippke, B., Comnick, J., Manriquez, C. (2007). An assessment of carbon pols, storage, and wood products market substitution using life-cycle analysis results. *Wood and Fiber Science* 37:140–148.

Ramankutty, N., Evan, A.T., Monfreda, C., Foley, J. (2008). Farming the plante: 1. Geographic distribution of global agricultural lands in the year 2000. *Global Biological Cycles* 22(1). doi 10.1029/2007GB002952.

Raven, J.A. and Falkowski, P.G. (1999). Oceanic sinks for atmospheric CO_2. *Plant, Cell, and Environment* 22(6):741–755.

Rost, B. and Reibessel, U.L.F. (2004). *Coccolithophores and the Biological Pump: Responses to Environmental Changes*. Berlin, Heidelberg: Springer.

Skog, K.E. (2008). Sequestration of carbon in harvested wood products for the United States. *Forest Products Journal* 58(6):56–72.

Skog, K.E. and Nicholson, G.A. (2000). Chapter 5: Carbon sequestration in wood and paper products. In: Joyce, L.A., Birdsey, R. (eds) *The Impact of Climate Change on American's Forests*. A Technical Document Supporting the 2000 USDA Forest Service. RPA Assessment. Fort Collins, CO: Rocky Mountain Research Station, USDA Forest Service.

Smith, J.E., Heath, L.S., Skog, K.E., Birdsey, R.A. (2006). *Methods for Calculating Forest Ecosystem and Harvested Carbon with Standard Estimates for Forest Types of the United States*. Newtown Square, PA: U.S. Department of Agriculture, Forest Service, Northeastern Research Station, 216.

Spellman, F.R. (2020). *The Science of Water*, 4th ed. Boca Raton, FL: CRC Press.

Spellman, F.R. (2023). *The Science of Ocean Pollution.* (In production). Boca Raton, FL: CRC Press.

Swift, R.S. (2001). Sequestration of carbon by soil. *Soil Science* 166(11):858–871.

The Royal Society (2009). Geoengineering the Climate: Science, Governance, and Uncertainty. Accessed 4/12/23 @ https://royalsociety.org/Geoengineering-the-climate/.

Winsley, P. (2007). Biochar and bioenergy production for climate change mitigation. *New Zealand Science Review* 64(5):5.

Woolf, D., Amonette, J.E., Street-Perrott, F.A., Lehmann, J., Joseph, S. (2010). Sustainable biochar to mitigate global climate change. *Nature Communications* 1(5):56.

5 The Color of Carbon

1 Gigaton (GT)=1 billion tons=1×10^{15} g=1 Petagram (Pg)

INTRODUCTION

If we were to ask the average person or woman on the street if they know anything about carbon, the standard answer would probably be along the lines of, "Carbon, yeah, that is the stuff I use in my barbecue grill." Others may be more precise and respond, "Yeah, carbon, which is that black charcoal stuff we use in out outdoor fire pits for barbecuing." Or they may reply that the carbon "is the black stuff on my fireplace chimney."

These are just a few of the answers explaining what carbon is; there seemed to be no limit to the number replies from several of the college upper level/grad students in my environmental health and science courses. And I was not surprised when they could define carbon right down to its extended atomic mass number of 12.0107 u. No problem, my students were the best, the brightest, smartest, highest IQ types, dedicated individuals—open-minded to the core, and wanting to learn.

So, again, and not surprised because the students I taught had at least 3 years of chemistry course backgrounds and a few worked part-time as lab technicians in local environmental laboratories. However, the truth be told when I asked my students about carbon what I was really doing was setting them up for a question that I did not think any—well, maybe are few would be knowledgeable about.

The question I put forward to the students in my classes was: "How many of you have heard about the carbon rainbow?"

Surprisingly (at least to me, anyway), several students raised their hands indicating that they had heard about the carbon rainbow. So, logically I asked: "can anyone in the class explain to all of us in the classroom … that is to those who have no idea what the carbon rainbow means—what it is all about?"

I did have a few respondents who were able to simply point out that carbon comes in many colors, the colors of the rainbow. And for the benefit of those who were not aware of this I presented the following information.

THE CARBON RAINBOW

Okay, getting down to the science of carbon, its physical characteristics, it can be simply stated that not all carbon is the same. In the grand scheme of all things classified in one form, specification or other scientists use color to classify carbon at different points in the carbon cycle. Carbon color is based on function, characteristics,

DOI: 10.1201/9781003432838-6

and location. What this really does is to allow us to create a more descriptive outline than customary labels, such as "organic" and "inorganic" tags. The point is that a full spectrum of color-based descriptions has emerged to describe the properties and distribution of organic carbon. According to USGS (2022), the carbon rainbow consists of eight colors. Blue, green and teal highlight the role of carbon in climate change mitigation via sequestration (Zinke, 2020). Black, brown, and red types impact Earth's heat balance or promote cyrospheric melting. Cyrospheric melting is an all-encompassing term that refers to those portions of Earth's surface where water is in solid form, including sea ice, lake ice, river ice, snow cover, glaciers, ice caps, and permafrost (frozen ground).

- **Purple**—carbon captured through the air or industrial emissions
- **Blue**—carbon stored in ocean plants and sediments
- **Teal**—carbon stored in freshwater and wetland environments
- **Green**—carbon stored in terrestrial plants
- **Black**—carbon released through the burning of fossil fuels
- **Gray**—carbon released through industrial emissions
- **Brown**—carbon released by incomplete combustion of organic matter
- **Red**—carbon released through biological particles on snow and ice that reduce albedo

Before focusing on blue carbon (blue carbon is the author's choice to focus on because it is one of the most important carbon types at the present time, but before doing so a brief introduction to the newest member of the other carbon colors is first presented in the next section), it is important to have a fundamental understanding of albedo. *Albedo* (the ratio between the light reflected from a surface and the total light falling on it; that is, incident to it) always has a value less than or equal to 1. An object with a high albedo, near 1, is very bright, while a body with a low albedo, near 0, is dark. For example, freshly fallen snow typically has an albedo that is between 75% and 90%; that is, 75%–95% of the solar radiation that is incident on snow is reflected. Thus, a surface cover such as clean snow can reflect solar radiation because of its high albedo. At the other extreme, the albedo of a rough, dark surface, such as a green forest, may be as low as 5%. The albedos of some common surfaces are listed in Table 5.1. The portion of insolation not reflected is absorbed by the Earth's surface, warming it. This means that Earth's albedo plays an important part in the Earth's radiation balance and influences the mean annual temperature and the climate on both local and global scales.

Red Carbon

I listed red carbon last on the color of the carbon rainbow colors because it is the newest color in the carbon spectrum. In the widest perspective, red carbon includes all living biological particles on snow and ice that reduce albedo to survive (Zinke, 2020). Although red carbon does not enjoy a long history of study or analysis, its impact on albedo is beginning to accelerate investigation.

TABLE 5.1
The Albedo of Some Surface Types in % Reflected

Surface	Albedo
Water (low sun)	10–100
Water (high sun)	3–10
Grass	16–26
Glacier ice	20–40
Deciduous forest	15–20
Coniferous forest	5–15
Old snow	40–70
Fresh snow	75–95
Sea ice	30–40
Blacktopped tarmac	5–10
Desert	25–30
Crops	15–25

THE BLUE CARBON SINK

Blue carbon refers to carbon dioxide that is absorbed from the atmosphere and stored in the ocean—blue-to-blue storage, but note that "Blue" refers to the watery nature of this storage. Most of the blue carbon is carbon dioxide that has dissolved directly into the ocean blue. A smaller, much smaller, amount is stored in underwater sediments, coastal vegetation, and soils.

Okay, for clarity, what is the amount, what's in the amount?

What is in the amount is DNA and proteins, also including ocean life from whales to phytoplankton.

Blue carbon is increasing in studies by climatologists and others.

Why?

Blue carbon, especially coastal blue carbon, is stored by saltwater ecosystems in their vegetation and soils. These saltwater ecosystems, in terms of total area, include salt marshes, mangroves, seagrass meadows and all have a small global footprint, but it is their deep, water-logged soils that can bury many times more carbon per acre than even a tropical rainforest.

Okay, as Voltaire stated, "if you wish to converse with me, please define your terms." So we will. Let's begin with salt marshes, what are they, exactly? Then the same question about mangroves and seagrass meadows.

- **Salt marshes**—are among the most productive and valuable ecosystems in the world. The vegetation in these coastal systems acts as a buffer between land and sea. This naturel buffer helps reduce flooding, erosion, and subsequent damage to property and infrastructure by slowing down the flow of water and dissipating waves during storms.

The impressive thing about salt marshes is that they provide essential food, refuge, and nursery habitat for commercially and recreationally important species. But that is just the beginning; they also improve water quality buy filtering runoff and absorbing excess nutrients. From the point of climate change, salt marshes decrease the effect of climate change by storing copious quantities of carbon dioxide from the atmosphere. It is understood (hopefully) that we need to preserve these important carbon sinks.

DID YOU KNOW?

Salt marshes can absorb up to 1.5 million gallons of floodwater—the equivalent of more than 2.25 Olympic-size swimming pools—per acre (USEPA, 2002).

- **Mangrove forests**—all eighty species of all mangrove trees grow in areas with low-oxygen soil, where low-moving waters allow fine sediments to accumulate. Note that mangrove forests only grow at tropical and subtropical latitudes near the equator because they cannot withstand freezing temperatures. One of the noteworthy features of the mangrove forests is their tangle of roots that make the trees appear to be standing on stilts above the water. This is important because of the daily rise and fall of tides; they get flooded at least twice per day. The roots also play the role of navigator to sediments in that they use the movement of tidal waters to settle out the sediments and build up muddy bottoms. Moreover, mangrove forests stabilize the coastline, reducing erosion from storm surges, currents, waves, and tides. The complex root system of mangroves also makes these forests attractive to fish and other organisms seeking food and shelter from predators.

DID YOU KNOW?

Mangrove forests can boost coastal resilience to storms, helping to protect 200 million people worldwide (Spalding et al., 2016).

- **Seagrass meadows—**a healthy seagrass meadow means a healthy water shed and to date more than 70 species of these aquatic plants, water filterers, and oxygen producers have been identified by scientists and they occupy every continent except for Antarctica. Seagrasses attract many species of fish and shellfish, some of which are only found in seagrass meadows. These meadows often serve as nursery areas for fish species to grow and mature. Seagrass plants are vital food sources for animal grazers including manatees, green sea turtles, and aquatic birds. The plants trap sand with their roots reducing erosion. Seagrasses act as buffers that absorb energy from waves.

DID YOU KNOW?

Seagrass meadows provide nursery, habitat, and feeding grounds for 20% of the world's largest fisheries, including cod, pollock, and hearing (Unsworth et al., 2019).

SIDE BAR 5.1—IS IT SEAWEED OR SEAGRASS?

Note that during a routine field trip to survey seagrasses in the shallow coastal waters of lower Chesapeake Bay in the Tidewater Region of Virginia with a group of college students in one of my environmental science courses I was initially surprised when a few of the students initially referred to seagrass as seaweed. They mistakenly described the waving fronds tickling their ankles as seaweed when they were within an underwater meadow of seagrass. It always amazes the students (and me) that the acres of seagrass we survey, while there is nothing special or exciting about them, basically insignificant. That is until these vast meadows are viewed from space—hardly insignificant. Moreover, they also makeup a powerful carbon storage system—and if carefully supported they could help lessen or rein in global carbon dioxide emissions responsible for worsening natural, cyclical climate change events. These extensive meadows take in, or sequester, copious amounts of carbon by trapping it temporarily in plant stems and over longer periods of time in sediments that accumulate where they grow.

▲ **Flux**—the transfer of carbon from one carbon pool to another.

WHITE KNIGHT VERSUS CLIMATE CHANGE

A White Knight is traditionally (and historically) a mythological figure who fits right in with today's situations when a heroic figure arrives on the scene to fights against evil or hostile takeovers in business in operation. In this instance and presentation our White Knight is blue, specifically Blue Carbon, a valuable tool (knight-errant) to the rescue in combating climate change by providing a carbon sink. Coastal wetlands can store far greater amounts of carbon they naturally release, and this trait makes them one of the world's most important natural "carbon sinks." Despite occupying less than 5% of global land area and less than 2% of the ocean, studies have shown that they store roughly 50% of carbon buried in ocean sediments (Duarte et al., 2005).

The bottom line: In this case whereby we are looking for some type of White Knight to aid in mitigating or combating climate change—to perform some heroics to save the planet—we might have to depend on a Blue Knight to aid the globe in conducting this.

It is a variety of chemical and physical factors that determine the capacity of coastal wetlands and their ability to capture and store carbon. The crucial element is the wet, low-oxygen conditions of tidally influenced and submerged soils characteristic of these ecosystem that slow the decay of plant and other organic material. Plants absorb carbon dioxide along with water from the air during photosynthesis and use the carbon to support their growth. Later, when the plants shed their leaves or roots or die, the carbon in that decaying organic matter becomes lodged in the soil. If the wetlands are healthy, they can keep carbon stored awry for millennia, providing a natural way to prevent it from being released into the atmosphere and contributing to climate change (Crooks et al., 2014). When coastal ecosystems degrade, however, the vast stores of carbon are released in the form of three major greenhouse gases: carbon dioxide, methane, and nitrous oxide (Moomaw et al., 2018). Worldwide, an estimated 450 million metric tons of carbon dioxide—equal to almost 100 million tons worth—is emitted from the destruction of coastal wetlands each year, accelerating planetary warming (Pendleton et al., 2021).

THE BOTTOM LINE

It is important to point out that coastal ecosystems' carbon-storing power is a double-edged sword because when they are disturbed or drained, they can release massive amounts of carbon dioxide back into the atmosphere. However, when restored and protected, they can become a valuable tool for offsetting carbon dioxide emissions. This is especially the case for island nations and developing nations whose greenhouse gas emissions are relatively low.

REFERENCES

Crooks, S. et al. (2014). Coastal Blue Carbon Opportunity Assessment for the Snohomish Estuary: The Climate Benefits of Estuary Restoration. Report by Environmental Science Associates, Western Washington University, EarthCorps, and Restore America's Estuaries. doi 10.13140/RG.2.1.1371.6568.

Duarte, D.M., Middelburg, J.J., Caraco, N. (2005). Major role of marine vegetation on oceanic carbon cycle. *Biogeosciences* 2(1):1–18.

Moomaw, W.R. et al. (2018). Wetlands in a changing climate: Science, policy and management. *Wetlands* 38(2):183–205.

Pendleton, L. et al. (2021). Estimating global blue carbon emissions from conversion and degradation of vegetated coastal ecosystems. *Plos One* 7(9):e43542.

Spalding, M.D., Brumbaugh, R.D., Landis, E. (2016). Atlas of Ocean Wealth (The Nature Conservancy). Accessed 4/19/23 @ https://oceanweatlh.org/wpcontent/uploads/2016/07/Atlas_of_Ocean_Wealth.pdf.

Unsworth, F.K.F., Nordlund, L.M., Cullen-Unsworth, L.C. (2019). Seagrass meadows support global fisheries production. *Conservation Letters* 12(1):e12566.

USEPA (2002). Functions and Value of Wetlands. Accessed 4/19/23 @ https://www.epa.gov/sites/production/files/2021-01/documents/functions_values_of_wetlands.pdf.

USGS (2022). Carbon Rainbow. Accessed 4/17/23 @ https://www.usgs.gov/media/images/carbon-rainbow.

Zinke, L. (2020). The colours of carbon. *Nature Reviews* 1:141.

6 Adsorption vs Absorption

1 kg Carbon (C) = 3.664 kg Carbon Dioxide (CO_2)

INTRODUCTION

Before we discuss carbon removal and various techniques used to remove, store, and sequester carbon, presented and described in Part 2—those especially dealing with forest capture and storage of carbon, it is important to be familiar and comfortable with terms and the definitions of these processes along with a discussion of adsorption and absorption.

ADSORPTION AND ABSORPTION AND ION EXCHANGE

Adsorption is the process by which one substance is attracted to and adheres to the surface of another substance, without penetrating its internal structure. In soil, adsorption works to bond (hold) a pollutant to the surface of a soil particle or mineral in such a way that the substance is only available or disperses slowly. Clay and highly organic materials (for example) tend to adsorb pesticides rather than absorb them. Pollutant adsorption takes place on mineral surface where defects in their crystalline structure result in imbalances of electrical charges on the mineral surface. Dissolved pollutant molecules and ions with charge imbalances are attracted to mineral surfaces that have an opposite charge imbalance. Adsorption occurs when ions, atoms, or molecules adhere to the surface. When a substance is adsorbed onto the surface it is called the *adsorbate*. The surface of the substance is called the *adsorbent*. Adsorption is an exothermic process because energy is released when the adsorbate sticks to the adsorbent. It is the surface area and temperature that control the rate of the process. Adsorption is promoted by low temperatures because particles with less thermal energy have less kinetic energy and are more likely to stick to surfaces from intermolecular forces and covalent bond formation.

Ion exchange is another step in the adsorption process whereby a dissolved pollutant substitutes itself for another chemical already adsorbed on the mineral surface. *Adsorption site density* involves the concentration of the sorptive surface available from the mineral and organic contents of soils. An increase in adsorption sites shows an increase in the ability of the soils to immobilize hydrocarbon compounds in the soil matrix.

Absorption involves mass transfer into another material and occurs when ions, atoms, or molecules pass into a bulky substance. These *absorbates* (particles) dissolve or diffuse into the *absorbent* material. A good example that most of us are familiar with is the action that takes place whenever we use a paper towel to pick up water or some other liquid substance. As we are know awareand have observed the water evenly saturates the paper. Absorption is an endothermic process in which absorption

DOI: 10.1201/9781003432838-7

TABLE 6.1
Adsorption and Absorption Differences

Adsorption	Absorption
Particles accumulate onto a substance surface	Accumulation of particles throughout another substance
Surface phenomenon	Bulk phenomenon
Exothermic process	Endothermic process
Effected by favored by low temperature	Not affected temperature
Steady increase in rate to meet equilibrium	Occurs at uniform rate
Surface differs from internal concentration	Eventually concentration is the same throughout the material

Source: Helmenstine, A. (2023). Science Note. Adsorption vs Absorption. Accessed 4/19/23 @ https://sciencenotes.org/adsorption-vs-absorption-differences-and-examples/.

occurs via diffusion (passively) or is facilitated diffusion or active transport. Several factors including pressure, surface area, and concentration affect the absorption rate.

Note that absorption and adsorption are confused; however, there are several differences between the two. The confusion arises whenever people think about adsorption and absorption where they consider the mass transfer of liquid particles onto (adsorption) and into (absorption) solids. The truth be told these processes can involve plasma, gases, liquids, or dissolved solids where the ions, atoms, or molecules are adsorbed or absorbed by liquids or solids (Helmenstine, 2023).

Both sorption processes share similarities, but they differ in many ways. Consider Table 6.1.

REFERENCE

Helmenstine, A. (2023). Science Note. *Adsorption vs Absorption.* Accessed 4/19/23 @ https://sciencenotes.org/adsorption-vs-absorption-differences-and-examples/

Part 2

Fundamentals of Carbon Capture and Sequestration

7 Carbon Capture and Sequestration

1 GtC = 3.664 billion tons; CO_2 = 3.664 $GtCO_2$

INTRODUCTION

Human activities, especially the burning of fossil fuels such as coal, oil, and gas, have caused a substantial increase in the concentration of carbon dioxide (CO_2) in the atmosphere. This increase in atmospheric CO_2—from about 280 to more than 380 parts per million (ppm) over the last 250 years … and currently at more than 410 ppm—is causing measurable global warming. Potential adverse impacts include sea-level rise; increased frequency and intensity of wildfires, floods, droughts, and tropical storms; changes in the amount, timing, and distribution of rain, snow, and runoff; and disturbance of coastal marine and other ecosystems. Rising atmospheric CO_2 is also increasing the absorption of CO_2 by seawater, causing the ocean to become more acidic, with potentially disruptive effects on marine plankton and coral reefs. Technically and economically feasible strategies are needed to mitigate the consequences of increased atmospheric CO_2.

THE 411 ON CARBON CAPTURE AND SEQUESTRATION

The reader might wonder what does carbon capture and sequestration (CCS) have to do with the environmental impact of climate change and with the use of renewable energy to mitigate potential damage caused by release of carbon dioxide into the environment? Renewable energy has two advantages: (1) it is a possible source of energy now and in the future (it is renewable and sustainable) and will be called on to replace nonrenewable hydrocarbon energy sources as they are depleted, and (2) renewable energy produces little or no waste products such as carbon dioxide or other chemical pollutants, so has a minimal impact on the environment. It is the latter of the two that is related to carbon capture and sequestration. That is, at the present time, and soon, we have and will continue to have an ongoing increase in atmospheric carbon dioxide. Many scientists agree that global climate change is occurring and that to prevent its most serious effects we must begin immediately to significantly reduce our greenhouse gas (GHG) emissions. As stated throughout this text, one major contributor to climate change is the release of the greenhouse gas carbon dioxide (CO_2). This points to the essence of the carbon capture and sequestration process: capture and sequester carbon dioxide. Further, to control atmospheric carbon dioxide will require deliberate mitigation with an approach the combines reducing emissions by utilizing renewable sources and by increasing capture and storage.

DOI: 10.1201/9781003432838-9

The term *carbon sequestration* is used to describe both natural and deliberate processes by which CO_2 is either removed from the atmosphere or diverted from emission sources and stored in the ocean, terrestrial environments (vegetation, soils, and sediments), and geologic formations. Before human-caused CO_2 emissions began, the natural processes that make up the global carbon cycle maintained a near balance between the uptake of CO_2 and its release back to the atmosphere. However, existing CO_2 uptake mechanisms (sometimes called CO_2 or carbon *sinks*) are insufficient to offset the accelerating pace of emissions related to human activities. Annual carbon emissions from burning fossil fuels in the United States are about 1.6 gigatons (billion metric tons), whereas annual uptake amounts are only about 0.5 gigatons, resulting in a net release of about 1.1 gigatons per year.

Scientists at the U.S. Geological Survey (USGS) and elsewhere are working to assess both the potential capacities and the potential limitations of the various forms of carbon sequestration and to evaluate their geologic, hydrologic, and ecological consequences. USGS is providing information needed by decision makers and resource managers to maximize carbon storage while minimizing undesirable impacts on humans and their physical and biological environments.

Important Point: Before moving on in this presentation, discussion, or whatever it is, it is important to note that there are two types of carbon sequestrations: geologic and biologic carbon sequestration. *Geologic carbon sequestration* is the process of storing carbon dioxide in underground geologic formations. Carbon dioxide is usually pressurized until it becomes a liquid, and then it is injected into porous rock formations in geologic basins. Because it is typically used later in the life of a producing oil well carbon storage is a method and sometimes a part of enhanced oil recovery (EOR)—known as tertiary recovery. In the EOR, the liquid carbon dioxide is injected into the oil-bearing formation to reduce the viscosity of the oil and allow it to flow more easily to the oil well. *Biologic carbon sequestration (aka Terrestrial Carbon Sequestration)* refers to storage of atmosphere carbon in vegetation, soils, woody products, and aquatic environments. Encouraging the growth of plants—particularly trees in our forests and elsewhere—advocates of biologic sequestration hope to remove carbon dioxide from the atmosphere.

The Bottom Line: This book focuses on biologic (natural) sequestration and storage of carbon in forests and the final part of this book is where this focus is placed—intentionally.

Terrestrial Carbon Sequestration

Terrestrial sequestration (as mentioned earlier sometimes termed "biological sequestration") is the removal of gaseous carbon dioxide from the atmosphere and binding it in living tissue by plants (especially trees). Terrestrial sequestration is typically accomplished through forest and soil conservation practices that enhance the storage of carbon (such as restoring and establishing new forests, wetlands, and grasslands) or reduce CO_2 emissions (such as reducing agricultural tillage and suppressing wildfires). In the United States, these practices are implemented to meet a variety of land-management objectives. Although the net terrestrial uptake fluxes offset about 30% of U.S. fossil fuel CO_2 emissions, only a small fraction of this uptake results from activities undertaken specifically to sequester carbon. The largest net uptake is due primarily to ongoing natural regrowth of forests that were harvested during the 19th and early 20th centuries.

Existing terrestrial carbon storage is susceptible to disturbances such as fire, disease, and change in climate and land use. Boreal forests, also known as taiga, and northern peatlands, which store nearly half the total terrestrial carbon in North American, are already experiencing substantial warming, resulting in large-scale thawing of permafrost and dramatic changes in aquatic and forest ecosystems. USGS scientists have estimated that at least 10 gigatons of soil carbon in Alaska are stored in organic soils that are extremely vulnerable to fire and decomposition under warming conditions.

The capacity of terrestrial ecosystems to sequester additional carbon is uncertain. An upper estimate of potential terrestrial sequestration in the United States might be the amount of carbon that would be accumulated if U.S. forests and soils were restored to their historic levels before they were depleted by logging and cultivation. These amounts (about 32 and 7 gigatons for forests and soils, respectively) are probably not attainable by deliberate sequestration because restoration on this scale would displace a large percentage of U.S. agriculture and disrupt many other present-day activities. Decisions about terrestrial carbon sequestration require careful considerations of priorities and tradeoffs among multiple resources. For example, converting farmlands to forests or wetlands may increase carbon sequestration, enhance wildlife habitat and water quality, and increase flood storage and recreational potential, but the loss of farmlands will decrease crop production. Converting existing conservation lands to intensive cultivation, while perhaps producing valuable corps (for example, biofuels), may diminish wildlife habitat, reduce water quality and supply, and increase CO_2 emissions. Scientists are working to determine the effects of climate and land-use change on potential carbon sequestration and ecosystem benefits, and to provide information about these effects for use in resource planning.

SIDEBAR 7.1 URBAN FORESTS AND CARBON SEQUESTRATION[1]

The urban environment presents important considerations for terrestrial carbon sequestration and global climate change. Over half of the world's population lives in urban areas (Population Reference Bureau, 2012). Because cities are denser and walkable, urban per capita emissions of greenhouse gases (GHGs) are almost always substantially lower than average per capita emissions for the counties in which they are located (The Cities Alliance, 2007). Urban areas are more likely than non-urban areas to have adequate emergency services, and so may be better equipped to provide critical assistance to residents in the case of climate-related stress and events such as heat waves, floods, storms, and disease outbreaks (Myers et al., 2013). However, cities are still major sources of GHG emissions (Dodman, 2009). Studies suggest that cities account for 40%–70% of all GHG emissions worldwide due to resource consumption

[1] From Safford et al., (2013). *Urban Forests and Climate Change*. USDA; Spellman, F.R. (2019). *Impact of Renewable Energy*. Boca Raton, FL: CRC Press.

and energy, infrastructure, and transportation demands (USEPA, 2009). Highly concentrated urban areas, especially in coastal regions and in developing countries, are disproportionately vulnerable to extreme weather and infectious disease.

The term "urban forest" refers to all trees within a densely populated area, including trees in parks, on streetways, and on private property. Although the composition, health, age, extent, and costs of urban forests vary considerably among different cities, all urban forests offer some common environmental, economic, and social benefits. Urban forests play a key role in climate change mitigation and adaptation. Active stewardship of a community's forestry assets can strengthen local resilience to climate change while creating more sustainable and desirable places to live. Trees in a community help to reduce air and water pollution, alter heating cooling costs, and increase real estate values. Trees can improve physical and mental health, strengthen social connections, and are associated with reduce crime rates. Trees, community gardens, and other green spaces get people outside, helping foster active living and neighborhood pride.

Like any forest, urban forests help mitigate climate change by capturing and storing atmospheric carbon dioxide during photosynthesis, and by influencing energy needs for heating and cooling buildings; trees typically reduce cooling costs but can increase or decrease winter heating use depending on their location around a building and whether they are evergreen or deciduous. In the contiguous United States alone, urban trees store over 708 million tons of carbon (approximately 12.6% of annual carbon dioxide emissions in the United States) and capture an additional 28.2 million tons of carbon (approximately 0.05% of annual emissions) per year (Nowak et al., 2013; USEPA, 2013). The value of urban carbon sequestration is substantial: approximately $2 billion per year, with a total current carbon storage value of over $50 billion (Nowak et al., 2010). Shading and reduction of wind speed by trees can help to reduce carbon emissions by reducing summer air conditioning and winter heating demand and, in turn, the level of emissions from supplying power plants (Nowak et al., 2010). Shading can also extend the useful life of street pavement by as much as ten years, thereby reducing emissions associated with the petroleum-intensive materials and operation of heavy equipment required to repave roads and haul away waste (McPherson and Muchnick, 2005). Establishing 100 million mature trees around residences in the United States would save an estimated $2 billion annually in reduced energy costs (Akbari et al., 1992; Population Reference Bureau, 2012). However, this level of tree planting would only offset less than 1% of United States emissions over a 50-year period (Nowak and Crane, 2002).

The sustainable use of wood, food, and other goods provided by the local urban forest may also help mitigate climate change by displacing imports associated with higher levels of carbon dioxide emitted during production and transport. Urban wood is a valuable and underutilized resource. At current utilization rates, forest products manufactured from felled urban trees are estimated to save several hundred million tons of CO_2 over a 30-year period. Furthermore, wood chips made from low-quality urban wood may be combusted for heat and/or power to displace an additional 2.1 million tons of fossil fuel emissions per year (Sherill and Bratkovitch, 2011).

Urban forests enable cities to better adapt to the effect of climate change on temperature patterns and weather events. Cities are generally warmer than their surroundings (typically by about 1°C –2°C, though this difference can be as high as 10°C under certain climatic conditions), meaning that average temperature increases caused by global warming are frequently amplified in urban areas (Bristow et al., 2012; Kovats and Akhtar, 2008). Urban forests help control this "heat island" effect by providing shade and by reducing urban albedo (the fraction of solar radiation reflected into the environment), and through cooling evapotranspiration (Romero-Lankao and Gratz, 2008; Bristow et al., 2012). Cities are also particularly susceptible to climate-related threats such as storms and flooding. Urban trees can help control runoff from these by catching rain in their canopies and increasing the infiltration rate of deposited precipitation. Reducing stormwater flow reduces stress on urban sewer systems by limiting the risk of hazardous combined sewer overflows (Fazio, 2010). Furthermore, well-maintained urban forests help buffer high winds, control erosion, and reduce drought (Nowak et al., 2010; Fazio, 2010; Cullington and Gye, 2010).

Urban forests provide critical social and cultural benefits that may strengthen community resilience to climate change. Street trees can hold spiritual value, promote social interaction, and contribute to a sense of place and family for residents (Dandy, 2010). Overall, forested urban areas appear to have potentially stronger and more stable communities (Dandy, 2010). Community stability is essential to the development of effective long-term sustainable strategies for addressing climate change (Williamson et al., 2010). For example, neighborhoods with stronger social networks are more likely to check on elderly and other vulnerable residents during heat waves and other emergencies (Klinenberg, 2002).

Urban forests help control the causes and consequences of climate-related threats. However, forests may also be negatively impacted by climate change. While it is true that carbon dioxide levels and water temperature may initially promote urban tree growth by accelerating photosynthesis, it is also true that too much warming in the absence of adequate water and nutrients stresses trees and retards future development (Tubby and Webber, 2010). Warmer winter temperatures increase the likelihood of winter kill, in which trees, responding to their altered environment, prematurely begin to circulate water and nutrients in their vascular tissue. If rapid cooling follows these unnatural warm periods, tissues will freeze, and trees will sustain injury or death.

Warmer winter temperatures favor many populations of tress pest and pathogen species normally kept at low levels by chilly winter temperatures (Tubby and Webber, 2010). Although climate change may reduce populations of some species, many others are better capable than their arboreal host to adapt to changing environments due to their short lifecycles and rapid evolutionary capacity (Cullington and Gye, 2010; Tubby and Webber, 2010). The consequences of these population changes are compounded by the fact that hot, dry environments enrich carbohydrate concentrations in tree foliage, making urban trees more attractive to pests and pathogens (Tubby and Webber, 2010).

Climate change alters water cycles in ways that impact urban forests. Increased winter precipitation puts urban forests at greater risk from physical damage due to increased snow and ice loading (Johnston, 2004). Increased summer evaporation and transpiration create water shortages often exacerbated by urban soil compaction and impermeable surfaces. More frequent and intense extreme weather events increase the likelihood of severe flooding, which may uproot trees and cause injury or death to tree root systems if waterlogged soils persist for prolonged periods (Johnston, 2004).

Cold regions especially may benefit from increased tourism, agricultural productivity, and ease of transport because of climate change. However, the potential positive implications of climate change are far eclipsed by the negative (Parry et al., 2007). Rising temperatures, increased pest and pathogen activity, and water cycle changes impose physiological stresses on urban forests that compromise forest ability to deliver ecosystem services that protect against climate change. Climate change will also continue to alter species ranges and regeneration rates, further affecting the health and composition of urban forests (Nowak, 2010; Ordonez et al., 2010). Proactive management is necessary to protect urban forests against climate-related threats, and to sustain desired urban forest structures for future generations.

GEOLOGIC CARBON SEQUESTRATION

Geologic sequestration begins with capturing carbon dioxide from the exhaust of fossil fuel power plants and other major sources. The captured carbon dioxide is piped 1–4 km below the land surface and injected into porous rock formations. Compared to the rates of terrestrial carbon uptake, geologic sequestration is currently used to store only lesser amounts of carbon per year. Much larger rates of sequestration are envisioned to take advantage of the potential permanence and capacity of geologic storage.

The permanence of geologic sequestration depends on the effectiveness of several carbon dioxide trapping mechanisms. After carbon dioxide is injected underground, it will rise buoyantly until it is trapped beneath an impermeable barrier, or seal. In principle, this physical trapping mechanism, which is identical to the natural geologic trapping of oil and gas, can retain carbon dioxide for thousands to millions of years. Some of the injected carbon dioxide will eventually dissolve in groundwater, and some may be trapped in the form of carbonate minerals formed by chemical reactions with the surrounding rock. All these processes are susceptible to change over time following carbon dioxide injection. Scientists are studying the permanence of these trapping mechanisms and developing methods to determine the potential for geologically sequestered carbon dioxide to leak back to the atmosphere.

The capacity for geologic carbon sequestration is constrained by the volume and distribution of potential storage sites. According to the U.S. Department of energy, the total storage capacity of physical traps associated with depleted oil and gas reservoirs in the United States is limited to about 38 gigatons of carbon and is geographically distributed in locations that are distant from most U.S. fossil fuel power plants. The potential U.S. storage capacity of deep porous rock formations that contain saline groundwater is much larger (estimated by the U.S. Department of Energy to be about 900–3,400 gigatons of carbon) and more widely distributed, but less is known about the effectiveness of trapping mechanisms at these sites. Unmineable coal beds have also been proposed for potential carbon dioxide storage, but more information is needed about the storage characteristics and impacts of carbon dioxide injection in these formations. Scientists are developing methods to refine estimates of the national capacity for geologic carbon sequestration.

To fully assess the potential for geologic carbon sequestration, economic costs and environmental risks must be considered. Infrastructure costs will depend on the locations of suitable storage sites. Environmental risks may include seismic disturbances, deformation of the land surface, contamination of portable water supplies, and adverse effects on ecosystems and human health. Many of these environmental risks and potential environmental impacts are discussed in the sections below.

POTENTIAL IMPACT OF TERRESTRIAL SEQUESTRATION

Potential environmental impacts associated with terrestrial sequestration include ground disturbance and the loss of soil resources due to erosion; equipment-related noise, visual impact, and air emissions; the disturbance of ecological, cultural, and paleontological resources, and conflicts with current or proposed land use.

Establishing and managing a terrestrial sequestration plot could involve ground clearing (removal of vegetative cover) to prepare the ground for planting, grading, vehicular traffic, and pedestrian traffic. Management could require the use of water for dust control and in some cases, water could be required to establish and maintain seeds, seedlings, or crops. The addition of soil additives like fertilizers and pesticides could have an impact on water quality. Equipment used to maintain a terrestrial sequestration plot could be a source of noise and air emissions and create a visual impact if frequent and conspicuous use were required.

Ecological, cultural, and paleontological resources could be impacted, especially if a terrestrial sequestration plot were going to replace an established ecological habitat or otherwise impact undisturbed land that hosts important cultural or paleontological resources. Impacts to land use could occur if there were conflicts with existing land-use plans; for example, if land zoned for future commercial or housing development is used to establish a forest sequestration plot.

Soil resources can also be impacted by terrestrial sequestration. The careful management of a sequestration plot should result in an improvement of soil resources, but poor management practices could adversely impact soils and the viability of the sequestration project. Practices like no-till cultivation and plant, crop rotation, and the use of cover crops, should result in the maintenance of soil organic materials and nutrients and an increase in the relative health of soil resources. Some management practices, however, could involve the use of hazardous materials like herbicides to kill a cover crop before planting the terrestrial sequestration crop.

POTENTIAL IMPACT OF GEOLOGIC SEQUESTRATION

The potential impacts of geologic sequestration, including the transportation of carbon, are discussed in this section. For this discussion, we have assumed that carbon capture would likely occur at a single power generating station. Because captured carbon may have to be transported for some distance away from the power station, transport, in general, has been evaluated. The significance of the impacts depends upon factors such as the number and size of transport pipelines and injection wells, the amount of land disturbed by drilling and transport activities, the amount of land occupied by facilities over the life of the sequestration project, the project's location with respect to other resources (e.g., wildlife use, distant to surface water bodies), and so forth.

THE BOTTOM LINE

The key point presented in this chapter that needs to be remembered is that there is a difference between carbon capture and geologic sequestration. Geologic sequestration involves injecting carbon dioxide deep underground where it stays permanently. Now terrestrial, or biologic, sequestration is where carbon is stored via agricultural and forestry practices.

The real bottom line: Current and future environmental scientists/engineers and other environmental professionals need to add carbon capture/sequestration and all that it entails to their professional toolboxes. Remember, there is nothing, absolutely nothing static about the environment and those who work to maintain it.

REFERENCES

Akbari, H., Davis, S., Dorsano, S., Huang, J., Winnett, S. (1992). *Cooling Our Communities: A Guidebook on Tree Planting and Light-Colored Surfacing*. Washington, DC: U.S. Environmental Protection Agency.

Bristow, R.S., Blackie, R., Brown, N. (2012). Parks and the Urban Heat Island: A Longitudinal study in Westfield, Massachusetts. In: Fisher, C.I., Watts Jr., C.E. (eds) *Proceedings of the 2010 Northeastern Recreation Research Symposium*. General Technical Report. NRS-P-94. Newtown Square, PA: U.S. Department of Agriculture, Forest Service, Northern Research Station, pp. 224–230.

Cullington, J. and Gye, J. (2010). Urban Forests: A Climate Adaptation Guide. Part of the BC Regional Adaption Collaborative (RAC). Ministry of Community, Sport, and Cultural Development, British Columbia, Canada.

Dandy, N. (2010). The Social and Cultural Values, and Governance, of Street Trees. Climate Change & Street Trees Project: Social Research Report. The Research Agency of the Forestry Commission, Dodman, London, England.

Dodman, D. (2009). Blaming cities for climate change? An analysis of urban greenhouse gas emissions inventories. *Environment and Urbanization* 21(1):185–201.

Fazio, J.R. (ed). (2010). *How Trees Can Retain Stormwater Runoff.* Tree City USA Bulletin No. 55. Arbor Day Foundation: Nebraska City, NE.

Fish Kill on March 8, 1977 (No. 3); and (4) investigation of Cedar River Flow Fluctuation on May 21, 1977. WDF. Internal memos. Olympia, WA.

Hamilton, W.S. (1983). Preventing cavitation damage to hydraulic structures. *International Water Power and Dam Construction* 35(12):48–53.

Johnston, M. (2004). Impacts and Adaptation for Climate Change in Urban Forests. In *Proceedings of 6th Canadian Urban Forest Conference*, Kelowna, BC.

Klinenberg, E. (2002). *Heat Wave: A Social Autopsy of Disaster in Chicago.* Chicago: University of Chicago Press.

Kovats, S. and Akhtar, R. (2008). Climate, climate change and human health in Asian cities. *Environment and Urbanization* 20:165–175.

McPherson, G. and Muchnick, J. (2005). Effects of street tree shade on asphalt concrete pavement performance. *Journal of Arboriculture* 31(6):303–310.

Myers, S.R., Branas, C.C., French, B.C., Kallan, M.L., Wiebe, D.J., Carr, H.G. (2013). Safety in Numbers: Are Major Cities the Safety Places in the United States? Annals of Emergency Management (Epub ahead of print). Accessed 4/16/2014 @ https://www.annemergmed.com/webfiles/images/jounrnals/ymem/FA-5548.pdf.

Nowak, D.J. (2010). Urban Biodiversity and Climate Change. In: Miller, N., Werner, P., Kelcey, J.G. (eds) *Urban Biodiversity and Design.* Hoboken, NJ: Wiley-Blackwell Publishing, pp. 101–117.

Nowak, D.J. et al. (2010). Sustaining America's Urban Trees and Forests: A Forests on the Edge Report. General Technical Report NRS-62. Newtown Square, PA: U.S. Department of Agriculture, Forest Service, Northern Research Stations.

Nowak, D.J. and Crane, D.E. (2002). Carbon storage and sequestration by urban trees in the USA. *Environmental Pollution* 116:381–389.

Nowak, D.J., Greenfield, E.J., Hoehn, R., LaPoint, E. (2013). Carbon storage and sequestration by trees in urban and community areas of the United States. *Environmental Pollution* 178: 229–236.

Ordonez, C., Dunker, P.N., Steenberg, J. (2010). Climate Change Mitigation and Adaptation in Urban Forests: A Framework for sustainable Urban Forest Management. Paper prepared for presentation at the 18th Commonwealth Forestry Conference, Edinburgh.

Parry, M.L. et al. (eds) (2007). Impacts, Adaptation, and Vulnerability. Contribution of Working Group II to the Fourth Assessment, Report of the Intergovernmental Panel on Climate Change. Cambridge, MA: Cambridge University Press.

Population Reference Bureau (2012). 2012 World Population Data Sheet. Accessed 4/16/2022 @ https://www.prb.org/pdf/2012-population-data-sheet_eng.pdf.

Romero-Lankao, P. and Gratz, D.M. (2008). Urban Areas and Climate Change: Review of Current Issues and Trends. Issues paper for the 2011 Global Report on Human Settlements.

Sherrill, S. and Bratkovitch, S. (2011). *Carbon and Carbon Dioxide Equivalent Sequestration in Urban Forest Products.* Minneapolis, MN: Dovetail Partners, Inc.

The Cities Alliance (2007). Liveable Cities: The Benefits of Urban Environmental Planning. Accessed 4/15/2022 @ https://www.citiesalliance.org/sites/cities alliance.org/files/CA_Docs/resources/cds/liveable/liveablecities_web_7dec07.pdf.

Tubby, K.V. and Webber, J.F. (2010). Pests and diseases threatening urban trees under a changing climate. *Forestry* 83(4):451–59.

USEPA (2009) Buildings and their Impact on the Environment: A Statistical Summary. Accessed 4/16/2022 @ https://www.epa.gov/greenbuilding/pubs/gbstats.pdf.

USEPA (2013) DRAFT Inventory of U.S. Greenhouse Gas Emissions and Sinks: 1990–2011. Accessed 4/16/2014 https://www.epa.gov/climatechange/Downloads/ghgemissions/US-GHG-Inventory-2011.

USGS (2008) *Carbon Sequestration to Mitigate Climate Change*. Washington, DC: US Geological Survey.

Williamson, T., Dubb, S., Alperovitz, G. (2010) *Climate Change, Community Stability, and the Next 150 Million Americans*. College Park, MD: The Democracy Collaboration.

Part 3

The Green Knight

8 Taller than the Trees

Note to the Reader: While it is a common practice to refer to the woods as the forests, the author chooses to use the term woods instead of forests (i.e., where the term woods fits), because having expended years exploring the "woods," I have learned that the term woods is best used when describing or discussing the biomass within—it is the biomass that stores the carbon. Besides, the term forest conveys a thought about trees only. However, the "woods" is an all-encompassing, wholistic term conveying thoughts about the dense, dark, that place filled with shrubs, plants, and, yes, trees—including animals. Including "Big Foot?" Anyway, I use the terms woods and forests where the term fits.

LARGEST CARBON SINK

The woods (aka forests) are the largest terrestrial carbon sink in the world. In the author's view, it is the woods that is the Green Knight of environmental preservation—our preservation (hopefully). The woods absorb carbon dioxide and store it as carbon (C) in soil and woody plants. The truth be told, in the United States, woods, wood products, and urban trees collectively offset annual carbon dioxide emissions by about 10%–15%. Woods health and conditions, management practices, disturbances such as fire, and wood harvesting and use all influence how much carbon is stored and released from woods over time (this is referred to as *carbon flux*). Tracking these dynamic processes over extensive landscapes is significantly important for understanding and employing the power of the woods to curb our greenhouse gas footprint.

Estimates from wood land uses and land use change are determined by the USDA Forest Service Forest Inventory and Analysis (FIA) program legislated by Congress to conduct the Nation's forest census. Local-, State-, and national-level policy makers use FIA estimates on the carbon content among forest types to estimate the carbon benefits from land management activities and inform climate change mitigation measures.

CARBON STORAGE

Woods ecosystems in the United States contained just under 60 gigatons of carbon in 2019; this weighs out at more than 6% of the global terrestrial biomass. Terrestrial biomass generally decreases markedly at each higher trophic level (plants, herbivores, carnivores). Examples of terrestrial producers are grasses, trees, and shrubs.

With trophic levels it is important to point out that all organisms, alive or dead, are potential sources of food for other organisms. All organisms that share the same general type of food in a food chain are said to be at the same *trophic level* (aka nourishment or feeding level). Since green plants use sunlight to produce food for animals,

DOI: 10.1201/9781003432838-11

they are called the *producers*, or the first trophic level. The herbivores, which eat plants directly, are called the second trophic level or the *primary consumers*. The carnivores are flesh-eating consumers; they include several trophic levels from the third on up. At each transfer, a large amount of energy (about 80%–90%) is lost as heat and waste. Thus, nature normally limits food chains to four or five links. In aquatic ecosystems, however, food chains are commonly longer than those on land. The aquatic food chain is longer because several predatory fish may be feeding on the plant consumers. Even so, the built-in inefficiency of the energy transfer process prevents development of extremely long food chains.

Only a few simple food chains are found in nature. Most simple food chains are interlocked. This interlocking of food chains forms a *food web*. Most ecosystems support a complex food web. A food web involves animals that do not feed on one trophic level. For example, humans feed on both plants and animals. An organism in a food web may occupy one or more trophic levels. The trophic level is determined by an organism's role in its community, not by its species. Food chains and webs help to explain how energy moves through an ecosystem.

An important trophic level of the food web is comprised of the *decomposers*. The decomposers feed on dead plants or animals and play a key role in recycling nutrients in the ecosystem. Simply, there is no waste in ecosystems. All organisms, dead or alive, are potential sources of food for other organisms.

It is important to note that terrestrial producers have a much higher biomass than the animals (including us to an extent) who consume them, including deer, zebras, and insects. The levels with the lowest biomass are the highest predators in the food chain such as wolves, foxes, and eagles.

Note that most of the carbon stored in woods ecosystem pools (95%), with the remainder as harvested wood products. It is important to remember that the trees and plants in the woods, even when they fall due to winds or human interference, still contain carbon and this carbon stays with fallen trees and shrubs until they rot. Also, the wood that is harvested to produce wood products, such as furniture retain certain amounts of carbon until the wood is destroyed by fire, rot, or abuse. Wood soils also store carbon in copious amounts (54%), followed by aboveground biomass (26%). Note that in active carbon removal live vegetation has the largest impact, accounting for more than 84% (0.5 metric tons of carbon per hectare per year) of the uptake (USDA, 2021).

DID YOU KNOW?

Recall that it was pointed out earlier that when we clear land, especially forests, we remove a dense growth of plants that had stored carbon in biomass—wood, stems, and leaves. Trees take carbon out of the atmosphere as they grow and when we clear the forest we eliminate this vital function—we eliminate plants that would otherwise take carbon out of the atmosphere as they grow. Our tendency is to replace the dense growth with crops or pasture, which store less carbon. This tendency is also negative in that we expose soil that vents carbon from decayed plant matter into the atmosphere. These land use changes are currently emitting almost a billion tons of carbon into the atmosphere per year. In 2004 it was estimated that in the United States alone more than 100 billion gallons of light fuel oil per year can be produced from waste and biomass. The source waste and biomass included in this estimation included municipal solid waste, municipal sewage sludge, Hazmat waste, agricultural crop waste, feedlot manure, plastics, tires, heavy oil or tar sands, forestry waste, restaurant grease, and biomass crops (switchgrass) which grow on idle land and cropland.

DID YOU KNOW?

Most states in the conterminous United States are carbon sinks, yet in recent years, several intermountain western states have become a source of carbon (USDA, 2021).

REFERENCE

USDA (2021). *Forest Carbon Status and Trends.* Washington, DC: U.S. Department of Agriculture.

9 Wood-Based Biomass
Heat Energy and Weight

INTRODUCTION

As a steppingstone to the advancement of knowledge about the importance of trees, underbrush, and soil (the woods) and about their huge role as carbon storage sinks, USDA (1979) points out that wise utilization of the forest resource relates to awareness of its value. This chapter explains the value of forests in storing carbon, but initially focuses on the basics to gain understanding of the value of trees and forest products besides their ability to store carbon. So, in this chapter, you are introduced to the value of forests products beyond their ability to store carbon, and we begin our discussion with the heat value of the biomass contained in forests—forest or woods are all about biomass content.

FUEL/HEAT VALUE OF FORESTS

The amount of heat energy that can be recovered from wood or bark determines its fuel value. The amount of recoverable heat energy varies with moisture content and chemical composition. Recoverable heat energy varies among tree species and even within a species. In this chapter, we provide a summary of information which may be used to estimate recoverable heat energy in wood or bark fuel, biomass weight considerations, biomass weight examples, and stand level biomass estimation.

Lower/Higher Heating Values

The *lower heating value* (LHV, also known as net calorific value) of a fuel is defined as the amount of heat released by combusting a specified quantity (initially at 25°C) and returning the temperature of the combustion products to 150°C, which assumes that the latent heat of vaporization of water in the reaction produces is not recovered. The LHVs are the useful calorific values in boiler combustion plants and are frequently used in Europe.

The *higher heating value* (HHV, also known as gross calorific value or gross energy) of a fuel is defined as the amount of heat released by a specified quantity (initially at 25°C) once it is combusted and the products have returned to a temperature of 25°C, which considers the latent heat of vaporization of water in the combustion products. The HHVs are derived only under laboratory conditions and are frequently used in the United States for solid fuels.

DOI: 10.1201/9781003432838-12

Key Lower and Higher Heating Values of Solid Fuels include the following:

- Btu = British thermal unit.
- The heating values in units of MJ/kg are converted form heating values in units of Btu/lb.
- For solid fuels, the heating values in units of BTU/lb are converted from the heating values in units of Btu/ton.

The lower and higher heating values of solid forest-based biomass fuels are listed in Table 9.1.

Table 9.2 attempts to capture the variation in reported heat content values (on a dry weight basis) in the US and European Literature based on values in the Phyllis database, the US DOE/EERE feedstock database, and selected literature sources. Metric values show both HHV and LHV since Europeans normally report the LHV (or net calorific values) of biomass fuels.

Effect of Fuel Moisture on Wood Heat Content

Because recently harvested wood fuels usually contain 30%–55% water, it is useful to understand the effect of moisture content on the heating value of wood fuels. Table 9.3 shows the effect of percent moisture content (MC) on the higher heating value as-fired (HHV-AF) of a wood sample starting at 8,500 Btu/lb (oven-dry).

Fuel moisture content is usually reported as the wet weight basis moisture content. Moisture content expressed on a wet weight basis (also called "green" or "as-fired" moisture content) is the decimal fraction of the fuel that consists of water. For example, a pound of wet wood fuel at 50% moisture content contains 0.50 pound of water and 0.50 pound of wood. Note that the wet weight basis differs from the total dry weight basis method of expressing moisture content which is more commonly used for describing moisture content of finished wood products. The dry weight basis is the ratio of the weight of water in wood to the oven-dry weight of the wood. The formulas used in the paper require the moisture content being expressed on the wet weigh basis (USDA, 1979).

TABLE 9.1
Lower and Higher Heating Values of Solid Fuels

Fuels	**Lower Heating Value (LHV)**			**Higher Heating Value (HHV)**		
Solid fuels	Btu/ton	Btu/lb	MJ/kg	Btu/ton	Btu/lb	MJ/kg
Farmed trees	16,811,000	8,406	19,551	17,703,170	8,852	20,589
Forest residue	13,243,490	6,622	15,402	14,164,160	7,082	16,473

Source: GREET (2008). *Transportation Fuel Cycle Analysis Model*, Greet 1.8b, developed by Argonne National Laboratory, Argonne, IL.

TABLE 9.2
Heat Content Ranges for Forest Biomass Fuels (Dry Weight Basis)

Fuel Type and Source	English		Metric			
	Higher Heating Value		Higher Heating Value		Lower Heating Value	
	Btu/lb	MBtu/ton	kJ/kg	MJ/kg	kJ/kg	MJ/kg
Woody Crops						
Black locust	8,409–8,582	16.8–17.2	19,547–19,948	19.5–19.9	18,464	18.5
Eucalyptus	8,174–8,432	16.3–16.9	19,000–19,599	19.0–19.6	17,963	18.0
Hybrid popular	8,183–8,491	16.4–17.0	19,022-19,737	19.0–19.7	17,700	17.7
Willow	7,983–8,497	16.0–17.0	18,556–19,750	18.6–19.7	16,734–18,419	16.7–18.4
Forest Residues						
Hardwood wood	8,017–8,920	16.0–17.5	18,635–20,734	18.6–20.7	–	–
Softwood wood	8,000–9,120	16.0–18.24	18,595–21,119	18.6–21.1	17,514–20,768	17.5–20.8

Sources: http://www1.eere.energy.gov/biomass/feedstock_databases.html. Retrieved 12/13/22; Jenkins, B. (1993). *Properties of Biomass*, Appendix to Biomass Energy Fundamentals, EPRI Report TR-102107; Jenkins, B.M., Baxter, L.L., Miles Jr., T.R., Miles (1998). *Combustion Properties of Biomass, Fuel Processing Technology* 54, pp. 17–46; Tillman, D. (1978). *Wood as an Energy Resource*. New York, Academic Press; Bushnell, D. (1989). *Biomass Fuel Characterization: Testing and Evaluating the Combustion Characteristics of Selected Biomass Fuels*, BPA report.

TABLE 9.3
The Effect of Fuel Moisture on Wood Heat Content

	Moisture Content (MC) Wet Basis										
(%)	0	15	20	25	30	35	40	45	50	55	60
	Higher Heating Value as-Fired (HHV-AF)										
Btus/lb	8,500	7,275	6,800	6,375	5,950	5,525	5,100	4,575	4,250	3,825	3,400

Sources: Borman, G.L. and Ragland, K.W. (1998). *Combustion Engineering*. New York: McGraw-Hill; Maker, T.M. (2004). *Wood-Chip Heating Systems: A Guide for Institutional and Commercial Biomass Installations*, Biomass Energy Resource Center; American Pulpwood Association, Southern Division Office (1980) *The Forester's Wood Energy Handbook.*

Moisture Content (MC) Wet and Dry Weight Basis Calculations

Moisture contents (MC) wet and dry weight basis are calculated as follows:

$$\text{MC (dry basis)} = 100\ (\text{wet weight} - \text{dry weight})/\text{dry weight} \tag{9.1}$$

$$\text{MC (wet basis)} = 100\ (\text{wet weight} - \text{dry weight})/\text{wet weight} \tag{9.2}$$

To convert MC wet basis to MC dry basis:

$$\text{MC (dry)} = 100 \times \text{MC(wet)}/100 - \text{MC(wet)} \tag{9.3}$$

To convert MC dry basis to MC wet basis:

$$\text{MC (wet)} = 100 \times \text{MC(dry)}/100 + \text{MC(dry)} \tag{9.4}$$

Some sources report heat contents of fuels "as-delivered" rather than a 0% moisture for practical reasons. Because most wood fuels have bone dry (oven-dry) heat contents in the range of 7,600– 9,600 Btu/lb (15,200,000–19,200,000 Btu/ton or 18–22 GJ/Mg), lower values will always mean that some moisture is included in the delivered fuel.

FORESTRY VOLUME UNIT TO BIOMASS WEIGHT CONSIDERATIONS

Biomass is frequently estimated from forestry inventory merchantable volume data, particularly for purposes of comparing regional and national estimates of aboveground biomass and carbon levels. Making such estimations can be done several ways but always involves the use of either conversion factors or biomass expansion factors (or both combined), which defines what is included in each category of volume or biomass units.

Total volume or biomass includes stem, bark, stump, branches, and foliage, especially if evergreen trees are being measured. When estimating biomass available for bioenergy, the foliage is not included, and the stump may or may not be appropriate to include depending on whether harvest occurs at the ground level or higher. Both conversion and expansion factors can be used together to translate directly between merchantable volumes per unit area and total biomass per unit area, as demonstrated by the simple volume to weight conversion process shown below.

Estimation of Biomass Weights from Forestry Volume Data

Equation 9.5 is for estimation of merchantable biomass from merchantable volume assuming that the specific gravity and moisture content are known, and the specific gravity basis corresponds to the moisture content of the volume involved (Briggs, 1994). Specific gravity (SG) is a critical element of the volume to biomass estimation equation. The SG content should correspond to the moisture content of the volume involved. SG varies considerably from species to species, differs for wood and bark, and is closely related to the moisture content as explained in graphs and tables in Briggs (1994). The wood specific gravity of species can be found in several references through the moisture content basis is not generally given. Briggs (1994) suggests that a moisture content of 12% is the standard upon which many wood property measurements are based.

$$\text{Weight} = (\text{volume})(\text{specific gravity})(\text{density of } H_2O)\left(1 + MC^{od}/100\right) \quad (9.5)$$

where

volume = expressed in cubic feet or cubic meters
density of water = 62.4 lb/ft^3 or 1000 kg/m^3
MC^{od} = oven-dry moisture content.

Example 9.1

Problem:

Weight of fiber in an oven-dry log of 44 ft^3 with a specific gravity of 0.40

Solution

Weight = (40 ft^3) (0.40) (62.4 lb ft^3) (1 + 0/100) = 1,098 lb or 9.549 dry ton

Biomass Expansion Factors (BEF)

Methods for estimating total aboveground dry biomass per unit area from growing stock volume data in the USDA Forest Service FIA database were described by Schroeder et al. (1997). The growing stock volume was limited to trees > than or equal to 12.7 cm diameter. It is highly recommended that the paper be studied for details of how the biomass expansion factors (BEF) for oak-hickory and beech-birch were developed.

STAND LEVEL BIOMASS ESTIMATION

At the individual field or stand level, biomass estimation is relatively straight forward, especially if being done for plantation grown trees that are relatively uniform in size and other characteristics. The procedure involves first developing a biomass equation that predicts individual tree biomass as a function of diameter at breast height (dbh), or of dbh plus height. Secondly, the equation parameters (dbh and height) need to be measured on a sufficiently large sample size to minimize variation around the mean values, and thirdly, the mean individual tree weight results are scaled to the area of interest based on percent survival or density information (trees per acre or hectare). Regression estimates are developed by directly sampling and weighing enough trees to cover the range of sizes being included in the estimation. They often take the form of:

$$\begin{aligned} &\ln Y \ (\text{weight in Kg}) = -\text{factor } 1 + \text{factor } 2 \times \ln X \\ &\left(\text{where } X \text{ is dbh or dbh}^2 + \text{height}/100\right) \end{aligned} \quad (9.6)$$

Regression equations can be found for many species in a wide range of literature. Examples of trees common to the Pacific Northwest are provided in Briggs (1994). The equations will differ depending on whether foliage or live branches are included, so care must be taken in interpreting the biomass data. For plantation trees grown on cropland or marginal cropland it is usually assumed that tops and branches are included in the equations, but that foliage is not. For trees harvested from forests on lower quality land, it is usually recommended that tops and branches should not be removed (PA Biomass, 2007) to maintain nutrient status and reduce erosion potential, and thus biomass equations should assume regressions based on the stem weight only.

DID YOU KNOW?

Regarding scaling and log rules associated with timber, the measurement of timber to be harvested (the cruise), timber cut and removal from the forest (scaling), timber not recovered in the harvesting process (waste processing), and the use of formulae or tables to estimate net yield for logs (log rule) are the basis of forest-based biomass harvesting operations.

BIOMASS EQUATIONS[1]

The intensity of forest utilization has increased in recent years because of whole-tree harvesting and the use of wood for energy. Estimating tree biomass (weight) based on parameters that are easily measured in the field is becoming a fundamental task in forestry and forest-based biomass technology. Traditionally, cubic-foot or board-foot volume of merchantable products, such as sawlogs or pulpwood, adequately described forest stands. However, as mentioned, the intensity of forest utilization has increased in recent years because of whole-tree harvesting and the use of wood for energy. All aboveground branches, leaves, bark, small trees, and trees of poor form or vigor are now commonly included in the harvested product and are listed as biomass of whole trees (WT) or individual components. With this increasing emphasis on complete tree utilization and use of wood as a source of energy, tables and equations have been developed to show the whole-tree biomass as weights of total trees and their components.

Numerous equations for estimating tree biomass from dry weight in kilograms and dbh (tree diameter at breast height in centimeters) have been developed by numerous researchers based on local and tree species. For example, Landis and Mogren (1994) developed an equation for estimating biomass of individual Engelmann spruce trees employing the following model:

$$Y + b_o + b_{1\,\text{dbh}}^{\;2} \tag{9.7}$$

[1] Based on material contained in USDA, Tritton, L.M. and Hornbeck, J.W. (1982) *Biomass Equation for Major Tree Species of the Northeast.* Washington, DC: United States Department of Agriculture.

where Y is tree component dry weight in kilograms and *dbh* is the tree diameter at breast height in centimeters.

Similar sets of equations have been developed for other species and locations. Examples of many common equations used in the Northeastern United States are presented in below. First, however, it is important to point out that regression equations are used to estimate tree biomass in both forestry and ecosystem studies. These equations are typically developed in the following way: samples of major tree species are chosen for study, selected dimensions of each tree are recorded, the tree is felled and weighed either whole or in pieces, and subsamples are oven-dried and weighed again to determine tree moisture content. (Tree green weights are converted to dry weights by using moisture content values.) Because biomass is related to tree dimensions, regression analysis is used to estimate the constants or regression coefficients required for the actual calculation of biomass. The resultant regression equations may be used to estimate the biomass by species, of all trees for which dimensional data are available. The equations shown below are of several different forms; they can be used to predict biomass (y) from dbh or dbh and height (x). The most common forms used are allometric, exponential, and quadratic.

Key to Abbreviations Used in Example Biomass Equations

Lf	leaf biomass
Tw	twig biomass
Br	branch biomass
DdBr	dead branch biomass
St	stem biomass
St+Br	steam and branch biomass but not foliage
Rt	root and stump biomass
WT	whole-tree biomass-all aboveground components including leaves, branches, and stems
CT	complete tree biomass-aboveground and below-ground components: whole tree plus roots and stump
LOG	logarithm to the base 10
LN	natural logarithm to the base e
wt	weight measure din pounds (lb), grams (gm), or kilogram (kg)
dbh	diameter at breast height (1.37 m) measured in inches (in), millimeters (mm), or centimeters (cm)

Tree Species/Biomass Example Equations

Biomass Equation 1 (Young et al., 1980)

Balsam fir (*Abies balsams*) WT: $\text{LN wt} = 0.5958 + 2.4017\ \text{LN dbh}$

Biomass Equation 2 (Young et al., 1980)

Red Maple (*Acer rubrum*) WT: $\text{LN wt} = 0.9392 + 2.3804\ \text{LN dbh}$

Biomass Equation 3 (Whittaker et al., 1974)

Sugar Maple (*Acer saccharum*) St: $\text{LOG wt} = 2.0877 + 2.3718\ \text{LOG dbh}$

Br: $\text{LOG wt} = 0.6266 + 2.9740 \text{ LOG dbh}$

DdBr: $\text{LOG wt} = 0.0444 + 2.2803 \text{ LOG dbh}$

Lf + Tw: $\text{LOG wt} = 1.0975 + 1.9329 \text{ LOG dbh}$

Biomass Equation 4 (Ribe, 1973)

Yellow Birch (*Betula alleghaniensis* Britt.)

Lf: $\text{LOG wt} = 1.9962 + 1.9683 \text{ LOG dbh}$

Br: $\text{LOG wt} = 2.5345 + 1.6179 \text{ LOG dbh}$

St: $\text{LOG wt} = 2.9670 + 2.5330 \text{ LOG dbh}$

Biomass Equation 5 (Brenneman et al., 1978)

Black birch (*Betula lenta*) WT: $\text{wt} = 1.6542 \text{ dby}^{2.6606}$

Biomass Equation 6 (Kinerson and Bartholomew, 1977)

Paper birch (*Betula papyrifera* Marsh)

St: $\text{LN wt} = 3.720 + 2.877 \text{ LN dbh}$

Br: $\text{LN wt} = -1.351 + 4.368 \text{ LN dbh}$

Biomass Equation 7 (Young et al., 1980)

Gray birch (*Betula populifolia* Marsh)

Wt: $\text{LN wt} = 1.0931 + 2.3146 \text{ LN dbh}$

Biomass Equation 8 (Wiant et al., 1977)

Hickory (*Carya* spp.) St + Br: $\text{wt} = 1.93378 \text{ dbh}^{2.62090}$

Biomass Equation 9 (Ribe, 1973)

Beech (*Fagus grandifolia* Ehrh.) Lf: $\text{LOG wt} = 2.0660 + 1.8089 \text{ LOG dbh}$

Br: $\text{LOG wt} = 2.5983 + 1.5402 \text{ LOG dbh}$

St: $\text{LOG wt} = 3.0692 + 2.4868 \text{ LOG dbh}$

Biomass Equation 10 (Brenneman et al., 1978)

White ash (*Fraxinus americana*) WT: $\text{wt} = 2.3626 \text{ dbh}^{2.4798}$

Biomass Equation 11 (MacLean and Wein, 1976)

Aspen (*Populus* spp.) WT: $\text{LOG wt} = -0.7891 + 2.0673 \text{ LOG dbh}$

Biomass Equation 12 (MacLean and Wein, 1976)

Spruce (*Picea* spp.) WT: $\text{LOG wt} = -0.2112 + 1.5639 \text{ LOG dbh}$

Biomass Equation 13 (Dunlap and Shipman, 1967)

Red pine (*Pinus resinosa* Ait.) St: $\text{wt} = -113.954 + 35.265 \text{ (dbh)}$

Biomass Equation 14 (Swank and Schreuder, 1974)

White pine (*Pinus strobus*) Lf: $\text{LN wt} = 3.051 + 2.1354 \text{ LN dbh}$

Br: $\text{LN wt} = 3.158 + \text{e}2.5328 \text{ LN dbh}$

St: $\text{LN wt} = -2.788 + 2.1338 \text{ LN dbh}$

Biomass Equation 15 (Hitchcock, 1978)

Yellow poplar (*Liriodendron tulipifera*)

St + Br: $\text{LOG wt} = 1.9167 + 0.7993 \text{ LOG } (\text{dbh}^2 \times \text{ht})$

Biomass Equation 16 (Young et al., 1980)

Pin cherry (*Prunus pensylvanica*) WT: $\text{LN wt} = 0.9758 + 2.1948 \text{ LN dbh}$

Biomass Equation 17 (Wiant et al., 1979)

Black cherry (*Prunus serotina* Ehrh.)

St + Br: $\text{wt} = 0.12968 \ (\text{dbh}^2 \times \text{ht})^{0.97028}$

Biomass Equation 18 (Reiners, 1972)

White oak (*Quercus alba*) Lf: $\text{LOG wt} = 2.1426 + 1.6684 \text{ LOG dbh}$

Biomass Equation 19 (Clark et al., 1980a)

Scarlet oak (*Quercus coccinea*) St + Br: $\text{wt} = 0.12161 \ (\text{dbh}^2 \times \text{ht})^{1.00031}$

Biomass Equation 20 (Wiant et al., 1979)

Chestnut oak (*Quercus prinus*) St + Br: $\text{wt} = 0.06834 \ (\text{dbh}^2 \times \text{t})^{1.06370}$

Biomass Equation 21 (Clark et al., 1980b)

Northern red oak (*Quercus rubra*) WT: $\text{wt} = 0.10987 \quad (\text{dbh}^2 \times \text{ht})^{1.00197}$

Biomass Equation 22 (Bridge, 1979)

Black oak (*Quercus velutina*) WT: $\text{LN wt} = -0.34052 + 2.65803) \text{ LN dbh}$

Biomass Equation 23 (Young et al., 1980)

Hemlock (*Tsuga canadensis*) WT: $\text{LN wt} = 0.6803 + 2.3617 \text{ LN dbh}$

Biomass Equation 24 (Monk et al., 1970)

General Hardwoods WT: LOG wt = 1.9757 + 2.5371 LOG dbh

Biomass Equation 25 (Monteith, 1979)

General Softwoods WT: $wt = 4.5966 - (0.2364 \times dbh) + (0.00411 \times dbh^2)$

REFERENCES AND RECOMMENDED READING

Brenneman, B.B., Frederick, D.J., Gardner, W.E., Schoenhofen, L.H., Marsh, P.L. (1978). Biomass of Species and Stands of West Virginia Hardwoods. In Pope, P.E. (ed.) *Proceedings Central Hardwood Forest Conference II.* West LaFayette, IN: Purdue University.

Bridge, J.A. (1979). *Fuelwood Production of Mixed Hardwoods on Mesic Sites in Rhode Island.* Kingston, RI: University of Rhode Island.

Briggs, D. (1994). *Forest Products Measurements and Conversion Factors*, Chapter 1. Seattle, WA: College of Forest Resources University of Washington.

Clark, A., III, Phillips, D.R., Hitchcock, H.C. (1980a). Predicted Weights and Volumes of Scarlet Oak Trees on the Tennessee Cumberland Plateau. U.S. Department of Agriculture, Forest Service. Research Paper SE-214.

Clark, A. III and Schroeder, J.G. (1980b). Biomass of Yellow Polar in Natural Stands in Western North Carolina. U.S. Department of Agriculture Forest Service Note NE-230.

Dunlap, W.H. and Shipman, R.D. (1967). *Density and Weight Prediction of Standing White Oak, Red Maple, and Red Pine.* University Park: Pennsylvania State University School of Forest Resources.

Hitchcock, III, H.C. (1978). Aboveground tree weight equations for hardwood seedlings and saplings. *TAPPI* 61(10):119–120.

Kinerson, R.S. and Bartholomew, I. (1977). Biomass Estimation Equations and Nutrient Composition of White Pine, White Birch, Red Maple, and Red Oak in New Hampshire. New Hampshire Agricultural Experiment Station. Research Report 62.

Landis, T.D. and Mogren, E.W. (1994). Tree strata biomass of subalpine spruce-fir stands in southwestern Colorado. *Forest Science* 21(1):9–14.

MacLean, D.A. and Wein, R.W. (1976). Biomass of jack pine and mixed hardwood stands in northeastern New Brunswick. *Canadian Journal of Forest Research* 6(4):441–447.

Monk, C.D., Child, G.I., Nicholson, S.A. (1970). Biomass, litter, and leaf surface area estimates of an oak-hickory forest. *Oikos* 21:138–141.

Monteith, D.B. (1979). *Whole-Tree Weight Table for New York.* Syracuse: Univeristy of New York.

NREL (2010). Learning About Renewable Energy. Accessed 12/14/10 @ https://www.nrel.gov/learning/re_bimomass.html.

PA Biomass (2007). Guidance on Harvesting Woody Biomass for Energy in Pennsylvania. Pennsylvania Department of Conservation and Natural Resources. Accessed 12/14/10 @ https://www.dcnr.state.pa.us/PA_Biomass_guidance_final.pdf.

Reiners, W.A. (1972). Structure and energetics of three Minnesota forests. *Ecological Monography* 42(1):71–94.

Ribe, J.H. (1973). *Puckerbrush Weight Tables.* Orono, ME: University of Maine.

Schroeder, P., Brown, S., Mo, J., Birdsey, R., Cieszewski, C. (1997). Biomass estimation of temperate broadleaf forests of the US using forest inventory data. *Forest Science* 43:424–434.

Swank, W.T. and Schreuder, H.T. (1974). Comparison of three methods of estimating surface area and biomass for a forest of young eastern white pine. *Forest Science* 20:91–100.

USDA (1979). *How to Estimate Recoverable Heat Energy in Wood or Bark Fuels.* Washington, DC: U.S. Department of Agriculture.

Whittaker, R.H., Bormann, F.H., Likens, G.E., Siccama, T.G. (1974). The Hubbard Book ecosystem study: Forest biomass and production. *Ecological Monographs* 4:233–254.

Wiant, H.V. Jr., Castaneda F., Sheetz, C.E., Colaninno, A, DeMoss, J.C. (1979). Equations for predicting weights of some Appalachian hardwoods. *WV Forestry Notes* 7:21–28.

Wiant, H.V. Jr., Sheetz, C.E., Colaninno, A., DeMoss, J.C., Castaneda, F. (1977). Tables and procedures for estimating weights of some Appalachian hardwoods. *WV Agricultural and Forestry Experiment Station Bulletin* 659T. 8 p.

Young, H.E., Ribe, J.H, Wainwright, K. (1980). *MR230: Weight Tables for Tree and Shrub Species in Maine.* Orono, ME: University of Maine: Life Science and Agriculture Experiment Station. 84 p.

10 Forest Tree Carbon

INTRODUCTION

Note: when describing forests and their ability to store carbon, in this book, we consider the entire system: the standing trees, the shrub layer, litterfall, the dead materials on the forest floor, and the soil (covered in Chapter 11).

Forests are huge carbon sinks—the Green Knights of protecting our environment. Annually, America's forests sequester over 800 million tons of carbon. Depending on the year, forests can store about 12% of the United States annual emissions. Forests store or sequester carbon in biomass contained in trees and soil (see Chapter 11). Photosynthesis pulls carbon out of the atmosphere to make sugar in trees, but they release carbon dioxide back into the atmosphere when they decompose. It is a cyclical process that captures and releases carbon and other gases within forests. Careful management of forests enables these cycles and enhances carbon capture.

When one studies carbon storage and sequestration it becomes quite clear almost right away that forests are the best natual mechanisms in capturing carbon. When they accomplish photosynthesis, they pull carbon dioxide out of the air, bind it up in sugar—to build wood, branches, and roots—and release oxygen. Because trees are composed of about 50% carbon, they are incredible carbon sinks that last for years as standing trees and continue to last for years after the tree dies. Note that even though trees store carbon they also release some carbon because of leaf fall and decomposition. Trees also give up carbon when their roots burn sugar to capture nutrients and water.

Healthy trees have a dynamic cycle in capturing and storing carbon. For example, consider the White Pine. This variety of trees lives a long life before disease, wind, or fire kills it. But when dead, not completely consumed by fire, the tree begins the decomposition phase of the cycle, slowly breaking down, but the rotten tree is still holding carbon from the atmosphere.

The variables influencing the amount of carbon a tree or stand of trees can capture and store depend on the trees' age, number of trees in the stand, and amount of physical space available. Young forest trees are excellent carbon capturers. Young forests have many trees that grow quickly and can capture carbon rapidly. Young trees compete for light, growing space, and resources, but when they die and decompose little carbon is released. The trees that survive and continue to grow and mature continue to sequester more carbon.

Mature forests are made up of a variety of plants and trees, which are medium to large, healthy and have very large root systems. As the trees age or mature they grow slower than the younger trees, but the amount of carbon stored by the older trees is greater. When the larger trees die, they are quickly replaced by younger trees that take advantage of the new space available. Because there are more trees growing than dying carbon capture and storage is enhanced.

DOI: 10.1201/9781003432838-13

In old growth forests it is the large trees that dominate and shade out small saplings, and thus the recruitment of young trees is nearly zero. The old growth forest carbon cycle is fixed and less dynamic. The plus side is that large, old trees contain carbon throughout the tree, from bud to root, slowly rotting logs contain carbon, and thick leaf litter and soil contain carbon.

FOREST BIOMASS SAMPLING[1]

> Most human decisions are made with incomplete knowledge. In daily life, a physician may diagnose disease from a single drop of blood or a microscopic section of tissue; a homemaker judges a watermelon by its "plug" or by the sound it emits when thumped; and amid a bewildering array of choices and claims we select toothpaste, insurance, vacation spots, mates, and careers with but a fragment of the total information necessary or desirable for complete understanding. All these we do with the ardent hope that the drop of blood, the melon plug, and the advertising claim give a reliable picture of the population they represent.
>
> —*Frank Freese, 1976*

Key Point: Monitoring, modeling, measuring, sampling, and surveying carbon in forests can be thought of in terms of carbon per area, and area. Simply, the change in forest carbon can be estimated by obtaining total forest carbon at two separate times, subtracting them, and dividing the by the years between the surveys to determine annul average change.

Freese points out above that partial knowledge is a normal state of doing business in many professions; the same can be said for the practice of forestry. The complete census is rare—the sample is commonplace. A forester must advertise timber sales with estimated volume, estimated grade yield and value, estimated cost, and estimated risk. The nurseryperson sows seed whose germination is estimated from a tiny fraction of the seed lot, and at harvest he or she estimates the seedling crop with sample counts in the nursery beds. Enterprising pulp companies, seeking a source of raw material in sawmill residue, may estimate the potential tonnage of chippable material by multiplying reported production by a set of conversion factors obtained at a few representative sawmills (Freese, 1976).

On the surface, and in many cases, it would seem better to measure and not to sample. However, there are several good reasons why sampling is often preferred. In the first place, complete measurement or enumeration may be impossible. That is, not all units in the population can be identified. For example, how does one accurately count each branch or twig on a tree? How do we figure out the extent of the tree's root system? How do we test the quality of every drop of water in a reservoir? How do we weigh every fish in a stream, or count all seedlings in a 1,000-bed nursery, enumerate all the egg masses in a turpentine beetle infestation, or measure the diameter and height of all merchantable trees in a 20,000-acre forest? Moreover, the nurseryperson might be somewhat better informed if he or she knew the germinative capacity of all

[1] This section is based on material in Spellman, F.R. (2011). *Forest-Based Biomass Energy*. Boca Raton, FL: CRC Press.

the seed to be sown, but the destructive nature of the germination test precludes testing every seed. For identical reasons, it is impossible to conduct tests that are destructive on every chainsaw without destroying every chainsaw. Likewise, it is impossible to measure the bending strength of all the timbers to be used in a bridge, the tearing strength of all the paper to be put into a book, or the grade of all the boards to be produced in a timber sale. If the tests were permitted, no seedlings would be produced, no bridges would be built, no books printed, and no stumpage sold. Clearly where testing is destructive some sort of sampling is inescapable. Obviously, the enormity of the counting task or the destructive effects of testing demand some sort of sampling procedure.

Sampling will frequently provide the essential information at a far lower cost and in less time than a complete enumeration. Surveying 100% of the lumber market is not going to provide information that is very useful to a seller if it takes 11 months to complete the job. In addition, it is often the case that sampling information may at times be more reliable than that obtained by a 100% inventory. There are several reasons why this might be true. With fewer observations to be made and more time available, measurement of the units in the sample can be and is more likely to be made with greater care. Moreover, a portion of the saving resulting from sampling could be used to buy better instruments and to employ or train higher caliber personnel. It is not hard to see that good measurements on 5% of the units in a population could provide more reliable information than sloppy measurements on 100% of the units.

The bottom line in sampling, to make it effective and accurate, is to obtain reliable data from the population sampled and to make certain inferences about that population are correct. How well the sampling is done depends on items such as the rule by which the sample is drawn, the care exercised in measurement, and the degree to which bias can be avoided (Avery and Burkhart, 2002).

TERMS AND CONCEPTS

The following terms required an expanded discussion beyond the scope of the Glossary:

1. **Inventory**—is the systematic acquisition and analysis of information needed to describe, characterize, or quantify vegetation. As might be expected, data for many different vegetation attributes can be collected. Inventories can be used not only for mapping and describing ecological sites, but also for determining ecological status, assessing the distribution and abundance of species, and establishing baseline data for monitoring studies.
2. **Population**—a population (used here in the structural, not biological, sense) is a complete collection of objects (usually called units) about which one wishes to make statistical inferences. Population units can be individual plants, points, plots, quadrats, or transects.
3. **Sampling unit**—is one of a set of objects in a sample that is drawn to make inferences about the population of those same objects. A collection of sampling units is a sample. Sampling units can be individual plants, points, plot, quadrats, or transects.

4. **Sample**—is a set of units selected from a population used to estimate something about the population (statisticians call this making inference about the population). To properly make inferences about the population, the units must be selected using some random procedure. The units selected are called sampling units.
5. **Sampling**—is a means by which inferences about a plan community can be made based on information from an examination of a small proportion of that community. The most complete way to determine the characteristics of a population is to conduct a complete enumeration of census. In a census, each individual unit in the population is sampled to provide the data for the aggregate. This process is both time-consuming and costly. If may also result in inaccurate values when individual sampling units are difficult to identify. Therefore, the best way to collect vegetation data is to sample a small subset of the population. If the population is uniform, sampling can be conducted anywhere in the population. However, most vegetation populations are not uniform. It is important that data be collected so that the sample represents the entire population. Sample design is an important consideration in collected representative data.
6. **Shrub characterization**—is addressed here because it is not covered in most of the techniques in this text. Shrub characterization is the collection of data on the shrub and tree component of a vegetation community. Attributes that could be important for shrub characterization are height, volume, foliage density, crown diameter, from class, age class, and total number of plants by species (density). Another important feature of shrub characterization is the collection of data on a vertical as well as horizontal plant. Canopy layering is almost important. The occurrence of individual species and the extent of canopy cover of each species is recorded in layers. The number of layers chosen should represent the herbaceous layer, the shrub layer, and the tree layers, though additional layers can be added if needed.
7. **Trend**—refers to the direction of change. Vegetation data are collected at different points in times on the same site and the results are then compared to detect a change. Trend is described as moving "towards meeting objectives," "away from meeting objectives, "not apparent," or "stated." Trend data are important in determining the effectiveness of on-the-ground management actions. Trend data indicated whether the rangeland is moving towards or away from specific objectives. The trend of a rangeland area may be judged by noting changes in vegetation attributes such as species composition, density, cover, production, and frequency. Trend data, along with actual use, authorized use, estimated use, utilization, climate, and other relevant data, are considered in evaluating activity plans.

SIMPLE RANDOM SAMPLING

Many statistical procedures assume *simple random sampling*—the fundamental selection method (not to be confused with random sampling). By this approach, a simple random sample is a subset of individuals (a sample) chosen from a larger set (a population). Everyone is chosen randomly and entirely by chance (must be free from deliberate choice), such that everyone has the same probability of being chosen at any stage during the sampling process, and each subset of k individuals has the same probably of being chosen for the sample as any other subset of k individuals (Yates et al., 2008).

In small populations and often in large ones, such sampling is typically done without replacement (i.e., one deliberately avoids choosing any member of the population more than once). Although simple random sampling can be conducted with replacement instead, this is less common and would normally be described more fully as simple random sampling with replacement. Sampling done without replacement is no longer independent, but still satisfies exchangeability, hence many results still hold. Further, for a small sample from a large population, sampling without replacement is approximately the same as sampling with replacement, since the odds of choosing the same sample twice is low.

Conceptually, simple random sampling is the simplest of the probability sampling techniques. Though simple in form and use, the goal, of course, of simple random sampling is to produce an unbiased random selection of individuals, which in the long run ensures that the sample represents the population.

Simple Random Sampling Methods[2]

It may be difficult to visualize giving every possible combination of n units an equal chance of appearing in a sample of size n, but it can be easily accomplished. It is only necessary to be sure that at any stage of the sampling, the selection of a particular unit is in no way influenced by the other units that have been selected. Stated differently, the selection of any given unit should be completely independent of the selection of all other units. One way to do this is to assign every unit in the population a number and then draw n numbers from a table of random digits. Alternately, the numbers can be written on some equal-sized disks or slips of paper which are placed in a bowl, thoroughly mixed, and then drawn one at a time. For units such as individual tree seeds or seedlings, the units themselves may be drawn at random.

As mentioned, the units may be selected with or without replacement. If selection is with replacement, each unit is allowed to appear in the sample as often as it is selected. In sampling without replacement, a particular unit is allowed to appear in the sample only once. Most forest sampling is without replacement.

Sample selection. The selection method and computations may be illustrated by the sampling of a 250-acre plantation. The objective of the survey was to estimate the mean cordwood volume per acre in trees more than 5 inches D.B.H. outside bark. The population and sample units were defined to be square quarter-acre plots with the unit value being the plot volume. The sample consisted of 25 units selected at random and without replacement.

The quarter-acre units were plotted on a map of the plantation and assigned numbers from 1 to 1,000. From a table of random digits, 25 three-digit numbers were selected to identify the units to be included in the sample (the number 000 was associated with the plot numbered 1,000). No unit was counted in the sample more than ounce. Units drawn a second time were rejected, and an alternative unit was randomly selected.

[2] Much of this section is Adapted from Freese, F. (1976). USDA *Elementary Forest Sampling*, reprint. Washington, DC: US Department of Agriculture.

The cordwood volumes measured on the 25 units were as follows:

7	10	7	4	7
8	8	8	7	5
2	6	9	7	8
6	7	11	8	8
7	3	8	7	7

$$\text{Total } = \overline{175}$$

Estimates. If the cordwood volume on the ith sampling unit is designated y_i, sampling unit is designated y_i, and the estimated mean volume $(\bar{y})$ per sampling unit is

$$\bar{y} = \frac{\sum_{i=1}^{n} y_i}{n} = \frac{7+8+2+\ldots+7}{25} = \frac{175}{25}$$

= 7 cords per quarter-acre plot.

The mean volume per acre would, of course, be 4 times the mean volume per quarter-acre plot, or 28 cords.

As there is a total of $N = 1{,}000$ quarter-acre units in the 250-acre plantation, the estimated total volume (Y) in the plantation would be

$$\hat{Y} = N\bar{y} = (1{,}000)(7) + 7{,}000 \text{ cords}$$

Alternatively,

$$\hat{Y} = (28 \text{ cords per acre})(250 \text{ acres}) = 7{,}000 \text{ cords}$$

Standard errors. A first step in computing the standard error of estimate is to make an estimate (s_y^2) of the variance of individual values of y.

$$s_y{}^2 = \frac{\sum_{i=1}^{n} y_i^2 - \frac{\left(\sum_{i=1}^{n} y_i\right)^2}{n}}{(n-1)} \tag{10.1}$$

In this example,

$$s_y{}^2 = \frac{\left(7^2 + 8^2 + \ldots + 7^2\right) - \frac{(175)^2}{25}}{(25-1)}$$

$$= \frac{1{,}317 - 1{,}225}{24} = 3.83333 \text{ cords}$$

When sampling is without replacement the standard error of the mean ($s_{\bar{y}}$) for a simple random sample is

$$s_{\bar{y}} = \sqrt{\frac{s_y^2}{n}\left(1 - \frac{n}{N}\right)} \tag{10.2}$$

where: N = total number of sample units in the entire population,
n = number of units in the sample

For the plantation survey,

$$s_{\bar{y}} = \sqrt{\frac{3.8333}{25}\left(1 - \frac{25}{1{,}000}\right)} = \sqrt{(.1533)(.975)}$$

$$= 0.337\text{cord}$$

This is the standard error for the mean per quarter-acre plot. By the rules for the expansion of variances and standard errors, the standard error for the mean volume per acre will be (4) (0.887) = 1.548 cords.

Similarly, the standard error for the estimated total volume ($s_{\bar{Y}}$) will be

$$s_{\bar{y}} = Ns_{\bar{y}} = (1{,}000)(.387) = 387 \text{ cords.}$$

Sampling with replacement. In the formula for the standard error of the mean, the term $\left(1 - \frac{n}{N}\right)$ is known as the finite-population correction or fpc. It is used when units are selected without replacement. If units are selected with replacement, the fpc is omitted and the formula for the standard error of the mean becomes

$$S_{\bar{y}} = \sqrt{\frac{S_y^2}{n}} \tag{10.3}$$

Even when sampling is without replacement the sampling fraction (n/N) may be extremely small, making the fpc very close to unity. If n/N is less than 0.05, the fpc is commonly ignored and the standard error computed from the shortened formula.

Confidence limits for large samples. By itself, the estimated mean of 28 cords per acre does not tell us very much. Had the sample consisted of only 2 observations we might conceivably have drawn the quarter-acre plots having only 2 and 3 cords, and the estimated mean would be 10 cords per acre. Or if we had selected the plots with 10 and 11 cords, the mean would be 42 cords per acre.

To make an estimate meaningful it is necessary to compute confidence limits that indicate the range within which we might expect (with some specified degree of confidence) to find the parameter. The 95% confidence limits for large samples are given by

$$\text{Estimate} \pm 2(\text{Standard Error of Estimate}) \tag{10.4}$$

Thus the mean volume per acre (28 cords) that had a standard error of 1.548 cords would have confidence limits of

$$28 \pm 2(1.548) = 24.90 \text{ to } 31.10 \text{ cords per acre.}$$

And the total volume of 7,000 cords that had a standard error of 387 cords would have 95% confidence limits of

$$7{,}000 \pm 2(387) = 6{,}226 \text{ to } 7{,}774 \text{ cords.}$$

Unless a 1-in-20 chance has occurred in sampling, the population mean volume per acre is somewhere between 24.9 and 31.1 cords, and the true total volume is between 6,226 and 7,774 cords.

Because of sampling variation, the 95% confidence limits will, on average, fail to include the parameter in 1 case out of 20. It must be emphasized, however, that these limits and the confidence statement take account of sampling variation only. They assume that the plot values are without measurement error and that the sampling and estimating procedures are unbiased and free of computational mistakes. If these basic assumptions are not valid, the estimates and confidence statements may be nothing more than a statistical hoax.

Confidence limits for small samples. Ordinarily, large-sample confidence limits are not appropriate for samples of less than 30 observations. For smaller samples, the proper procedure depends on the distribution of the unit values in the parent population, a subject that is beyond the scope of this book. Fortunately, many forest measurements follow the bell-shaped normal distribution, or a distribution that can be made nearly normal by transformation of the variable.

For samples of any size from normally distributed populations, Student's t value can be used to compute confidence limits. The general formula is

$$\text{Estimate} \pm (t)\ (\text{Standard Error of Estimate})$$

The values of t have been tabulated. The value of t to be used depends on the degree of confidence, desired and on the size of the sample. For 95% confidence limits, the t values are taken from the column for a probability of .05. For 99% confidence limits, the t value would come from the .01 probability column, Within the specified columns, the appropriate t for a simple random sample of n observations is found in the row for $(n - 1)$ df's (degrees of freedom). For a simple random sample of 25 observations the t value for computing the 95% confidence limits will be found in the .05 column and the 24 df row. This value is 2.064. Thus, for the plantation survey that showed a mean per-acre volume of 28 cords and a standard error of the mean of 1.548 cords, the small-sample 95% confidence limits would be

$$28 \pm (2{,}064)(1.548) = 24.80 \text{ to } 31.20 \text{ cords}$$

The same t value is used for computing the 95% confidence limits on the total volume. As the estimated total was 7,000 cores with a standard error of 387 cores, the 95% confidence limits are

$$7{,}000 \pm (2.064)(387) = 6{,}201 \text{ to } 7{,}799 \text{ cords}$$

DID YOU KNOW?

In this book the expression "degrees of freedom" refers to a parameter the distribution of Student's *t*. When a tabular value of *t* is required, the number of degrees of freedom (df's) must be specified. The expression is not easily explained in nonstatistical language. One definition is that the df's are equal to the number of observations in a sample minus the number of independently estimated parameters used in calculating the sample variance. Thus, in a simple random sample of *n* observations, the only estimated parameter needed in calculating the sample variance is the mean (x), and so the df's would be $(n - 1)$.

Size of sample. In the example illustrating simple random sampling, 25 units were selected. But why 25? Why not 100? Or 10? All too often the number depends on the sampler's view of what looks about right. But there is a somewhat more objective solution. That is to take only the number of observations needed to give the desired precision.

In planning the plantation survey, we could have stated that unless a 1-in-2 change occurs we would like our sample estimate of the mean to be within $\pm E$ cords of the population mean.

As the small-sample confidence limits are computed as $\bar{y} \pm t(s_{\bar{y}})$, this is equivalent to saying that we want

$$t\left(s_{\bar{y}}\right) = E \tag{10.5}$$

For a simple random sample,

$$s_{\bar{y}} = \sqrt{\frac{s_y{}^2}{n}\left(1 - \frac{n}{N}\right)} \tag{10.6}$$

Substituting for $s_{\bar{y}}$ in the first equation, we get

$$(t)\sqrt{\frac{s_y{}^2}{n}\left(1 - \frac{n}{N}\right)} = E \tag{10.7}$$

Rewritten in terms of the sample size (n), this becomes

$$n = \frac{1}{\dfrac{E^2}{t^2 sy_2} + \dfrac{1}{N}} \tag{10.8}$$

To solve this relationship for n, we must have some estimate (Sy^2) of the population variance. Sometimes the information is available from previous surveys. In the illustration, we found $Sy^2 = 3.88$, a value which might be taken as a representative of the variation among quarter-acre plots in this or similar populations. In the absence of this information, a small preliminary survey might be made to obtain an estimate of the variance. When, as often happens, neither of these solutions is feasible, a very crude estimate can be made from the relationship:

$$Sy^2 = \left(\frac{R}{4}\right)^2 \tag{10.9}$$

where:

R = estimated range from the smallest to the largest unit value likely to be encountered in sampling. For the plantation survey we might estimate the smallest y-value on quarter-acre plots to be 1 cord and the largest to be 10 cords. As the range is 9, the estimated variance would be

$$S_{y^2} = \left(\frac{9}{4}\right)^2 = 5.06$$

This approximation procedure should be used only when no other estimate of the variance is available.

Having specified a value of E and obtained an estimate of the variance, the last piece of information we need is the value of t_2. Here we hit somewhat of a snag. To use t we must know the number of degrees of freedom. But the number of df's must be $(n - 1)$ and n is not known and cannot be determined without knowing t.

An iterative solution will give us what we need, and it is not as difficult as it sounds. The procedure is to guess at a value of n, use the guessed value to get the degrees of freedom for t, and then substitute the appropriate t value in the sample-size formula to solve for a first approximation of n. Selecting a new n somewhere between the guessed value and the first approximation, but closer to the latter, we compute a second approximation. The process is repeated until successive values of n are the same or only slightly different. Three trials usually suffice.

To illustrate the process, suppose that in planning the plantation survey we had specified that, barring a 1-in-100 change, we would like the estimate to be within 3.0 cords of the true mean volume per acre. This is equivalent to E = 0.75 cord per quarter-acre. From previous experience, we estimate the population variance among quarter-acre plots to be $s_y^2 = 4$, and we know that there is a total of N = 1,000 units in the population. To solve for n, this information is substituted in the sample-size formula 6.7.

$$n = \frac{1}{\frac{(0.75)^2}{(t^2)(4)} + \frac{1}{1{,}000}}$$

We will have to use the t value for the .01 probability level, but we do not know how many degrees of freedom t will have without knowing n. As a first guess, we can try n = 61; then the value of t with 60 degrees of freedom at the .01 probability level is t = 2.66. Thus, the first approximation will be

$$n_1 = \frac{1}{\dfrac{(0.75)^2}{(2.66^2)(4)} + \dfrac{1}{1{,}000}} = \frac{1}{\dfrac{.5625}{(7.0756)(4)} + \dfrac{1}{1{,}000}}$$

$$= 47.9$$

A second-guessed value for n would be somewhere between 61 and 48, but closer to the computed value. We might test $n=51$, for which the value of t (50 df's) at the 0.1 level is about 2.68, whence

$$n_2 = \frac{1}{\dfrac{.5625}{(7.184)(4)} + \dfrac{1}{1{,}000}}$$

$$= 48.6$$

The desired value is somewhere between 51 and 48.6 but much closer to the latter. Because the estimated sample size is, at best, only a good approximation, it is rather futile to strain on the computation of n. In this case we would probably settle on $n=50$, a value that could have been easily guessed after the first approximation was computed.

If the sampling fraction n/N is likely to be small (say, less than 0.05), the finite population may be ignored in the estimation of sample size and the formula simplified to

$$n = \frac{t^2 S_{y^2}}{E^2}$$

This formula is also appropriate in sampling with replacement. In the previous example the simplified formula gives an estimated sample size of $n=51$.

The short formula is frequently used to get a first approximation of n. Then, if the sample size indicated by the short formula is a considerable proportion (say over 10%) of the number of units in the population and sampling will be without replacement, the estimated sample size is recomputed with the long formula.

Effect of plot size. —In estimating sample size, the effect of plot size and the scale of the unit values on variance must be kept in mind. In the plantation survey a plot size of one-quarter acre was selected and the variance among plot volumes was estimated to be $s^2=4$. This is the variance among volumes per quarter-acre. Because the desired precision was expressed on a per-acre basis it was necessary to modify the precision specification of s^2 to get them on the same scale. In the example, s^2 was used without change and the desired precision was divided by 4 to put it on a quarter-acre basis. The same result could have been obtained by leaving the specified precision unchanged and putting the variance on a per acre basis. The same result could have been obtained by leaving the specified precision unchanged and putting the variance on a per-acre basis. Since the quarter-acre volumes should be multiplied by 4 to put them on a per-acre basis, the variance of quarter-acre volumes should be multiplied by 16. (Remember: If x is a variable with variance s^2, then the variance of a variable $z=k$ is k^2s^2.)

Plot size has an additional effect on variance. At the same scale of measurement, small plots will almost always be more variable than large ones. The variance in volume per acre on quarter-acre plots would be somewhat larger than the variance in volume per acre on half-acre pots, but slightly smaller than the variance in volume per acre of fifth-acre plots. Unfortunately, the relation of plot size to variance changes from one population to another. Large plots tend to have a smaller variance because they average out the effect of flumping and holes. In very uniform populations, changes in plot size have negligible effect on variance. In nonuniform populations the relationship of plot size to variance will depend on how the sizes of clumps and holes compare to the plot sizes. Experience is the best guide as to the effect of changing plot size on variance. Where neither experience nor advice is available, a very rough approximation can be obtained by the rule:

$$s_2^2 = s_1^2\sqrt{P_1 / P_2} \tag{10.10}$$

Thus, if the variance in cordwood volume *per acre* on quarter-acre plots is $s_1{}^2 = 61$, the variance in cordwood volume *per acre* on tenth-acre plots will be roughly

$$61\sqrt{0.25 / 0.10} = 96$$

The same results will be obtained without worry about the scale of measurement if the squared coefficients of variation (C^2) are used in place of the variances. The formula would then be

$$C_2^2 = C_1^2\sqrt{P_1 / P_2} \tag{10.11}$$

Example 10.1

Problem:

A survey is to be made to estimate the mean board-foot volume per acre in a 200-acre tract. Barring a 1-in-20 chance, we would like the estimate to be within 500 board feet of the population mean. Sample plots will be one-fifth acre. A survey in a similar tract showed the standard deviation among quarter-acre plot volumes to be 520 board feet. What size sample will be needed?

Solution:

The variance among quarter-acre plot volumes in $520^2 = 270{,}400$. For quarter-acre volumes expressed on a per-acre basis, the variance would be

$$s_1^2 = (4^2)(270{,}400) = 4{,}326{,}400$$

The estimated variance among fifth-acre plot volumes expressed on a per-acre basis would then be

$$s_2^2 = s_1^2 \sqrt{\frac{P_1}{P_2}} = 4,326,400 \sqrt{\frac{0.25}{0.20}}$$

$$= (4,326,400)\ (1.118)$$

$$= 4,836,915$$

The population size is $N = 1,000$ fifth-acre plots.
If as a first guess $n = 61$, the t value at the .05 level with 60 degrees of freedom is 2.00. The first computed approximation of n is

$$n_1 = \frac{1}{\dfrac{(500)^2}{(4)(4,836,915)} + \dfrac{1}{1,000}} = 71.8$$

The correct solution is between 61 and 71.8 but much closer to the computed value. Repeated trials will give values between 71.0 and 71.8. The sample size (n) must be an integral value and, because 71 is too small, a sample of $n = 72$ observations would be required for the desired precision.

STRATIFIED RANDOM SAMPLING[3]

Stratified random sampling refers to a sampling method that has a population that consists of N elements. This N population consists of data compiled in forest inventory work. The purpose of stratification is to reduce the variation within the forest subdivision and increase the precision of the population estimate, increasing the usefulness of the sample. The population is divided into H groups (stratifications combined based on similarity of some characteristics), called *strata*. Each element of the population can be assigned to one, and only one, *strata*. The number of observations within each stratum N_h is known, and $N = N_1 + N_2 + N_3 + \ldots + N_{H-1} + N_H$. A probability sample is drawn from each stratum. Stratified random sampling in forest inventory has the following advantages over simple random sampling:

1. A stratified sample can provide greater precision than a simple random sample of the same size.
2. A stratified sample uses a smaller sample to ensure greater precision; this often saves money.
3. A stratified sample can guard against an "unrepresentative" sample (e.g., an all-pine tree sample from a mixed stand population).
4. The researcher can ensure that he or she obtains sufficient sample points to support a separate analysis of any subgroup.

On the other hand, the disadvantages of stratification are that it may require more administrative effort than a simple random sample. Also, the size of each stratum

[3] Based on Freese, F. (1976). USDA.

must be known, and that the sampling units must be taken in each stratum if an estimate for that stratum is needed.

In sampling a forest, we might set up strata corresponding to the major timber types, make separate sample estimates of each type, and then combine the type data to give an estimate for the entire population. If the variation among units within types is less than the variation among units that are not in the same type, the population estimate will be more precise than if sampling had been at random over the entire population.

Example 10.2

The sampling and computational procedures can be illustrated with data from a cruise made to estimate the mean cubic-foot volume per acre on an 800-acre forest. On aerial photographs the tract was divided into three strata corresponding to the three major forest types: pine, bottom-land hardwoods, and upland hardwoods. The boundaries and total acreage of each type were known. Ten one-acre plots were selected at random and without replacement in each stratum.

Stratum		Observations			
I.	Pine	570	510	600	
		640	590	780	Total = 6.100
		480	670	700	
		560			
II.	Bottom-land hardwoods	520	630	810	
		710	760	580	Total = 7.370
		770	890	860	
		840			
III.	Upland hardwoods	420	540	320	
		210	180	270	Total = 3.040
		290	260	200	
		350			

Estimates. The first step in estimating the population mean per unit is to compute the mean ($\overline{y}h$) for each stratum. The procedure is the same as for the mean of a simple random sample.

$\overline{y}_{I} = 6{,}100 / 10 = 610$ cubic feet per acre for the pine type

$\overline{y}_{II} = 7{,}370 / 10 = 737$ cubic feet per acre for bottom-land hardwoods

$\overline{y}_{III} = 3{,}040 / 10 = 304$ cubic feet per acre for upland hardwoods.

The mean of a stratified sample ($\overline{y}_{st}$) is then computed by

$$\overline{y}_{st} = \frac{\sum_{h=1}^{L} N_h \overline{y}_h}{N}$$

where:

L = The number of strata

N_h = The total size (number of units) of stratum h ($h=1, \ldots, L$)

N = The total number of units in all strata

$$\left(N = \sum_{h=1}^{L} N_h \right)$$

If the strata sizes are

$$\begin{aligned} &1.\text{Pine} &&= 320 \text{ acres} = N_{\text{I}} \\ &2.\text{Bottom-landhardwoods} &&= 140 \text{ acres} = N_{\text{II}} \\ &3.\text{Upland hardwoods} &&= 340 \text{ acres} = N_{\text{III}} \\ & \qquad\qquad \text{Total} &&= \overline{800 \text{ acres}} = N \end{aligned}$$

Then the estimate of the population means is

$$\begin{aligned} \overline{y}_{\text{st}} &= \frac{(320)(610)+(140)(737)+(340)(304)}{800} \\ &= 502.175 \text{ cubic feet per acre} \end{aligned}$$

For the estimate of the population total ($\hat{Y}_{\text{st}}$), simply omit the divisor N.

$$\hat{Y}_{\text{st}} = \sum_{h=1}^{L} N_h \overline{y_h} = 320(610)+140(737)+340(304) = 401,740$$

Alternatively,

$$Y_{\text{st}} = N\overline{y}_{\text{st}} = 800(502.175) = 401,740$$

Standard errors. —To determine standard errors, it is first necessary to obtain the estimated variance among individuals within each stratum ($s_h^{\,2}$). These variances are computed in the same manner as the variance of a simple random sample. Thus, the variance within Stratum I (Pine) is

$$\begin{aligned} s_{\text{I}}^{\,2} &= \frac{\left(570^2+640^2+\ldots+700^2\right)-\frac{(6100)^2}{10}}{(10-1)} \\ &= \frac{3{,}794{,}000-3{,}721{,}000}{9} \\ &= 8111.1111 \end{aligned}$$

Similarly,

$$s_{\text{II}}^2 = 15{,}556.6667$$

$$s_{\text{III}}{}^2 = 12{,}204.4444$$

From these values we fined the standard error of the mean of a stratified random sample ($s_{\bar{y}\text{st}}$) by the formula

$$s_{\bar{y}\text{st}} = \sqrt{\frac{1}{N^2} \sum_{h=1}^{L} \left[\frac{N_h^2 s_h^2}{n_h} \left(1 - \frac{n_h}{N_h} \right) \right]}$$

where:

$$n_h = \text{Number of units observed in stratum } h.$$

For the timber cruising example we would have

$$s_{\bar{y}st} = \sqrt{\frac{1}{800^2} \left[\frac{(320)^2 (8111.1111)}{10} \left(1 - \frac{10}{320} \right) + \ldots + \frac{(340)^2 (12{,}204.4444)}{10} \left(1 - \frac{10}{340} \right) \right]}$$

$$= \sqrt{383.920659}$$

$$= 19.594$$

As a rough rule we can say that unless a 1-in-20 chance has occurred, the population mean is included in the range

$$\bar{y}_{\text{st}} \pm 2\left(s_{\bar{y}\text{st}}\right) = 502.175 \pm 2(19.594)$$

$$= 463 \text{ to } 541$$

If sampling is with replacement or if the sampling fraction within a particular stratum (n_h/N_h) is small, we can omit the finite-population correction $\left(1 - n_h / N_h\right)$ for that stratum when calculating the standard error.

The population total being estimated by $\hat{Y}_{\text{st}} = N\bar{y}_{\text{st}}$ the standard error of $\hat{Y}_{\text{st}}$, the standard error of $\hat{Y}_{\text{st}}$ is simply

$$S_{\hat{Y}\text{st}} = Ns_{\bar{y}\text{st}} = 800(19.594) = 15{,}675$$

WORTH REPEATING!

Stratified random sampling offers two primary advantages over simple random sampling. First, it provides separate estimates of the mean and variance of each stratum. Second, for a given sampling intensity, it often gives more precise estimates of the population parameters than would a simple random sample of the same size. For this latter advantage, however, it is necessary that the strata be set up so that the variability among unit values within the strata is less than the variability among units that are not in the same stratum. Some of the drawbacks are that each unit in the population must be assigned to one and only one stratum, that the size of each stratum must be known, and that a sample must be taken in each stratum. The most common barrier to the use of stratified random sampling is lack of knowledge of the strata sizes. If the sampling fractions are small in each stratum, it is not necessary to know the exact strata sizes; the population mean, and its standard error can be computed from the relative sizes.

Estimation of Number of Sampling Units

Assuming we have decided on a total sample size of n observations, how do we know how many of these observations to make in each stratum? To estimate the number of sampling units needed, it is necessary to have preliminary information of the variability of the strata in the population and to choose an allowable error and probability level. With this information, two common solutions to this problem are known as proportional and optimum allocation.

Proportional allocation. —In this procedure the proportion of the sample that is selected in the hth stratum is made equal to the proportion of all units in the population which fall in that stratum. If a stratum contains half of the units in the population, half of the sample observations would be made in that stratum. In equation form, if the total number of sample unit is the be n, then for proportional allocation the number to be observed in stratum h is

$$n_h = \left(N_h / N\right) n$$

In Example 10.2, the 30 sample observations were divided equally among the strata. For proportional allocation we would have used

$$n_{\text{I}} = \left(N_{\text{I}}/N\right) n = (320/800)30 = 12$$

$$n_{\text{II}} = (140/800)30 = 5.25 \text{ or } 5$$

$$n_{\text{III}} = (340/800)30 = 12.75 \text{ or } 13$$

Optimum allocation. —In optimum allocation the observations are allocated to the strata to give the smallest standard error possible with a total of n observations. If we wish to get the most precise estimate of the population mean for the expenditure of money, optimum allocation should be used. Note that this allocation can be done either if the costs of sampling in all strata are equal or if they differ. For a sample of size n, the number of observations (n_h) to be made in stratum h under optimum allocation is

$$N_h = \left(\frac{N_h s_h}{\sum_{h=1}^{L} N_h s_h} \right)^n$$

In terms of Example 10.2, the value of $N_h s_h$ for each stratum is

$$\begin{aligned}
N_{\text{I}s\text{I}} &= 320\sqrt{8111.1111} = 320(90.06) &&= 28{,}819.20 \\
N_{\text{II}s\text{II}} &= 140\sqrt{15{,}556.6667} = 140(124.73) &&= 17{,}462.20 \\
N_{\text{III}s\text{III}} &= 340\sqrt{12{,}204.4444} = 340(110.47) &&= 37{,}559.80
\end{aligned}$$

$$\text{Total} = \overline{83{,}841.20} = \sum_{h=1}^{\text{III}} N_h s_h$$

Applying these values in the formula, we would get

$$n_{\text{I}} = \left(\frac{28{,}819.20}{83{,}841.20} \right) 30 = 10.3 \text{ or } 10$$

$$n_{\text{II}} = \left(\frac{17{,}462.20}{83{,}841.20} \right) 30 = 6.2 \text{ or } 6$$

$$n_{\text{III}} = \left(\frac{37{,}559.80}{83{,}841.20} \right) 30 = 13.4 \text{ or } 14$$

Here optimum allocation is not much different from proportional allocation. Sometimes the difference is great.

Optimum Allocation with Varying Sampling Costs

Optimum allocation as described above assumes that the sampling cost per unit is the same in all strata. When sampling costs vary from one stratum to another, the allocation giving the most information per dollar is

$$N_h = \left(\frac{\frac{N_h s_h}{\sqrt{c_h}}}{\sum \left(\frac{N_h sh}{c_h} \right)} \right)^n$$

where:

c_h = Cost per sampling unit in stratum h.

The best way to allocate a sample among the various strata depends on the primary objectives of the survey and our information about the population. One of the two forms of optimum allocation is preferable if the objective is to get the most precise estimate of the population mean for a given cost. If we want separate estimates for each stratum and the overall estimate is of secondary importance, we may want to sample heavily in the strata having high-value material. Then we would ignore both optimum and proportional allocation and place our observations to give the degree of precision desired for the strata.

We can't, of course, use optimum allocation without having some idea about the variability within the various strata. The appropriate measure of variability within the stratum is the standard deviation (not the standard error), but we need not know the exact standard deviation (s_h) for each stratum. In place of actual s_h values, we can use relative values. In Example 10.2, if we had known that the standard deviations for the strata were about in the proportions $s_I:s_{II}:s_{III}=9:12:11$, we could have used these values and obtained about the same allocation. Where optimum allocation is indicated but nothing is known about the strata standard deviations, proportional allocation is often very satisfactory.

NOTE OF CAUTION!

In some situations the optimum allocation formula will indicate that the number of units (n_h) to be selected in a stratum is larger than the stratum (N_h) itself. The common procedure then is to sample all units in the stratum and to recompute the total sample size (n) needed to obtain the desired precision.

Sample Size in Stratified Random Sampling

Overall sample (n) size needed to achieve a desired degree of precision at a specified probability level can be computed; however, the exact form of the sample-size formula varies somewhat depending on the method of allocating the sample to the strata. To estimate the total size of sample (n) needed in a stratified random sample, the following pieces of information are required:

- A statement of the desired size of the standard error of the mean. This will be symbolized by D.
- A reasonably good estimate of the variance (s_h^2) or standard deviation (s_h) among individuals within each stratum.
- The method of sample allocation. If the choice is optimum allocation with varying sampling costs, the sampling cost per unit for each stratum must also be known.

Given this hard-to-come-by information, we can estimate the size of sample (n) with these formulae:

For equal samples in each of the L strata,

$$n = \frac{L\sum_{h=1}^{L} N_h^2 s_h^2}{N^2D^2 + \sum_{h=1}^{L} N_h s_h^2}$$

For proportional allocation,

$$n = \frac{N\sum_{h=1}^{L} N_h^2 s_h^2}{N^2D^2 + \sum_{h=1}^{L} N_h s_h^2}$$

For optimum allocation with equal sampling costs among strata,

$$n = \frac{\left(\sum_{h=1}^{L} N_h s_h\right)^2}{N^2D^2 + \sum_{h=1}^{L} N_h s_h^2}$$

For optimum allocation with varying sampling costs among strata,

$$n = \frac{\left(\sum_{h=1}^{L} N_h S_h \sqrt{c_h}\right)\left(\sum_{h=1}^{L} \frac{N_h S_h}{\sqrt{c_h}}\right)}{N^2D^2 + \sum_{h=1}^{L} N_h S_h^2}$$

When the sampling fractions n_h/N_h are likely to be very small for all strata or when sampling will be with replacement, the second term of the denominators of the above formulae

$\left(\sum_{h=1}^{L} N_h S_h{}^2\right)$ may be omitted only N^2D^2.

If the optimum allocation formula indicates a sample (n_h) greater than the total number of units (N_h) in a particular stratum, n_h is usually made equal to N_h; i.e., all units in that stratum are observed. The previously estimated sample size (n) should then be dropped and the total sample size (n') and allocation for the remaining strata recomputed omitting the N_h and s_h values for the offending stratum but leaving N and D unchanged.

Example 10.3

Problem:

Assume a population of 4 strata with sizes (N_h) and estimated variances sh^2 as follows:

Stratum	N_h	sh^2	s_h	N_bS_n	$N_hS_n^2$
1…………	200	400	20	4,000	80,000
2…………	100	900	30	3,000	90,000
3…………	400	400	20	8,000	160,000
4…………	20	19,600	140	2,800	392,000
	----			--------	----------
	$N = 720$			17,800	722,000

Solution:

With optimum allocation (same sampling cost per unit in all strata), the number of observations to estimate the population mean with a standard error of $D=1$ is

$$n = \frac{(17{,}800)^2}{\left(720^2\right)\left(1^2\right) + 722{,}000} = 255.4 \text{ or } 256$$

The allocation of these observations according to the optimum formula would be

$$n_1 = \left(\frac{4{,}000}{17{,}800}\right)256 = 57.5 \text{ or } 58$$

$$n_2 = \left(\frac{3{,}000}{17{,}800}\right)256 = 43.1 \text{ or } 43$$

$$n_3 = \left(\frac{8{,}000}{17{,}800}\right)256 = 115.1 \text{ or } 115$$

$$n_4 = \left(\frac{2{,}800}{17{,}800}\right)256 = 40.3$$

The number of units allocated to the fourth stratum is greater than the total size of the stratum. Thus every unit in this stratum would be selected ($n_4 = N_4 = 20$) and the sample size for the first three strata recomputed. From these three strata,

$$\Sigma = N_hSh = 15{,}000$$
$$\Sigma = N_hsh^2 = 330{,}000$$

Hence,

$$n' = \frac{(15,000)^2}{(720^2)(1^2) + 330,000} = 265$$

And the allocation of these observations among the three strata would be

$$n_1 = \left(\frac{4,000}{15,000}\right)265 = 70.7 \text{ or } 71$$

$$n_2 = \left(\frac{3,000}{15,000}\right)265 = 53.0 \text{ or } 53$$

$$n_3 = \left(\frac{8,000}{15,000}\right)265 = 141.3 \text{ or } 141$$

REGRESSION ESTIMATION[4]

Freese (1976) points out *regression estimators*, like stratification, were developed to increase the precision or efficiency of sampling by making use of the supplementary information about the population being studied. If we have exact knowledge of the basal area (i.e., cross section of a tree at breast height; used to determine percent stocking) of a stand of timber, the relationship between volume and basal area may help us to improve out estimate of stand volume. The sample data provides information on the volume-basal area relationship, which is then applied to the known basal area, giving a volume estimate that may better or cheaper than would be obtained by sampling volume alone.

Note that in this method, a *regression coefficient* is used to adjust an estimate of the mean volume of sampling units. The regression coefficient indicates the average change in volume per unit change in area between the sampling units in the sample and the population. The total number of sampling units in the population and their average size along with the total size of the forest area must be known.

Example 10.4

Problem:

Suppose a 100% inventory of a 200-acre pine stand indicates a basal area of 84 square feet per acre in trees 3.6 inches in dbh and larger. Assume further that on 20 random plots, each one-fifth acre in size, measurements were made of the basal area (x) and volume (y) per acre.

[4] Based on Freese, F. (1976). USDA.

Basal area per acre (x) (ft²)	Volume per acre (y) (ft³)
88	1,680
72	1,460
80	1,590
96	1,880
64	1,240
48	1,060
76	1,500
85	1,620
93	1,880
110	2,140
88	1,840
80	1,630
82	1,560
76	1,560
86	1,610
73	1,370
79	1,490
85	1,710
84	1,600
75	1,440
Total.......... 1,620	31,860
Mean..............81	1,593

Some values and parameters that will be needed later are

$n = 20$ $\qquad \sum xy = 2,635,500$

$\sum y = 31,860$ $\qquad \sum x = 1,620$

$\bar{y} = 1,593$ $\qquad \bar{x} = 81$

$\sum y^2 = 51,822,600$ $\qquad \sum x^2 = 134,210$

SP = corrected sum of cross products of x and y

SS = is the corrected sum of squares of x

$$SS_y = \sum y^2 - \frac{\left(\sum y\right)^2}{n} 51,822,600 - \frac{(31,860)^2}{20} = 1,069,620$$

$$s_y^2 - \frac{ss_y}{(n-1)} = \frac{1,069,620}{19} = 56,295.79$$

$$SS_x = \sum x^2 - \frac{\left(\sum x\right)^2}{n} = 134,210 - \frac{(1,620)^2}{20} = 2,990$$

$$SP_{xy} = \sum xy - \frac{\left(\sum x\right)\left(\sum y\right)}{n} = 2{,}635{,}500 - \frac{(1{,}620)(31{,}860)}{20} = 54{,}840$$

N = total number of fifth-acre plots in the population (=1,000)

The relationship between y and x may take one of several forms, but here we will assume that it is a straight line. The equation for the line can be estimated from

$$\bar{y}_R = \bar{y} + b(X - \bar{x})$$

Where:
$\bar{y}_x$ = The mean value of y as estimated from X (a specified value of the variable X)
$\bar{y}$ = The sample mean of y (= 1,593)
$\bar{x}$ = The sample mean of x (= 81)
b = The linear regression coefficient of y on x.
$\bar{y}_R$ = Linear regression estimate
For the linear regression estimator used here, the value of the regression coefficient is estimated by

$$b = \frac{SP_{xy}}{SS_x} = \frac{54{,}840}{2{,}990} = 18.84$$

Thus, the equation would be

$$\begin{aligned}\bar{y}_R &= 1{,}593 + 18.34(X - 81) \\ &= 107.46 + 18.34X\end{aligned}$$

To estimate the mean volume per acre for the tract we substitute for X the known mean basal area per acre.

$$\bar{y}_R = 107.46 + 18.34(84) = 1{,}648 \text{ cubic feet per acre}$$

Standard error. —The standard errors for simple random sampling and stratified random sampling can be estimated. To obtain the standard error for a regression estimator, we need an estimate of the variability of the individual y-values about the regression of y on x. A measure of this variability is the standard deviation from regression ($s_{y.s}$) which is computed by

$$\begin{aligned}s_{y \cdot s} &= \sqrt{\frac{SS_k - \frac{\left(SP_{xy}\right)^2}{SS_x}}{(n-2)}} \\ &= \sqrt{\frac{1{,}069{,}620 - \frac{(54{,}840)2}{2{,}990}}{(20\ -\ 2)}} \\ &= 59.53\end{aligned}$$

The symbol $s_{y.x}$ bears a strong resemblance to the covariance symbol (S_{yx}) with which it must not be confused.

Having the standard deviation from regression, the standard error of $\bar{y}_R$ is

$$S_{\bar{y}R} = S_{y \cdot s}\sqrt{\left(\frac{1}{n}+\frac{(X-\bar{x})^2}{\text{SS}_x}\right)\left(1-\frac{n}{N}\right)}$$

$$= 59.53\sqrt{\left(\frac{1}{20}+\frac{(84-81)^2}{2{,}990}\right)\left(1-\frac{20}{1{,}000}\right)}$$

$$= 13.57$$

With such a small sampling fraction ($n/N = 0.02$), the finite-population correction ($1-n/N$) could have been ignored, and the standard error would be 13.71.

Family of Regression Estimators. The regression procedure in the above example is valid only if certain conditions are met. One of these is, of course, that we know the population mean for the supplementary variable (x). As will be shown in the next section (Double Sampling), an estimate of the population mean can often be substituted.

Another condition is that the relationship of y to x must be reasonably close to a straight line within the range of x values for which y will be estimated. If the relationship departs very greatly from a straight line, our estimate of the mean value of y will not be reliable. Often a curvilinear function is more appropriate.

A third condition is that the variance of y about its mean should be the same at all levels of x. This condition is difficult to evaluate with the amount of data usually available. Ordinarily the question is answered from our knowledge of the population or by making special studies of the variability of y. If we know the way in which the variance changes with changes in the level of x a weighted regression procedure may be used.

Thus, the linear regression estimator that has been described is just one of many related procedures that enable us to increase our sampling efficiency by making use of supplementary information about the population. Two other members of this family are the ratio-of-means estimator and the mean-of-ratios estimator.

Ratio estimation. —The *ratio-of-means estimator* is appropriate when the relationship of y to x is in the form of a straight line passing through the origin and when the standard deviation of y at any given level of x is proportional to the square root of x. The ratio estimate (y_R) of mean $\bar{y}$ is

$$y_R = \hat{R}X$$

where:

R=The ratio of means obtained from the sample

$$= \frac{y}{\bar{x}} \text{ or } \frac{\sum y}{\sum x}$$

X = The known population mean of x.
For large samples (generally taken as $n > 30$; Cochran, 1953), the stand error of the ratio-of-means estimator can be approximated by

$$S_{\bar{y}R} = \frac{s_y^2 + \hat{R}^2 s_x^2 - 2\hat{R}s_{xy}}{n}(1 - n/N)$$

where:
s_y^2 = The estimated variance of y.
s_x^2 = The estimated variance of x.
s_{xy} = The estimated covariance of x and y.

Example 10.5

Problem:

Assume that for a population of $N = 400$ units, the population mean of x is known to be 62 and that from this population a sample of $n = 10$ units is selected. The y and x values of these 10 units are found to be

Observation	*y*	*X*
1	8	62
2	13	81
3	5	40
4	6	46
5	19	123
6	9	74
7	8	52
8	11	96
9	5	36
10	12	70
	–	–
Total	96	680

Solution:

From this sample the ratio-of-means is

$$\hat{R} = 9.6/68 = 0.141$$

The ratio-of-means estimator is then

$$\bar{y}_R = \hat{R}X = 0.141(62) = 8.742$$

To compute the standard error of the mean we will need the variances of y and x and the covariance. These values are computed by the standard formulae for a sample random sample, and thus,

$$S_y{}^2 = \frac{\left(8^2 + 13^2 + \ldots + 12^2\right) - (96)^2/10}{(10-1)} = 18.7111$$

$$s_x^2 = \frac{\left(62^2 + 81^2 + \ldots + 70^2\right) - (680)^2/10}{(10-1)} = 733.5556$$

$$Sxy = \frac{(8)(62) + (13)(81) + \ldots + (12)(70) - \frac{(96)(680)}{10}}{(10-1)}$$

$$= 110.2222$$

Substituting these values in the formula for the standard error of the mean gives

$$S_{\bar{y}R} = \frac{\sqrt{\dfrac{(18.7111) + \left(0.141^2\right)(733.5556) - 2(0.141)(110.222)}{10}}}{(1\ -\ 10/400)}$$

$$= \sqrt{.215690}$$

$$=\ 0.464$$

Means-of-ratios estimator. This estimator is appropriate when the relation of y to x is in the form of a straight line passing through the origin and the standard deviation of y at a given level of x is proportional to x (rather than to $\sqrt{x}$). The ratio (r_i) of y_i to x_i is computed for each pair of sample observations. Then the estimated mean of y for the population is

$$\bar{y}_R = \hat{R}X$$

where: $\hat{R}$ = the mean of the individual ratios (r_i), i.e.,

$$\hat{R} = \frac{\sum_{i=1}^{n} r_i}{n}$$

To compute the standard error of this estimate we must first obtain a measure ($s_r{}^2$) of the variability of the individual ratios (r_i) about their mean

$$s_r^2 = \frac{\sum_{n}^{i=1} r_i^2 - \frac{\left(\sum_{i=1}^{n} r_i\right)^2}{n}}{(n-1)}$$

The standard error for the mean-of-ratios estimator or mean y is then

$$s_{\bar{y}R} = X\sqrt{\frac{s_{r^2}}{n}\left(1 - n/N\right)}$$

Example 10.6

Problem:

A set of $n = 10$ observations is taken from a population of $N = 100$ units having a mean x value of 40:

Observations	Y_i	X_i	r_i
1	36	18	2.00
2	95	48	1.98
3	108	46	2.35
4	172	74	2.32
5	126	58	2.17
6	58	26	2.23
7	123	60	2.05
8	98	51	1.92
9	54	25	2.00
	Total		21.18

Solution:

The sample mean-of-ratios is

$$\hat{R} = \frac{21.18}{10} = 2.118$$

And this is used to obtain the mean-of-ratios estimator

$$\bar{y}_R = \hat{R}X = 2.118(40) = 84.72$$

The variance of the individual ratios is

$$s_{r^2} = \frac{\left(2.00^2 + 1.98^2 + \ldots + 2.00^2\right) - \frac{(21.18)^2}{10}}{(10-1)} = 0.022484$$

Thus, the standard error of the mean-of-ratios estimator is

$$S_{\bar{y}R} = 40\sqrt{\frac{0.022484}{10}(1-10/100)}$$
$$= 1.799$$

DOUBLE SAMPLING

Double sampling permits the use of regression estimators when the population mean or total of the supplementary variable is unknown. A large sample is taken to obtain a good estimate of the mean or total for the supplementary variable (x). On a subsample of the units in this large sample, the y-values are also measured to provide an estimate of the relationship of y to x. The large sample mean or total or x is then applied to the fitted relationship to obtain an estimate, or the population mean or total of y.

Example 10.7

Problem:

In updating a forest inventory in 1950 a sample of 200 quarter-acre plots in an 800-acre forest showed a mean volume of 372 cubic feet per plot (1,488 cubic feet per acre). A subsample of 40 plots, selected at random from the 200 plots, was marked for remeasurement in 1955. The relationship of the subsample was applied to the 1950 volume to obtain a regression estimate of the 1955 volume. The subsample was as follows:

1955	1950
Volume (y)	Volume (x)
370	280
290	240
520	410
490	360
530	390
330	220
310	270
400	340
450	360
430	360
460	400
480	380

430	350
500	390
640	480
660	520
490	400
510	430
270	230
380	270
420	330
530	390
550	430
550	460
520	400
420	390
490	340
500	420
610	470
460	350
430	340
510	380
450	370
380	300
430	290
460	340
490	370
560	440
580	480
540	420
Total 18,820	14,790
Mean 470.50	369.75

Solution:

$$\sum y^2 = 9{,}157{,}400$$

$$\sum x^2 = 5{,}661{,}300$$

$$\sum xy = 7{,}186{,}300$$

A plotting of the 40 pairs of plot values on coordinate paper suggested that the variability of *y* was the same at all levels of *x* and that the relationship of *y* to *x* was linear. The estimator selected based on this information was the linear regression

$y_{Rd} = \bar{a} + bX$. Values needed to compute the linear regression estimate and its standard error were as follows:

Large-sample data (indicated by the subscript 1):

n_1 = Number of observations in large sample = 200

N = Number of sample units in population = 3,200

$\overline{x_1}$ = Small sample means of x = 372

Small-sample data (indicated by the subscript 2):

n_2 = Number of observations in subsample = 40

$\overline{y}_2$ = Small sample mean of y = 470.50

$\overline{x}_2$ = Small sample mean of x = 369.75

$$SS_y = \left(\sum y^2 - \left(\sum y\right)^2 / n_2\right) = \left(9{,}157{,}400 - (18{,}820)^2/40\right) = 302{,}590.0$$

$$SS_x = \left(\sum x^2 - \left(\sum x\right)^2 / n_2\right)\left(5{,}661{,}300 - (14{,}790)^2/40\right) = 192{,}697.5$$

$$SP_{xy} = \left(\sum xy - \left(\sum x\right)\left(\sum y\right) / n_2\right) = (7{,}186{,}300 - (18{,}820)(14{,}700)/40)$$

$$= 227{,}605.0$$

$$sy^2 = SS_y/(n_2 - 1) = 302{,}590 / 40 - 1 = 7{,}758.72$$

The regression coefficient (b) and the squared standard deviation from regression ($s_{y.x}$) are

$$b = SP_{xy}/SS_x = 227{,}605.0/192{,}697.5 = 1.18$$

$$s_{y \cdot x}{}^2 = \frac{\left(SS_y - \left(SP_{xy}\right)^2/SS_x\right)}{(n_2 - 2)} = \frac{\left(302{,}590.0 - (227{,}605.0)^2/192{,}697.5\right)}{40 - 2} = 888.2617$$

And the regression equation is

$$\begin{aligned} \overline{y}_{Rd} &= \overline{y}_2 = b\left(X - \overline{x}_2\right) \\ &= 470.50 + 1.18(X - 369.75) \\ &= 34.2 + 1.18X \end{aligned}$$

Substituting the 1950 mean volume (372 cubic feet) for X gives the regression estimate of the 1955 volume.

$$\overline{y}_{xd} = 34.2 + 1.18(372) = 473.16 \text{ cubic feet per plot (excluding standard error)}$$

SAMPLING PROTOCOLS AND VEGETATION ATTRIBUTES[5]

By now you should understand that attempting to count every live, down-tree or down-wood material, and/or snag in an entire forest (of more than 10,000+ acres, for example) is not fun, not feasible, not practicable, not time or cost effective, and just not likely to be conducted at all. Instead, sampling is preferred. Why sample? We sample because it is not as arduous undertaking and is not as time-consuming as a total count; simply, it is feasible and if conducted properly provides an accurate representative sample to work with. This sample to "work with" provides the sampler/researcher with valuable information that is unbiased and empirically derived, which aids in the analysis of fuel loads for forest fires, carbon pools, and the wildlife habitat of the ecosystem sampled.

Note: Although this text dedicates most of its space to describing the author's modified version of US Forest Service FIREMON (Fire Effects Monitoring and Inventory System—designed to characterize changes in ecosystem attributes over time) sampling methods (the author has used this modified method extensively over the years in various forest locations throughout the Cascade Mountain Range of Washington State), it is important to discuss other sampling protocols and methods used in forest inventory, sampling and management. Keep in mind, however, that the same vegetation attributes and many of the following sampling methods are also used in the author's modified version of FIREMON.

VEGETATION ATTRIBUTES

When we are sampling forest biomass we are in fact measuring. Measuring what? Well, for our purposes in this text we are measuring the features of interest (attributes). A forest biomass inventory is where we take measurements to obtain dimensional or physiological information about the resource. For example, as stated above we could be interested in quantifying the amount of carbon sequestered (carbon pool) by a forest each year or the amount of water used by surrounding undergrowth in a semiarid landscape.

A typical forest inventory normally focuses on assessing the volume or value of standing trees to use for different forest products. A forest biomass inventory is somewhat different than the typical forest inventory in that it can be defined as the process involved in the estimation of any forest feature such as tree height, grass cover, and so forth as precisely as available time, money, personnel, etc. can permit. Again, we are talking about measuring the features or interest. Often, in practice, we refer to these features as attributes or plant attributes. Generally, there are only 6 attributes that are measured. A typical forest inventory normally focuses on assessing the volume or value of standing trees to use for different forest products. Vegetation or plant attributes are characteristics of vegetation or plants that can be measured or quantified referring to how many, how much, or what kind of plant species are present. The most used attributes are listed and described in the following:

[5] Material in this section is based on USDA Forest Service General Technical Report RMRS_GTR-164-CD. *Integrated Sampling Guide* (2006); *Sampling Vegetation Attributes*. Bureau of Land Management (BLM), USDA 1999. BLM/RS/ST-96/002+1730. Washington, DC: US Dept of Interior.

1. **Frequency**—(was the plant present or not?) This is one of the easiest and fastest methods available for monitoring vegetation. It describes the abundance and distribution of species and is useful to detect changes in a plant community over time.

   Frequency has been used to determine rangeland condition but only limited work has been done in most communities. This makes the interpretation difficult. The literature has discussed the relationship between density and frequency but his relation is only consistent with randomly distributed plants (Greig-Smith, 1983).

   Frequency is the number of times a species is present in each number of sampling units. It is usually expressed as a percentage.

   Advantages and Limitations

   - Frequency s highly influenced by the size and shape of the quadrat used. Quadrats or nested quadrats are the most common measurement use; however, point sampling and step-point methods have also been used to estimate frequency. The size and shape of a quadrat needed to adequately determine frequency depends on the distribution, number, and size of the plant species.
   - To determine change, the frequency of a species must generally be at least 20% and nor greater than 80%. Frequency comparison must be made with quadrats of the same size and shape. While change can be detected with frequency, the extent to which the vegetation community has changed cannot be determined.
   - High repeatability is obtainable.
   - Frequency is highly sensitive to changes resulting from seedling establishment. Seedlings present one year may not be persistent the following year. This situation is problematic if data is collected only every few years. It is less of a problem if seedlings are recorded separately.
   - Frequency is also very sensitive to changes in pattern of distribution in sample area.
   - Rooted frequency data is less sensitive to fluctuations in climatic and biotic influences.
   - Interpretation of changes in frequency is difficult because of the inability to determine the vegetation attribute that changed. Frequency cannot tell which of three parameters has changed: canopy cover, density, or pattern of distribution.

     Appropriate Use of Frequency for Rangeland Monitoring

     If the primary reason for collecting frequency data is to demonstrate that a change in vegetation has occurred, then on most sites the frequency method can accomplish the task with statistical evidence more rapidly and at less cost than any other method that is currently available (Hironaka, 1985).

     Frequency should not be the only data collected if time and money are available. Additional information on ground cover, plant cover, and other vegetation and site data would contribute to a better understanding of the changes that have occurred (Hironaka, 1985).

West (1985) noted the following limitations: "Because of the greater risk of misjudging a downward than upward trend, frequency may provide the easiest early warning of undesirable changes in key or indicator species. However, because frequency data are so dependent on quadrat size and sensitive to non-random dispersion patterns that prevail on rangelands, managers are fooling themselves if they calculate percentage composition from frequency data and try to compare different sites at the same time or the same site over time in terms of total species composition. This is because the numbers derived for frequency sampling are unique to the choice of sample size, shape, number, and placement. For variables of over and weight, accuracy is mostly what is affected by these choices and the variable can be conceived independently of the sampling protocol" (p. 97).

2. **Cover**—(how much space did they cover?) Cover is an important vegetation and hydrologic characteristic. It can be used in many ways to determine the contribution of each species to a plant community. Cover is also important in determining the proper hydrologic function of a site. This characteristic is very sensitive to biotic and edaphic forces (i.e., soil factors). For watershed stability, some have tried to use a standard soil cover, but research has shown each edaphic site has its own potential cover.

   Cover is generally referred to as the percentage of ground surface covered by vegetation. However, numerous definitions exist. It can be expressed in absolute terms (square meters/hectares) but is most often expressed as a percentage. The objective being measured will determine the definition and type of cover measured.

   - Vegetation cover is the total cover of vegetation on a site.
   - Foliar cover is the area of ground covered by the vertical projection of the aerial portions of the plans. Small openings in the canopy and intraspecific overlap are excluded.
   - Canopy cover is the area of ground covered by the vertical projection of the outermost perimeter of the natural spread of foliage of plants. Small openings within the canopy are included. It may exceed 100%.
   - Basal cover is the area of ground surface occupied by the basal portion of the plants.
   - Ground cover is the cover of plants, litter, rocks, and gravel on a site.

     Advantage and Limitations

     - Ground cover is most often used to determine the watershed stability of the site, but comparisons between sites are difficult to interpret because of the different potentials associated with each ecological site.
     - Vegetation cover is a component of ground cover and is often sensitive to climatic fluctuations that can cause errors in interpretation. Canopy cover and foliar cover are components of vegetation cover and are the most sensitive to climatic and biotic factors. This is particularly true with herbaceous vegetation.

- Overlapping canopy cover often creates problems, particularly in mixed communities. If species composition is to be determined, the canopy of each species is counted regardless of any overlap with other species. If watershed characteristics are the objective, only the uppermost canopy is generally counted.
- For trend comparisons in herbaceous pant communities, basal cover is generally considered to be the most stable. It does not vary as much due to climatic fluctuation or current-year grazing.

3. **Density**—(how many plants were there and how close are individual plants to one another?) Density has been used dot describe characteristics of plant communities. However, comparisons can only be based on similar life forms and size. Therefore density is rarely used as a measurement by itself when describing plant communities. For example, the importance of a particular species to a community is very different if there are 1,000 annual plants per acre versus 1,000 shrubs per acre. Density was synonymous with cover in earlier literature.

   Advantages and Limitations

   - Density is useful in monitoring threatened and endangered species or other special status plants because it samples the number of individuals per unit area.
   - Density is useful when comparing similar life forms (annuals to annuals, shrubs to shrubs) that are approximately the same size. For trend measurements, this parameter is used to determine if the number of individuals of a specific species is increasing or decreasing.
   - The problem with using density is being able to identify individuals and comparing individuals of varied sizes. It is often hard to identify individuals of plants that are capable of vegetative reproduction (e.g., rhizomatous plants like western wheatgrass or Gambles oak). Comparisons of bunchgrass plants to rhizomatous plants are often meaningless because of these problems. Similar problems occur when looking at the density of shrubs of different growth forms or comparing seedlings to mature plants. Density on rhizomatous or stoloniferous (i.e., grows horizontally above the ground) plants is determined by counting the number of stems instead of the number of individuals. Seedling density is directly related to environmental conditions and can often be interpreted erroneously as a positive or negative trend measurement. Because of these limitations, density has generally been used with shrubs and not herbaceous vegetation. Seedlings and mature plants should be recorded separately.

   If the individuals can be identified, density measurements are repeatable over time because there is small observer error. The type of vegetation and distribution will dictate the technique used to obtain the density measurements. In homogenous plant communities, which are rare, square quadrats have been recommended, while heterogeneous communities should be sampled with rectangular or line strip quadrats. Plotless methods have also been developed for widely dispersed plants.

4. **Biomass (Production)**—(how much did the plants weigh?) Many believe that the relative production of varied species in a plant community is the best measure of these species' roles in the ecosystems.

   The terminology associated with vegetation is normally related to production.

   - Gross primary production is the total amount of organic material produced, both above ground and below ground.
   - Biomass is the total weight of living organisms I the ecosystem, including plants and animals.
   - Standing crop is the amount of plant biomass present above ground at any given point.
   - Peak standing crop is the greatest amount of plant biomass above ground present during a given year.
   - Total forage is the total herbaceous and woody palatable plant biomass available to herbivores.
   - Allocated forage is the difference of desired amount of residual material subtracted from the total forage.
   - Browse is the portion of woody plant biomass accessible to herbivores.

     Advantages and Limitations

     - Biomass and gross primary production are rarely used in rangeland trend studies because it is impractical to obtain the measurements below ground. In addition, the animal portion of biomass is rarely obtainable.
     - Standing crop and peak standing crop are the measurements most often used in trend studies. Peak standing crops are generally measured at the end of the growing season. However, varied species reach their peak standing crop at different times. This can be a significant problem in mixed plant communities.
     - Often, the greater the diversity of plant species or growth patterns, the larger the error if only one measurement is made.
     - Other problems associated with the use of plant biomass are that fluctuations in climate and biotic influences can alter the estimates. When dealing with large ungulates, enclosures are generally required to measure this parameter. Several authors have suggested that approximately 25% of the peak standing crop is consumed by insects or trampled; this is rarely discussed in most trend studies.
     - Collecting production data also tends to be time and labor intensive. Cover and frequency have been used to estimate plant biomass in some species.

5. **Structure**— (how tall were the plants and how were branches and leaves arranged?) Structure of vegetation primarily looks at how the vegetation is arranged in a three-dimensional space. The primary use for structure measurements is to help evaluate a vegetation community's value in providing habitat for associated wildlife species.

Vegetation is measured in layers on vertical planes. Measurements generally look at the vertical distribution by either estimating the cover of each layer or by measuring the height of the vegetation.

Advantages and Limitations

- Structure data provide information that is useful in describing the suitability of the sites for screening and escape cover, which are important for wildlife. Methods used to collect these data are quick, allowing for numerous samples to be obtained over relatively large areas. Methods that use visual obstruction techniques to evaluate vegetation height have little observer bias. Those techniques that estimate cover require more training to reduce observer bias. Structure is rarely used by itself when describing trends.

6. **Species composition**—(what kind of plant was it?) Composition is a calculated attribute rather than one that is directly collected in the field. It is the proportion of various plant species in relation to the total of a given area. It may be expressed in terms of relative over, relative density, relative weight, etc.

Composition has been used extensively to describe ecological sites and to evaluate rangeland condition.

To calculate composition, the individual value (eight, density, percent cover) of a species or group of species is divided by the total value of the entire population.

Advantages and Limitations

- Quadrats, point sampling, and step-point methods can all be used to calculate composition.
- The repeatability of determining composition depends on the attribute collected and the method used.
- Sensitivity to change is dependent on the attribute used to calculate composition. For instance, if plant biomass is used to calculate composition, the values can vary with climatic conditions and the timing of climate events (precipitation, frost-free period, etc.). Composition based on basal cover, on the other hand, would be relatively stable.
- Composition allows the comparison of vegetation communities at various locations within the same ecological sites.

MATRIX OF MONITORING TECHNIQUES/VEGETATION ATTRIBUTES AND MONITORING METHODS

Matrix of Monitoring Techniques and Vegetation Attributes

The following is a matrix of commonly used monitoring techniques and vegetation attributes. The X indicates that this is the primary attribute that the technique collects. Some techniques have the capability of collected other attributes; the ● indicates the secondary attribute that can be collected or calculated.

Method	Frequency	Cover	Density	Biomass	Structure	Composition
Frequency	X	●				
Dry-weight Rank	●			●		X*
Daubenmire	●	X				●
Line Intercept		X				●
Step Point		X				●
Point Intercept		X				●
Density			X			●
Double-Weight Sampling				X		●
Harvest				X		●
Comparative Yield				X		●
Cover Board		X			X	
Robel Pole				●	X	

* Species composition is calculated using production data. Frequency date should not be used to calculate species composition.

Monitoring Methods

1. **Frequency methods**—Pace Frequency, Quadrat Frequency, and Nested Frequency Methods: All three methods consist of observing quadrats along transects, with quadrats systematically located at specified intervals along each transect. The only differences in these techniques are the size and configuration of the quadrat frames and the layout of the transect. The following vegetation attributes are monitored with this method:
   - Frequency.
   - Basal cover and general cover categories (including litter).
   - Reproduction of key species (if seedling data are collected).

   It is important to establish a photo plot and take both close-up and general view photographs. This allows the portrayal of resource values, conditions, and furnishes visual evidence of vegetation and soil changes over time.

   Note: Close-up and general view photographs should be taken for each of the methods described for the following methods.

   This method is applicable to a wide variety of vegetation types and is suited for ruses with grasses, forbs, and shrubs.

   Advantages and Disadvantages

   - Frequency sampling is highly objective, repeatable, rapid, and simple to perform, and it involves a minimum number of decisions. Decisions are limited to identifying species and determining whether species are rooted within the quadrats (presence or absence).

- Frequency data can be collected in different-sized quadrats with the use of the nested frame. When a plant of a particular species occurs within a plot, it also occurs in all the successively large plots. Frequency of occurrence for various size plots can be analyzed even though frequency is recorded for only one size plot. This eliminates problems with comparing frequency data from different plot sizes. Use of the nested plot configuration improves the change of selecting a proper size plot for frequency sampling.
- Cover data can also be collected at the same time frequency data is gathered. However, cover data collected in this manner will greatly overestimate cover; unless the tines are honed to a fine point, observer bias will come into play. Another limitation is that the use of one size quadrat will likely result in values falling outside the optimum frequency range (greater than 20 percent to less than 80%) for some of the species of interest.

2. **Dry-weight rank method**—This method is used to determine species composition. It consist of observing various quadrats and ranking the three species which contribute the most weight in the quadrat.

   This method has been tested in a wide variety of vegetation types and is generally considered suitable for grassland/small shrub types or understory communities of large shrub or tree communities. It does not work well on large shrubs and trees.

   Advantages and Disadvantages

   - One advantage of the Dry-Weight Rank Method is that many samples can be obtained very quickly. Another advantage is that it deals with estimate of production, which allows for better interpretation of the data to make management decisions. It can be done in conjunction with frequency, canopy cover, or comparative yield methods. Because it is easier to rank the top three species in a quadrat, there is less observer bias.
   - The limitation with this technique is that, by itself, it will not give a reliable estimate of plant standing crop, and it assumes that there are few empty quadrats. In many large shrub or sparse desert communities, a high percentage of quadrats are empty or have only one species present. The quadrat size required to address these concerns is often impractical.

3. **Daubenmire method**—this method consists of systematically placing a 20- × 50-cm quadrat frame along a tope on permanently located transects. The following vegetation attributes are monitored using the Daubenmire method:

   - Canopy cover
   - Frequency
   - Composition of canopy cover

   This method is applicable to a wide variety of vegetation types if the plants do not exceed waist height.

Advantages and Disadvantages

- This method is relatively simple and rapid to use. A limitation is that there can be substantial changes in canopy cover of herbaceous species between years because of climatic conditions, with no relationship to the effects of management. In general, quadrats are not recommended for estimating cover (Floyd and Anderson, 1987). This method cannot be used to calculate rooted frequency.

4. **Line intercept method**—this method consists of horizontal, linear measurements of plant intercepts along the course of the line (tape). It is designed for measuring grass or grass-like plants, forbs, shrubs, and trees. The following vegetation attributes are monitored with this method:
   - Foliar and basal cover
   - Composition (by cover)

   This method is ideally suited for semiarid bunchgrass-shrub vegetation types.

   Advantages and Disadvantages

   - This method is best suited where the boundaries of plant growth are relatively easy to determine. It can be adapted to sampling varying densities and types of vegetation. It is not well adapted, however, for estimating cover on single-stemmed species, dense grass land situations, litter, or gravel less than ½ inch in diameter. It is best suited to estimating cover on shrubs.

5. **Step-point method**—This method involves making observations along a transect at specified intervals, using a pin to record cover "hits." It measures cover for individual species, total cover, and species composition by cover.

   This method is best suited for use with grasses and fobs, as well as low shrubs. The greater the structure of the community, the more difficult it becomes to determine "hits" due to parallax, observer bias, wind, etc. This method is good for an initial overview of an area not yet subjected to intensive monitoring.

   Advantages and Limitations

   - This method is relatively simple and easy to use if careful consideration is given to the vegetation type to which it is applied. It is suitable for measuring major characteristics of the ground and vegetation cover of an area. Large areas can easily be sampled, particularly if the cover is reasonably uniform. It is possible to collect a considerable number of samples within a relatively short time.

     A limitation of this method is that there can be extreme variation in the data collected among examiners when sample sizes are small. Tall or armored vegetation reduces the ability to pace in a straight line, and the offset for obstructions described in the procedures adds bias to the data collection by avoiding certain components to the community. Another limitation is that less predominant plant species may not be hit on the transects and therefore do not show up in the study records. The literature contains numerous studies utilizing point-intercept procedures that required point densities ranging from 300 to 39,000 to adequately sample for minor species. One major consideration in the use of this method is to assure that a sharpened pin is used and that only the point is used to record "hits." Pins have finite diameters and therefore overestimate cover (Goodall, 1952). Another limitation of this method is that statistical analysis of the data is suspect unless two and preferable more transects are run per site.

6. **Point-intercept method-sighting devices, pin frames, and point frames**—this method consists of employing sighting device or pin/point frame along a set of transects to arrive at an estimate of cover. It measures cover for individual species, total cover, and species composition by cover.

This method is suited to all vegetation types less than about 1.5 m in height. This is because sighting devices and pin/point frames require the observer to look down on the vegetation from above in a vertical line with the ground. If the sighting device allows upward viewing, the method can also be used to estimate the canopy cover of large shrubs and trees.

Advantages and Limitations

Point interception measurements are highly repeatable and lead to more precise measurements than cover estimates using quadrats. The method is more efficient than line intercept techniques, at least for herbaceous vegetation, and it is the best method of determining ground cover and the cover of the more dominant species. Given the choice between sighting devices and pin/point frames, the optical sighing device is preferable.

A limitation of point-intercept sampling is the difficulty in picking up the minor species in the community without using a very large number of points. In addition, wind will increase the time required to complete a study because of the need to view a stationary plant.

One limitation that is specific to the use of point frames is that a given number of points grouped in frames gives less precise estimates of cover than the same number of points distributed individually (Goodall, 1952; Greig-Smith, 1983). In fact, single-pin measurements require only one-third as many points as when point frames are used (Bonham, 1989). Another problem with frames is that they overestimated the cover of large or clumped plants because the same plant is intercepted by different points on the same frame (Bonham, 1989). This problem is overcome with the method described here by treating the frames as the sampling units (rather than using the individual points as sampling units). However, this approach doesn't change the fact that more points must be read than when the points are independent.

Use of a pin frame device (as opposed to a grid frame made of crossing strings) will result in overestimation of cover because the pins have finite diameter. The use of a sharpened pin will greatly reduce overestimation when only the point of the pin is used to record a hit or a miss.

7. **Cover board method**—this method uses a profile board or density board to estimate the vertical area of a board covered by vegetation from a specified distance away. This technique is designed to evaluate changes in the vegetation structure over time. Quantifying the vegetation structure for statistical comparison was described by Nudds (1977). The following vegetation attributes are monitored using this method
   - Vertical cover
   - Structure

     This method is applicable to a wide variety of vegetation types. It should be used with those that show potential for changes, such as woody riparian vegetation.

     Advantages and Disadvantages

     The Cover Board technique is a fast and easily duplicated procedure. The size of the board can be modified to meet the purpose of the study.
8. **Density method**—density is the number of individuals of a species in each unit of area (e.g., plants/m^2). The term consequently refers to the closeness of individual plans to one another. For rhizomatous and other species for which the delineation of separate individual plants is difficult, density can also mean the number of stems, inflorescences, culm groups, or the plant parts per unit area.

   This method has wide applicability and is suited for use with grasses, forbs, shrubs, and trees.

   Advantages and Disadvantages
   - Generally, the density of mature perennial plant is not affected as much by annual variations in precipitation as are other vegetation attributes such as canopy cover of herbage production.
   - Density is a quantitative and absolute attribute.
   - Density is sensitive to changes in the adult population caused by long-term climatic conditions or resource uses.
   - Density provides useful information on seedling emergence, survival, and mortality.
   - Sampling is often quick and easy with certain lifeforms (e.g., trees, shrubs, bunchgrasses).
   - Plant communities on the same ecological sites can be compared using density estimates on specific species or lifeforms.
   - Density can be useful in estimating plant response to management actions.
   - It can often be difficult to delineate an individual, especially when sampling and forming plants (stoloniferous or rhizomatous plants) and multi-stemmed grasses of closely spaced shrubs. Although in these cases a surrogate plant part (e.g., upright stems, inflorescences, and culm groups) can be counted, the usefulness of such estimates is limited to the biological significance of changes in these surrogates.

- Sampling may be slow and tedious in dense populations; this also raises the risk of non-sampling errors.
- There is no single quadrat size and shape that will efficiently and adequately sample all species and life forms. For this reason, density estimations are usually limited to one or a few key species.

9. **Double-weight sampling**—this technique has been referred to by some as the Calibrated Weight Estimate method. The objective of this method is to determine the amount of current-year above-ground vegetation production on a defined area. The following vegetation are monitored
   - Peak standing crop, which is the above-ground annual production of each plant species
   - Species composition by weight

     This method can be used in a wide variety of vegetation types. It is best suited to grasslands and desert shrubs. It can also be used in large shrub and tree communities, but the difficulties increase.

     Advantages and Limitations

     – Double-weight sampling measures the attribute historically used to determine capabilities of an ecosystem.
     – It provides the basic data currently used for determining ecological status.
     – Seasonal and annual fluctuations in climate can influence plant biomass.
     – Measurements can be time-consuming.
     – Current year's growth can be hard to separate from previous years' growth.
     – Accurate measurements require collecting production data at peak production periods, which are usually short, or using utilization and phenology adjustment factors.
     – Green weights require conversion to air-dry weights.
     – In most areas, the variability in production between quadrats and the accuracy of estimating production within individual quadrats requires the sampling of large numbers of quadrats to detect reasonable levels of change.

10. **Harvest method**—the concept of this method is to determine the amount of current-year above-ground vegetation production on a defined area. The following vegetation attributes are monitored:
    - Peak standing crop, which is the above-ground annual production of each plant species
    - Species composition by weight

      This method can be used in a wide variety of vegetation types. If is best suited for grasslands and desert shrubs. It is not well suited to large shrub and tree communities.

Advantages and Limitations

- The harvest method measures the attribute historically used to determine the capabilities of an ecosystem.
- It provides the basic data currently used for determining ecological status.
- Seasonal and annual fluctuations in climate can influence plant biomass.
- Measurements can be time-consuming.
- Current year's growth can be hard to separate from previous years' growth.
- Accurate measurements require collecting production data at peak production periods which, are usually short, or using utilization and phenology adjustment factors.
- Green weights require conversion to air-dry weights.
- In most area, the variability in production between quadrats requires the sampling of large numbers of quadrats to detect reasonable levels of change.

11. **Comparative yield method**—this method is used to estimate total standing crops or production of a site. The total production in a sample quadrat is compared to one of five reference quadrats; relative ranks are recorded rather than estimating the weight directly.

    This method works best for herbaceous vegetation but can also be used successfully with small shrubs and half-shrubs. As with most production estimates, the comparative yield method can be used to compare relative production between different sites.

    Advantages and Limitations

    - The advantage of the comparative yield method is that a large sample can be obtained quickly. Total production is evaluated, so clipping calibration on a species basis is not needed. The process of developing reference quadrats for ranking purposes reduces both sampling and training time. This technique can be done in conjunction with the frequency, canopy cover, or day weigh rank methods. Identification of individual species is not required.
    - Large shrub communities are not well suited for this technique. If used in conjunction with other techniques (frequency and dry-weight rank), the quadrat size may need to be different. This technique can detect only substantial changes in production.

12. **Visual obstruction method-Robel Pole**—this method is used for determining standing plant biomass on an area. It has primarily been used to determine the quality of nesting cover for birds on the Great Plains and is commonly referred to as the Robel Pole Method. This method is applicable to other ecosystems throughout the western United States where height and vertical obstruction of cover are important. The following vegetation attributes are monitored using this method:

    - Vertical cover
    - Production
    - Structure

The Robel Pole Method is most effective in upland and riparian areas where perennial grasses, forbs, and shrubs less than 4 feet tall are the predominant species.

Advantages and Disadvantages

- Robel Pole measurements are simple, quick, and accurate. This method can be used to monitor height and density of standing vegetation over large areas quickly. Statistical reliability improves because numerous measurements can be taken in a relatively brief time. Limitations of the method may stem from infrequent application in a variety of rangeland ecosystems. While the Robel Pole Method has been used with remarkable success on the Great Plains, there needs to be more research in a variety of plant communities.

THE BOTTOM LINE

This chapter has been presented to introduce the reader to the intricacies involved in attempting to determine the total biomass available in various forest stands. Although very brief and basic the point has been made both that forest biomass is a huge carbon sink, but it is also difficult an exact the total biomass available.

CITED REFERENCES AND RECOMMENDED READING

Avery, T.E. and Burkhart, H.E. (2002). *Forest Measurements*, 5th ed. New York: McGraw-Hill.

Bailey, A.W. and Poulton, C.E. (1968). Plant communities and environmental relationships in a portion of the Tillamook Burn, northwest Oregon. *Ecology* 49:114.

Barbour, M.G., Burk, J.H., Pitts, W.D. (1987). *Terrestrial Plant Ecology*. San Francisco: The Benjamin/Cummings Publish Co. Inc., pp. 182–208.

Blackman, G.E. (1935). A study of statistical methods of the distribution of species in grassland associations. *Annals of Botany* 49:749–777.

Bonham, C.D. (1989). *Measurements for Terrestrial Vegetation*. New York: John Wiley and Sons.

Braun-Blanquet, J. (1965). *Plant Sociology: The Study of Plant Communities*. London: Hafner.

Brown, J.K. (1974). Handbook for Inventorying Downed Wood Material. General Technical Report INT-16, USDA Forest Service, Intermountain Forest & Range Experiment Station, Ogden, Utah, 24 p.

Busing, R., Rimar, K., Solte, K.W., Stohlgren, T.J. (1999). *Forest Health Monitoring: Vegetation Pilot Field Methods Guide: Vegetation Diversity and Structure Down Wood Debris Fuel Loading*. Research Triangle Park, NC: National Forest Health Monitoring Program.

Canfield, R.H. (1941). Application of the line interception method in sampling range vegetation. *Journal of Forestry* 39:388–394.

Cochran, W.G. (1953). *Sampling Techniques*. New York: Wiley.

Cole, W.P. (1977). *Using the UTM Grid System to Record Historic Sites*. Washington, DC: U.S. Department of Interior.

Cook, C.W. and Stubbendieck, J. (1986). *Range Research: Basic Problems and Techniques*. Denver, CO: Society for Range Management.

Coulloudon, B., Podborny, P., Eshelman, K., Rasmussen, A., Gianola, J., Robels, B., Habich, N., Shaver, P., Hughes, L., Spehar, J., Johnson, C., Willoughby, J., Pellant, M. (1999). *Sampling Vegetation Attributes*. Washington, DC: BLM, Technical Reference 1734-04, 164 p.

Daubenmire, R.R. (1959). A canopy-coverage method. *Northwest Science* 33:43–64.

Despain, D.W., Ogden, P.R., Smith, E.L. (1991). Plant frequency sampling for monitoring rangelands. In: Ruyle, G.B. (ed.) *Some Methods of Monitoring Rangelands and Other Natural Area Vegetation*. Extension Report 9043, University of Arizona, College of Agriculture, Tucson, AZ.

Diggle, P.J. (1975). Robust density estimation using distance methods. *Biometrika* 62:43–64.

Dilworth, J.R. (1989). *Log Scaling and Timber Cruising*. Corvallis, OR: Oregon State University.

Dilworth, J.R and Bell, J.F. (1973). *Variable probability sampling: Variable plot and 3P*. Corvallis OR: Oregon State University.

Elzinga, C.I., Salzer, D.W., Willoughby, J.W. (1998). *Measuring and Monitoring Plant Populations*. Washington, DC: BLM Technical Reference 1730-1.

Evans, R.A. and Love, R.M. (1957). The step-point method of sampling: A practical tool in range research. *Journal of Range Management* 10:208–212.

Floyd, D.A. and Anderson, J.E. (1987). A comparisons of three methods for estimating plant cover. *Journal of Ecology* 75:229–245.

Freese, F. (1976). *Elementary Forest Sampling. Agriculture Handbook No. 232*. Washington, DC: U.S. Department of Agriculture.

Goebel, C., DeBano, L.F., Lloyd, R.D. (1958). A new method of determining forage cover and production on desert shrub vegetation. *Journal of Range Management* 11:244–246.

Goldsmith, F.B., Harrison, C.M, Morton, A.J. (1986). Description and analysis of vegetation in methods. In: Moore, P.D., Chapman, S.B. (eds) *Plant Ecology*. New York: Blackwell Scientific Publications, pp. 437–424.

Goodall, D.W. (1952). Some considerations I the use of pint quadrats for the analysis of vegetation. *Australian Journal of Biological Sciences* 5:1–41.

Greig-Smith, P. (1983). *Quantitative Plant Ecology*. Berkeley, CA: University of California Press, pp. 1–53.

Hanley, T.A. (1978). A comparison of the line-interception and quadrat estimation methods of determining shrub canopy coverage. *Journal of Range Management*. 31:60–62.

Hansen, H.C. (1962). *Dictionary of Ecology*. New York: Bonanza Books, Crown Publishers, Inc., 382 p.

Hironoak, M. (1985). Frequency Approaches to Monitor Rangeland Vegetation Symposium on Use of Frequency and for Rangeland Monitoring. In: Krueger, W.C., Chairman (ed.) Proceedings of 38th Annual Meeting Society for Range Management, Feb 1985. Salt Lake City, UT, pp. 84-86.

Husch, B., Miller, C.I., Beers, T.W. (1982). *Forest Mensuration*. New York: John Wiley and Sons.

Hyder, D.N., Bement, R.E., Remmenga, E.E., Terwilliger, Jr., C. (1965). Frequency sampling of blue gramma range. *Journal of Range Management* 18:90–94.

Jasmer, G.E., Holechek, J. (1984). Determining Grazing Intensity on Rangelands. *Journal of Soil and Water Conservation* 39(1):32–35.

Jensen, M.E., Hann, W.H., Keane, R.E., Caratti, J. (1994). ECODATA-A Multiresource Database and Analysis System for Ecosystem Description and Analysis. In: Jensen, M.E., Bourgeron, P.S. (eds.) *Eastside Forest Ecosystem Health Assessment, Volume II: Ecosystem Management: Principles and Applications*. General Technical Report GTR-PNW-318. Portland, OR: U.S. Department of Agriculture, Forest Service, pp. 203–216

Krebs, C.J. (1989). *Ecological Methodology*. New York: Harper & Row.

Lapin, L.L. (1993). *Statistics for Modern Business Decisions*, 6th ed. Orlando, FL: Dryden Press

Laycock, W.A. (1987). *Setting Objectives and Picking Appropriate Methods for Monitoring Vegetation on Rangelands. Rangeland Monitoring Workshop Proceedings*. Golden, CO: U.S. Department of Interior. Bureau of Land Management.

Lucas, H.A. and Seber, G.A.F. (1977). Estimating coverage and particle density using the line intercept method. *Biometricka* 64:618–622.

Lutes, D.C. and Keane, R.E. (2006). In: Lutes, et al., 2006. *FIREMON: Fire Effects Monitoring and Inventory System.* General Technical Report RMRS-GTR-164-CD. Fort Collins, FO: U.S. Department of Agriculture, Forest Service, Rocky Mountain Research Station.

Kennedy, K.A. and Addision, P.A. (1987). Some considerations for the use of visual estimates of plant cover in biomonitoring. *Journal of Ecology* 75:151–157.

Muir, S. and McClaran, M.P. (1997). Rangeland Inventory, Monitoring, and Evaluation. Accessed 03/28/11 @ https://rangelandswest.arid.arizonaedu/rangelandswest/az.

Nudds, T.D. (1977). Quantifying the vegetative structure of wildlife cover. *Wildlife Society Bulletin* 5:113–117.

Pechanec, J.F. and Pickford, G.D. (1937). A weight-estimate method for the determination of range or pasture protection. *Agronomy Journal* 29:894–904.

Pound, P. and Clements, F.E. (1989). A method of determining the abundance of secondary species. *Minnesota Botanical Studies* 2:19–24.

Riser, P.G. (1984). *Methods for Inventory and Monitoring of Vegetation, Litter, and Soil Surface Condition. Developing Strategies for Rangeland Monitoring.* Washington, DC: National Research Council National Academy of Sciences.

Robel, R.J., Briggs, J.N., Dayton, A.D., Hulbert, L.C. (1970). Relationships between visual obstruction measurements and weight of grassland vegetation. *Journal of Range Management* 23:295.

Salzer, D. (1994). An Introduction to Sampling and Sampling Design for Vegetation Monitoring. Unpublished Papers Prepared for Bureau of Land Management Training/ Course 173–5. Phoenix, AZ: BLM Training Center.

Schreuder, H.T. Gregoire, T.G., Wood, G.B. (1993). *Sampling Methods for Multiresource Forest Inventory.* New York NY: John Wiley and Sons.

Spalinger, D.E. (1980). *Vegetation Changes on Eight Selected Deer Ranges in Nevada Over a 15-Year Period.* Reno: Nevada State Office Bureau of Land Management.

Shivers, B.D. and Borders, B.E. (1996). *Sampling Techniques for Forest Resource Inventory.* New York NY: John Wiley and Sons.

Tansley, A.G. and Chipp, T.F. (eds.) (1926). *Aims and Methods in Study of Vegetation.* London: British Empire Vegetation Committee and the Crown Agents for the Colonies, 383 p.

USDA, Soil Conservation Service. (1976). *National Range Handbook.* Washington, DC: USDA

USFS (2003a). Fact Sheet: *Tree Mortality.* Accessed 03/24/11 @ www.fia.fs.fed.us.

USFS (2003b). Fact Sheet: *Tree Growth.* Accessed 03/24/11 @ www.fia.fs.fed.us.

USFS (2003c). Fact Sheet: *Soil Quality Indicator.* Accessed 03/25/11 @ www.fia.fs.fed.us.

USFS (2003d). Fact Sheet: *Down Wood Materials Indicator.* Accessed 03/25/11 @ www.fia.fs.fed.us.

USFS (2003e). Fact Sheet: *Vegetation Indicator.* Accessed 03/26/11 @ www.fia.fs.fed.us.

USFS (2005). Fact Sheet: *Lichen Communities Indicator.* Accessed 03/24/11 www.fia.fs.fed.us.

USFS (2009). U.S. Forest Resource Facts and Historical Trends. Accessed 03/26/11 @ https://fia.fs/fed/us/

Van Dyne, G.M., Vogel, W.G., Fisser, H.G. (1963). Influence of small plot size and shape on range herbage production estimates. *Ecology* 44:746–759.

Van Wagner, C.E. (1982). Practical Aspects of the Line Intersect Method. Canadian Forest Service Information Report. PI-X-12.

Warren, W.G. and Olsen, P.F. (1964). A live intersect technique for assessing logging wastes. *Forest Service* 10:267–276.

West, N.E. (1985). Shortcomings of Plant Frequency-Based Methods for Range Condition and Trend. In: Krueger, W.C., Chairman (ed.) Proceedings of 38th Annual Meeting Society for Range Mangement, Feb. 1985. Salt Lake City, Utah, pp. 87–90.

Winkworth, R.E. (1955). The use of point quadrates for the analysis of heathland. *Australian Journal of Botany* 3:68–81.

Whysong, G.L. and Brady, W.W. (1987). Frequency sampling and type II errors. *Journal of Range Management* 40:172–174.

Yates, D.S., Moore, D.S., Starnes, D.S. (2008). *The Practice of Statistics*, 3rd ed. New York: Freeman.

11 Soil Carbon

INTRODUCTION

A significant sink within a forest stand is its soil. Soil organic matter (SOM) is a pervasive material composed of carbon (C) and other elements. It includes the O horizon (or organic horizon, made up mostly of organic matters such as leaf litter and decomposed plant material), senesced plant materials within the mineral soil matrix, dead organisms (including macro- and microorganisms), microbial and root exudates (i.e., oozes), and organic materials adhering to mineral surfaces.

Simply, and directly to the point, any discussion or consideration of forest carbon storage must include soil. Moreover, the truth be told, the carbon stored in the soils of temperate forest ecosystems is often greater than the amount stored aboveground in living and dead biomass. Thus, before continuing with this important chapter and to enable better understanding of the topics discussed in follow-up chapters, it is important to first present and discuss the basics of soil science. And because this is a science book this chapter includes metrics involved with soils.

DID YOU KNOW?

The total stock of soil organic carbon varies for forest types in the United States, with large stocks occurring where a forest type is common.

SOIL BASICS[1]

We take soil for granted. It has always been there, right?—with the implied corollary that it will always be there—right? But where does soil come from?

Of course, soil was formed, and in a never-ending process, it is still being formed. However, soil formation is a slow process—one at work over the course of millennia, as mountains are worn away to dust through bare rock succession.

Any activity, human or natural, that exposes rock to air begins the process. Through the agents of physical and chemical weathering, through extremes of heat and cold, through storms and earthquake and entropy, bare rock is gradually worn away. As its exterior structures are exposed and weakened, plant life appears to speed the process along.

Lichens cover the bare rock first, growing on the rock's surface, etching it with mild acids and collecting a thin film of soil that is trapped against the rock and clings. This changes the conditions of growth so much that the lichens can no longer survive and are replaced by mosses.

[1] Based on material from F.R. Spellman (2020) *The Science of Environmental Pollution*. Boca Raton, FL: CRC Press.

DOI: 10.1201/9781003432838-14

The mosses establish themselves in the soil trapped and enriched by the lichens and collect even more soil. They hold moisture to the surface of the rock, setting up another change in environmental conditions.

DID YOU KNOW?

Based on personal observation during the past 60 years, I have noticed and pointed out to my college level students that depauperization of lichens in eastern hardwood forest areas; my theory is that this is probably the result of air pollution. U.S. National Park Service (2017) points out that "lichens are a paradox … durable enough to grow on the bark of trees and bare rock, yet sensitive to pollution and air quality." Over time, sensitive may be replaced by pollution-tolerant species. The point herein is that the species of lichen present in a location and the concentration of pollutants in those lichens can tell us a lot about air quality.

Well established mosses hold enough soil to allow herbaceous plant seeds to invade the rock. Grasses and small flowering plants move in, sending out fine root systems that hold more soil and moisture, and work their way into minute fissures in the rock's surface. Increased organisms join the increasingly complex community.

Weedy shrubs are the next invaders, with heavier root systems that find their way into every crevice. Each stage of succession affects the decay of the rock's surface and adds its own organic material to the mix. Over the course of time, mountains are worn away, eaten away to soil, as time, plants, weather, and extremes of weather work on them.

The parent material, the rock, becomes smaller and weaker as the years, decades, centuries, and millennia go by, creating the rich, varied, and valuable mineral resource we call soil.

DID YOU KNOW?

Note that the amount of carbon stored in soils depends on a variety of factors, including carbon inputs from vegetation, carbon losses from decomposition and biodegradation, soil physicochemical characteristics, and climate variables like temperature and precipitation (D'Amore, 2015).

Soil: What is it?

Perhaps no term causes more confusion in communication between various groups of average persons, soil scientists, soil engineers, and earth scientists than the word soil itself. In simple terms, soil can be defined as the topmost layer of decomposed rock and organic matter which usually contains air, moisture, and nutrients, and can therefore support life. Most people would have little difficulty in understanding and accepting this simple definition. Then why are various groups confused on the exact meaning of the word soil? Quite simply, confusion reigns because soil is not simple—it is quite complex. In addition, the term soil has different meanings to diverse groups (like pollution, the exact definition of soil is a personal judgment call). Let us look at how some of these different groups view soil.

The average person: seldom gives soil a first or second thought. Why should they?—soil is not that big a deal—that important—it doesn't impact their lives, pay their bills, and feed the bulldog, right?

Not exactly. Not directly.

The average person seldom thinks about soil as soil. He or she may think of soil in terms of dirt, but hardly ever as soil. Why is this? Having said the obvious about the confusion between soil and dirt, let us clear up this confusion.

First, soil is not dirt. Dirt is misplaced soil—soil where we do not want it, contaminating our hands, clothes, automobiles, tracked in on the floor. Dirt we try to clean up, and to keep out our living environments.

Secondly, soil is too special to be called dirt. Why? Because soil is mysterious, and whether we realize it or not, it is essential to our existence. Because we think of it as common, we relegate soil to an ignoble position. As our usual course of action, we degrade it, abuse it, throw it away, contaminate it, ignore it—we treat it like dirt, and only feces hold a lower status. Soil deserves better.

Why?

Again, because soil is not dirt—how can it be? It is not filth, or grime, or squalor. Instead soil is clay, air, water, sand, loam, organic detritus of former lifeforms, and most important, the amended fabric of Earth itself; if water is Earth's blood and air is Earth's breath, then soil is its flesh, bone, and marrow—simply put, soil is the substance that most life depends on.

Soil scientists: (or pedologists) a group interested in soils as a medium for plant growth. Their focus is on the upper meter or so beneath the land surface (this is known as the weathering zone, which contains the organic-rich material that supports plant growth) directly above the unconsolidated parent material. Soil scientists have developed a classification system for soils based on the physical, chemical, and biological properties that can be observed and measured in the soil.

Soils engineers: typically soils specialists who look at soil as a medium that can be excavated using tools. Soils engineers are not concerned with the plant-growing potential of a soil, but rather are concerned with a particular soil's ability to support a load. They attempt to determine (through examination and testing) a soil's particle size, particle-size distribution, and the plasticity of the soil.

Earth scientists: (or geologists) have a view that typically falls between pedologists and soils engineers—they are interested in soils and the weathering processes as past indicators of climatic conditions, and in relation to the geologic formation of useful materials ranging from clay deposits to metallic ores.

Would you like to gain new understanding of soil? Take yourself out to a plowed farm field somewhere. Reach down and pick a handful of soil and look at it—really look at it closely. What are you holding in your hand? Look at the two descriptions of that handful of soil that follow, and you may gain a better understanding of what soil is and why it is critically important to us all.

1. A handful of soil is alive, a delicate living organism—as lively as an army of migrating caribou and as fascinating as a flock of egrets. Literally teeming with life of incomparable forms, soil deserves to be classified as an independent ecosystem, or more correctly stated as many ecosystems as possible.
2. When we reach down and pick up a handful of soil, exposing Earth's stark bedrock surface, it should remind us (and maybe startle some of us) to the realization that without its thin living soil layer, Earth is a planet as lifeless as our own moon (Spellman, 1999).

In our attempt to define soil, to differentiate soil from dirt, in our trying to make the point that soil is important, critical, and vital to all of us, that it is arguably the most valuable of all mineral resources on Earth, we would be hard-pressed to do so in a more succinct but complete fashion than to quote E. L. Konigsburg, who states, "Soil is the working layer of the earth."

Definitions of Key Terms

Every branch of science, including soil science, has its own language. The terminology used herein is as different from that of astrophysics as that of astronomy is from botany. To work even at the edge of soil science, soil pollution, and soil pollution remediation, you must acquire familiarity with the vocabulary used in this text.

ablation till—a super glacial coarse-grained sediment or till, accumulating as the sub adjacent ice melts and drains away, finally deposited on the exhumed sub glacial surface.

absorption—movement of ions and water into the plant roots because of either metabolic processes by the root (active absorption), or because of diffusion along a gradient (passive absorption).

acid rain—atmospheric precipitation with pH values less than about 5.6, the acidity being due to inorganic acids such as nitric and sulfuric that is formed when oxides of nitrogen and sulfur are emitted into the atmosphere.

acid soil—a soil with a pH value of <7.0 or neutral. Soils may be naturally acid from their rocky origin, by leaching, or may become acid from decaying leaves or from soil additives such as aluminum sulfate (alum). Acid soils can be neutralized by the addition of lime products.

actinomycetes—a group of organisms intermediate between the bacteria and the true fungi that usually produce a characteristic branched mycelium. Includes many (but not all) organisms belonging to the order of Actinomycetales.

adhesion—molecular attraction that holds the surfaces of two substances (e.g., water and sand particles) in contact.

adsorption—the attraction of ions or compounds to the surface of a solid.

aeration, soil—the process by which air in the soil is replaced by air from the atmosphere. In a well-aerated soil, the soil air is similar in composition to the atmosphere above the soil. Poorly aerated soil usually contains more carbon dioxide and correspondingly less oxygen than the atmosphere above the soil.

aerobic—growing only in the presence of molecular oxygen, as aerobic organisms.

aggregates (soil)—soil structural units of various shapes, composed of mineral and organic material, formed by natural processes, and having a range of stabilities.

agronomy—a specialization of agriculture concerned with the theory and practice of field crop production and soil management. The scientific management of land.

air capacity—percentage of soil volume occupied by air spaces or pores.

air porosity—the proportion of the bulk volume of soil that is filled with air at any given time or under a given condition, such as a specified moisture potential, usually the large pores.

alkali—a substance capable of liberating hydroxide ions in water, measured by a pH of more than 7.0, and possessing caustic properties; it can neutralize hydrogen ions, with which it reacts to form a salt and water, and is an important agent in rock weathering.

alluvium—a general term for unconsolidated, granular sediments deposited by rivers.

amendment, soil—any substance other than fertilizers (such as compost, sulfur, gypsum, lime, and sawdust) used to alter the chemical or physical properties of a soil, generally to make it more productive.

ammonification—the production of ammonia and ammonium-nitrogen through the decomposition of organic nitrogen compounds in soil organic matter.

anaerobic—without molecular oxygen.

anion an atom which has gained one or more negatively charged electrons and is thus itself negatively charged.

aspect (of slopes)—the direction that a slope faces with respect to the sun.

assimilation—the taking up of plant nutrients and their transformation into actual plant tissues.

atterburg limits—water contents of fine-grained soils at different states of consistency.

autotrophs—plants and microorganisms capable of synthesizing organic compounds from inorganic materials by either photosynthesis or oxidation reactions.

available water—the portion of water in a soil that can be readily absorbed by plant roots. The amount of water released between the field capacity and the permanent wilting point.

bedrock—the solid rock underlying soils and the regolith in depths ranging from zero (where exposed to erosion) to several hundred feet.

biological function—the role played by a chemical compound or a system of chemical compounds in living organisms.

biomass—the total weight of living biological organisms within a specified unit (area, community, population).

biome—a major ecological community extending over large areas.

blow-out—a deflation depression, eroded by wind from the face of a vegetated dune.

breccia—a rock composed of coarse angular fragments cemented together.

calcareous soil—containing sufficient calcium carbonate (often with magnesium carbonate) to effervesce visibly when treated with hydrochloric acid.

caliche—a layer near the surface, cemented by secondary carbonates of calcium or magnesium precipitated from the soil solution. It may occur as a soft, thin soil horizon, as a hard, thick bed just beneath the solum, or as a surface layer exposed to erosion.

capillary water—held within the capillary pores of soils, mostly available to plants.

catena—the sequences of soils which occupy a slope transect, from the topographic divide to the bottom of the adjacent valley.

cation—an atom which has lost one or more negatively charged electrons and is thus itself positively charged.

chelate—(Greek, claw) a complex organic compound containing a central metallic ion surrounded by organic chemical groups.

class, soil—a group of soils having a definite range in a particular property such as acidity, degree of slope, texture, structure, land-use capability, degree of erosion, or drainage.

clay—a soil separate consisting of particles <0.0002 mm in equivalent diameter.

cohesion—holding together: force holding a solid or liquid together, owing to attraction between like molecules. Decreases with rise in temperature.

colloidal—matter of very fine particle size.

convection—a process of heat transfer in a fluid involving the movement of substantial volumes of the fluid concerned. Convection is very important in the atmosphere, and to a lesser extent, in the oceans.

denitrification—the biochemical reduction of nitrate or nitrite to gaseous nitrogen, either as molecular nitrogen or as an oxide of nitrogen.

detritus—debris from dead plants and animals.

diffusion—the movement of atoms in a gaseous mixture, or ions in a solution, primarily because of their own random motion.

drainage—the removal of excess water, both surface and subsurface, from plants. All plants (except aquatics) will die if exposed to an excess of water.

duff—the matted, partly decomposed organic surface layer of forest soils.

erosion—the wearing away of the land surface by running water, wind, ice, or other geological agents, including such processes as gravitational creep.

eutrophication—a process of lake aging whereby aquatic plants are abundant and waters are deficient in oxygen. The process is usually accelerated by enrichment of waters with surface runoff containing nitrogen and phosphorus.

evapotranspiration—the combined loss of water from a given area, during a specified period, by evaporation from the soil surface and by transpiration from plants.

exfoliation—mechanical or physical weathering that involves the disintegration and removal of successive layers of rock mass.

fertility, soil—the quality of a soil that enables it to provide essential chemical elements in quantities and proportions for the growth of specified plants.

fixation—the transformation in soil of a plant nutrient from an available to an unavailable state.

fluvial—deposits of parent materials laid down by rivers or streams.

friable—a soil consistency term pertaining to the ease of crumbling of soils.

heaving—the partial lifting of plants, buildings, roadways, and fence posts, etc., out of the ground, because of freezing and thawing of the surface soil during the winter.

heterotroph—an organism capable of deriving energy for life processes only from the decomposition of organic compounds, and incapable of using inorganic compounds as sole sources of energy or for organic synthesis.

horizon, soil—a layer of soil, approximately parallel to the soil surface, differing in properties and characteristics from adjacent layers below or above it.

humus—stable fraction of the soil organic matter (usually dark in color) remaining after the major portions of added plant and animal residues have decomposed.

hydration—the incorporation of water into the chemical composition of a mineral, converting it from an anhydrous to a hydrous form; the term is also applied to a form of weathering in which hydration swelling creates tensile stress within a rock mass.

hydraulic conductivity—the rate at which water can move through a soil.

hydrolysis—the reaction between water and a compound (commonly a salt). The hydroxyl from the water combines with the anion from the compound undergoing hydrolysis to form a base; the hydrogen ion from the water combines with the cation from the compound to form an acid.

hygroscopic coefficient—the amount of moisture in a dry soil when it is in equilibrium with some standard relative humidity nears a saturated atmosphere (about 98%), expressed in terms of percentage based on oven-dry soil.

infiltration—the downward entry of water into the soil.

ions—atoms which have lost or gained one or more negatively charged electrons.

land classification—the arrangement of land units into various categories based upon the properties of the land and its suitability for some particular purpose.

leaching—the removal of materials in solution from the soil by percolating waters.

Liebig's law—the growth and reproduction of an organism are determined by the nutrient substance (oxygen, carbon dioxide, calcium, etc.) that is available in minimum quantity with respect to organic needs, the limiting factor.

loam—the textural-class name for soil having moderate amounts of sand, silt, and clay.

loess—an accumulation of wind-blown dust (silt) which may have undergone mild digenesis.

marl—an earthy deposit consisting mainly of calcium carbonate, usually mixed with clay. Marl is used for liming acid soils. It is slower-acting than most lime products used for this purpose.

mineralization—the conversion of an element from an organic form to an inorganic state because of microbial decomposition.

nitrogen fixation—the biological conversion of elemental nitrogen (N_2) to organic combinations, or to forms readily utilized in biological processes.

osmosis—the movement of a liquid across a membrane from a region of high concentration to a region of low concentration. Water and nutrients move into roots independently.

oxidation—the loss of electrons by a substance.

parent material—the unconsolidated and chemically weathered mineral or organic matter from which the solum of soils is developed by Pedogenic processes.

ped—a unit of soil structure such as an aggregate, crumb, prism, block, or granule, formed by natural processes.

pedogenic/pedagogical process—any process associated with the formation and development of soil.

pH—the degree of acidity or alkalinity of the soil. Also referred to as soil reaction, this measurement is based on the pH scale where 7.0 is neutral—values from 0.0 to 7.0 are acid and values from 7.0 to 14.0 are alkaline. The pH of soil is determined by a simple chemical test where a sensitive indicator solution is added directly to a soil sample in a test tube.

photosynthesis—the process by which green leaves of plants, in the presence of sunlight, manufacture their own needed materials from carbon dioxide in the air, and water and minerals taken from the soil.

porosity, soil—the volume percentage of the total bulk not occupied by solid particles.

profile, soil—a vertical section of the soil through all its horizons and extending into the parent material.

reduction—the gain of electrons, and therefore the loss of positive valence charge by a substance.

regolith—the unconsolidated mantle of weathered rock and soil material on the earth's surface; loose earth materials above solid rock.

rock—the material that forms the essential part of the earth's solid crust, including loose incoherent masses such as sand and gravel, as well as solid masses of granite and limestone.

rock cycle—the global geological cycling of lithospheric and crustal rocks from their igneous origins through all any stages of alteration, deformation, resorption, and reformation.

runoff—the portion of the precipitation on an area that is discharged from the area through stream channels.

salinization—the process of accumulation of salts in soil.

sand—a soil particle between 0.05 and 2.0 mm in diameter; a soil textural class.

silt—a soil separate consisting of particles between 0.05 and 0.002 mm in equivalent diameter. A soil textural class.

slope—the degree of deviation of a surface from horizontal, measured in a numerical ratio, percent, or degrees.

soil—an assemblage of loose and normally stratified granular minerogenic and biogenic debris at the land surface, it is the supporting medium for the growth of plants.

soil air—the soil atmosphere; the gaseous phase of the soil, being that volume not occupied by soil or liquid.

soil horizon—a layer of soil, approximately parallel to the soil surface, with distinct characteristics produced by soil-forming processes. These characteristics form the basis for systematic classification of soils.

soil profile—a vertical section of the soil from the surface through all its horizons, including C horizons.

soil structure—the combination or arrangement of primary soil particles into secondary particles, units, or peds. These secondary units may be, but usually are not, arranged in the profile in such a manner as to give a distinctive characteristic pattern. The secondary units are characterized and classified based on size, shape, and degree of distinctness into classes, types, and grades, respectively.

soil texture—the relative proportion of the various soil separates in a soil.

soluble—will dissolve easily in water.

solum— (plural sola) the upper and most weathered part of the soil profile; the A, E, and B horizons.

subsoil—that part of the soil below the plow layer.

till—unstratified glacial drift deposited directly by the ice and consisting of clay, sand, gravel, and boulders intermingled in any proportion.

tilth—the physical condition of soil as related to its ease of tillage, fitness as a seedbed, and its impedance to seedling emergence and root penetration.

topsoil—the layer of soil moved in cultivation.

weathering—all physical and chemical changes produced in rocks, at or near the earth's surface, by atmospheric agents.

ABOUT SOIL

> Weekend gardeners tend to think of soil as the first few inches below the Earth's surface—the thin layer that needs to be weeded and that provides a firm foundation for plants. But the soil extends from the surface down to the Earth's hard rocky crust. It is a zone of transition and, as in many of nature's transition zones; the soil is the site of important chemical and physical processes. In addition, because plants need soil to grow, it is arguably the most valuable of all the mineral resources on Earth.
>
> *—Macmillan Publishing Company, p. 30, 1992*

Before we begin a journey that takes us through the territory that is soil, and examine soil from micro to macro levels, we need to stop for a moment and discuss why, beyond the obvious reason, soil is so important to us—to our environment—to our very survival. Is soil really that big a deal? Is it that important? Do we need to even think about soil?

Yes, yes, and yes. Soil is all these things and more.

DID YOU KNOW?

Through various stabilization mechanisms. The formation of soil C feeds a major reservoir (sink) of global C.

FUNCTIONS OF SOIL

We normally think of or relate soil to our backyards, to farms, to forests, or to a regional watershed. We think of soil as the substance upon which plants grow. This is generally about as far as the average person's thoughts extend concerning soil and its usefulness to us. Soils play other roles, though. They have five main functions important to us: (1) soil is a medium for plant growth; (2) soils regulate our water supplies; (3) soils are recyclers of raw materials; (4) soils provide a habitat for organisms; and (5) soils are used as an engineering medium.

Let us take a closer look at each of the functions of soil.

Soil: A Plant Growth Medium

We are all aware of the primary function of soil: soil serves as a plant growth medium; it is responsible for plant growth (a function that becomes more important with each passing day as Earth's population continues to grow). However, while true that soil is a medium for plant growth (thus critical to maintaining life), let us also point out that soil is alive as well. Soil exists in paradox: we depend on soil for life, and at the same time, soil depends on life; it's very origin, its maintenance, and its true nature are intimately tied to living plants and animals. What does this mean? Let us look at how one renowned environmental writer, whose elegant prose brought this point to the forefront, explained this paradox.

> "The soil community . . . consists of a web of interwoven lives, each in some way related to the others—the living creatures depending on the soil, but the soil in turn is a vital element of the earth only so long as this community within it flourishes."
>
> —*Rachel Carson, Silent Spring, p. 56, 1962*

What Rachel Carson said is important—the meaning clear. To us, she points out what the soil might say to us if it could— "Don't kill off the life within me and I will do the best I can to provide life that will help to sustain your life." What we have here is a tradeoff—one vitally important to both to soil and to ourselves. Remember that most of Earth's people are tillers of the soil—and the soil is their source of life—and those soil tillers provide food for us all.

As a plant growth medium, soil provides vital resources and performs essential functions for the plant. To grow in soil, plants must have water and nutrients—soil provides these. To grow and to sustain growth, a plant must have a root system—soil provides pore spaces for roots. To grow and maintain growth, a plant's roots must have oxygen for respiration and carbon dioxide exchange and ultimate diffusion out of the soil—soil provides the air and pore spaces (the soil's ventilation system) for this. To continue to grow, a plant must have support—soil provides this support.

If a plant seed is planted in a soil and is exposed to the proper amount of sunlight, for growth to occur, the soil must provide nutrients through a root system that has space to grow, a continuous stream of water (it requires about 500 g of water to produce 1 g of dry plant material) for root nutrient transport and plant cooling, and a pathway for both oxygen and carbon dioxide transfer. Just as important, soil water provides the plant with its normal fullness or tension (turgor) it needs to stand—the structural support it needs to face the sun for photosynthesis to occur.

As well as the functions stated above, soil is also an important moderator of temperature fluctuations. If you have ever dug in a garden on a sizzling summer day, you probably noticed that the soil was warmer (even hot) on the surface, but much cooler just a few inches below the surface.

Soil: Regulator of Water Supplies

When we walk on land, few of us probably realize that we are walking across a bridge. This bridge (in many areas) transports us across a veritable ocean of water below us, deep—or not so deep—under the surface of the earth.

Consider what happens to rain. Where does the rainwater go? Some, falling directly over water bodies, become part of the water body again. But an enormous amount falls on land. Where does it go? Some of the water, obviously, runs off—always following the path of least resistance. In modern communities, stormwater runoff is a hot topic. Cities have taken giant steps to try and control runoff—to send it where it can be properly handled, to prevent flooding.

Let's take a closer look at precipitation and the "sinks" it "pours" into, and then relate this usually natural operation to soil water. We begin with surface water, and then move onto that ocean of water below the soil's surface: groundwater.

Surface water (water on the earth's surface as opposed to subsurface water—groundwater) is mostly a product of precipitation—rain, snow, sleet, or hail. Surface water is exposed or open to the atmosphere, and results from the movement of water on and just under the earth's surface (overland flow). This overland flow is the same thing as surface runoff, which is the amount of rainfall which passes over the earth's surface. Specific sources of surface water include rivers, streams, lakes, impoundments, shallow wells, rain catchments, and tundra ponds or meskegs (peat bogs).

Most surface water is the result of surface runoff. The amount and flow rate of surface runoff is highly variable. This variability stems from two main factors: (1) human interference (influences) and (2) natural conditions. In some cases, surface water runs quickly off land. Generally this is undesirable (from a water resources standpoint) because it does not provide enough time for water to infiltrate into the ground and recharge groundwater aquifers. Other problems associated with quick surface water runoff are erosion and flooding. Probably the only good thing that can be said about surface water that quickly runs off land is that it does not have enough time (normally) to become contaminated with high mineral content. Surface water running slowly off land may be expected to have all the opposite effects.

Surface water travels over the land to what amounts to a predetermined destination. What factors influence how surface water moves? Surface water's journey over the face of the earth typically begins at its drainage basin, sometimes referred to as its drainage area, catchment, and/or watershed. For a groundwater source, this is known as the recharge area—the area from which precipitation flows into an underground water source.

A surface water drainage basin is usually an area measured in square miles, acres, or sections, and if a city takes water from a surface water source, how large (and what lies within) the drainage basin is essential information for the assessment of water quality.

We know that water doesn't run uphill. Instead, surface water runoff (like the flow of electricity), follows along the path of least resistance. Water within a drainage basin will naturally (by the geological formation of the area) is shunted toward one primary watercourse (a river, stream, creek, and brook) unless some manufactured distribution system diverts the flow.

Numerous factors directly influence the surface water's flow over land. The principal factors are:

- **Rainfall duration**. Length of the rainstorm affects the amount of runoff. Even a light, gentle rain will eventually saturate the soil if it lasts long enough. Once the saturated soil can absorb no more water, rainfall builds up on the surface and begins to flow as runoff.
- **Rainfall intensity.** The harder and faster it rains, the more quickly soil becomes saturated. With hard rains, the surface inches of soil quickly become inundated, and with short, hard storms, most of the rainfall may end up as surface runoff, because the moisture is carried away before significant amounts of water are absorbed into the earth.
- **Soil moisture**. Obviously, if the soil is already laden with water from previous rains, the saturation point will be reached sooner than if the soil was dry. Frozen soil also inhibits water absorption: up to 100% of snow melts or rainfall on frozen soil will end up as runoff because frozen ground is impervious.
- **Soil composition**. Runoff amount is directly affected by soil composition. Hard rock surfaces will shed all rainfall, obviously, but so will soils with heavy clay composition. Clay soils possess small void spaces that swell when wet. When the void spaces close, they form a barrier that does not allow additional absorption or infiltration. On the opposite end of the spectrum, coarse sand allows easy water flow-through, even in a torrential downpour.
- **Vegetation cover**. Runoff is limited by ground cover. Roots of vegetation and pine needles, pinecones, leaves, and branches create a porous layer (sheet of decaying natural organic substances) above the soil. This porous "organic" sheet (ground cover) readily allows water into the soil. Vegetation and organic waste also act as a cover to protect the soil from hard, driving rains. Hard rains can compact bare soils, close off void spaces, and increase runoff. Vegetation and ground cover work to maintain the soil's infiltration and water-holding capacity. Note that vegetation and groundcover also reduce evaporation of soil moisture.
- **Ground slope.** Flat land water flow is usually so slow that substantial amounts of rainfall can infiltrate the ground. Gravity works against infiltration on steeply sloping ground where up to 80% of rainfall may become surface runoff.
- **Human influences**. Various human activities have a definite impact on surface water runoff. Most human activities tend to increase the rate of water flow. For example, canals and ditches are usually constructed to provide steady flow, and agricultural activities generally remove ground cover that would work to retard the runoff rate. On the opposite extreme, manufactured dams are generally built to retard the flow of runoff. Human habitations, with their paved streets, tarmac, paved parking lots, and buildings create surface runoff potential, since so many surfaces are impervious to infiltration. All these surfaces hasten the flow of water, and they also increase the possibility of flooding, often with devastating results. Because

of urban increases in runoff, a whole new field (industry) has developed: Stormwater management. Paving over natural surface acreage has another serious side effect. Without enough area available for water to infiltrate the ground and percolate through the soil to eventually reach and replenish—recharge—groundwater sources, those sources may eventually fail, with devastating impact on local water supply.

DID YOU KNOW?

Soils with high soil C tend to have high nutrient content, promoting growth of trees and foliage. About 99% soil nitrogen (N) is found within soil organic matter (SOM). In addition, SOM provides much of the cation exchange capacity essential for making nutrients available to plant roots.

Now let us shift gears and look at groundwater.

Groundwater? What does groundwater have to do with carbon in soil?

Carbon has a lot to do with and basically works with infiltrating water (aboveground precipitation and water flow) and groundwater. Consider, for example, that proper management of farms, ranches, and public land leads to "increased soil carbon" which increases soil water holding capacity and increases hydrologic benefits such as increase baseflows and aquifer recharge, decreased flooding and erosion, and reduced climate-related water deficits.

So, groundwater is included herein because within the huge carbon sink, soil, carbon has a role beyond being sequestered, stored, or captured. Water falling to the ground as precipitation normally follows three courses. Some runs off directly to rivers and streams, some infiltrate to ground reservoirs, and the rest evaporates or transpires through vegetation. The water in the ground (groundwater) is "invisible," and may be thought of as a temporary natural reservoir (ASTM, 1969). Almost all groundwater is in constant motion toward rivers or other surface water bodies.

Groundwater is defined as water below the earth's crust, but above a depth of 2,500 ft. Thus, if water is located between the earth's crust and the 2,500-ft level, it is considered usable (potable) fresh water. In the United States, it is estimated "that at least 50% of total available freshwater storage is in underground aquifers" (Kemmer, 1979, p. 17).

We are concerned with the amount of water retained in the soil to ensure plant life and growth. Having said this, recall that earlier we stated that producing 1 g of dry plant material requires about five hundred grams of water. Note that about five grams of this water becomes an integral part of the plant. Unless rainfall is frequent, you do not have to be a rocket scientist to figure out that the ability of soil to hold water against the force of gravity is very important—thus, one of the vital functions of soil is to regulate the water supply to plants.

Soils: Recyclers of Raw Materials

Can you imagine what it would be like to step out into the open air and be hit by a stench that you could not only smell, but could almost reach out and grab you (like the situation we had in the cave earlier—but worse)? You look out upon the cluttered fields in front of your domicile and see nothing but stack upon stack upon stack of the sources of this horrible, putrefied, gagging stench. We are talking about plant and animal remains and waste (mountains of it), reaching toward the sky. "Impossible," you say. Well, thankfully (in most cases) you are right. However, if it were not for the power of the soil to recycle waste products, then this scene or something like it is imaginable and even possible—but of course it would be impossible because there would be no life to die and to stack up anywhere, for that matter.

Soil is a recycler—probably the premier recycler on Earth. The simple fact is that if it were not for soil's incredible recycling ability, plants and animals would have run out of nourishment long ago. Soil is recycled in other ways. For example, consider the geochemical cycles (i.e., the chemical interactions between soil, water, air, and life on earth) in which soil plays a significant role.

Soil possesses the incomparable ability and capacity to assimilate great quantities of organic wastes and turn them into beneficial organic matter (humus), then to convert the nutrients in the wastes to forms that can be utilized by plants and animals. In turn, the soil returns carbon to the atmosphere as carbon dioxide, where it again will eventually become part of living organisms through photosynthesis. Soil performs several different recycling functions—most of them good, some of them not so good.

Not so good? Yes. Consider one recycling function of soil that may not be so good. Soils have the capacity to accumulate substantial amounts of carbon as soil organic matter which can have a major impact on global change such as greenhouse effect.

Soil: Habitat for Soil Organisms

> Life not only formed the soil, but other living things of incredible abundance and diversity now exist within it; if this were not so the soil would be a dead and sterile thing.
>
> —*Carson, p. 53, 1962*

One thing is certain; most soils are not dead and sterile things. The fact is a handful of soil is an ecosystem. It may contain up to billions of organisms, belonging to thousands of species. Let's look at Table 11.1, which lists a few (very few) of these organisms.

DID YOU KNOW?

By increasing nutrient and water availability, soils with high soil C and organic matter support increased growth of forests, rangelands, and wildlands, leading to increased uptake of carbon dioxide (CO_2).

Obviously, communities of living organisms inhabit the soil. What is not so obvious is that they are as complex and intrinsically valuable as are those organisms that roam the land surface and waters of Earth.

Soil: An Engineering Medium We usually think of soil as being firm and solid—"solid ground" . . . terra firma. As solid ground (when it is), soil is usually a good substrate upon which to build highways and structures. We say "usually" true because in

TABLE 11.1
Soil Organisms (A Representative Sample)

Microorganisms (Protists)
Bacteria
Fungi
Actinomycetes
Algae
Protozoa
Nonarthropod animals
Nematodes
Earthworms and potworms
Arthropod animals
Springtails
Mites
Millipedes and centipedes
Harvestman
Ants
Diplopoda
Diptera
Crustacea
Vertebrates
Mice, moles, voles
Rabbits, gophers, squirrels

some ways this is untrue. Not all soils are firm and solid—some are not as stable as others. While construction of buildings and highways may be suitable in one location on one type of soil, it may be unsuitable in another location with different soil. To construct structurally sound, stable (and therefore reliable) highways and buildings, construction on soils and with soil materials requires knowledge of the diversity of soil properties—which really means that a knowledge of the engineering properties of soils is required.

Note that working with manufactured building materials that have been "engineered" to withstand certain stresses and forces is much different than working with natural soil materials, even though engineers have the same concerns about soils as they do with manufactured building materials (concrete and steel). It is much more difficult to make these predictions or determinations for soil's ability to resist compression, to remain in place, its bearing strength, shear strength, and stability, than it is to make the same determinations for manufactured building materials.

BACK TO SOIL BASICS

Okay, earlier we defined soil. Let's take another look and a bit different explanation of what soil is and is all about. Any fundamental discussion about soil should begin with a definition of what soil is. The word soil is derived through Old French from the Latin solum, which means floor or ground. A more concise definition is made difficult by the great diversity of soils throughout the globe. However, here is a generalized definition from the Soil Survey Staff, Science Society of America (1975):

Soil is unconsolidated mineral matter on the surface of the earth that has been subjected to and influenced by genetic and environmental factors of parent material, climate, macro- and microorganisms, and topography, all acting over a period and producing a product—soil—that differs from the material from which it is derived in many physical, chemical, and biological properties, and characteristics.

Engineers might define soil by saying that soil occupies the unconsolidated mantle of weathered rock making up the loose materials on the Earth's surface; commonly known as the regolith.

Soil can be described as a three-phase system, composed of a solid, liquid, and gaseous phase.

Note: This phase relationship is important in dealing with soil pollution, because each of the three phases of soil are in equilibrium with the atmosphere, and with rivers, lakes, and the oceans. Thus, the fate and transport of pollutants are influenced by each of these components.

Soil is also commonly described as a mixture of air, water, mineral matter, and organic matter; the relative proportions of these four components greatly influence the productivity of soils. The interface (where the regolith meets the atmosphere) of these materials that make up soil is what concerns us here.

Keep in mind that the four major ingredients that make up soil are not mixed or blended like cake batter. Instead, pore spaces (vital to air and water circulation, providing space for roots to grow and microscopic organisms to live) are a major (and critically important) constituent of soil. Without sufficient pore space, soil would be too compacted to be productive. Ideally, the pore space will be divided roughly equally between water and air, with about one-quarter of the soil volume consisting of air and one-quarter consisting of water. The relative proportions of air and water in a soil typically fluctuate significantly as water is added and lost. Compared to surface soils, subsoil's tend to contain less total pore space, less organic matter, and a larger proportion of micropores which tend to be filled with water.

Let's take a closer look at the four major components (the air, water, mineral matter, and organic matter) that make up soil.

Soil air circulates through soil pores in the same way air circulates through a ventilation system. Only when the pores (the ventilation ducts) become blocked by water or other substances does the air fail to circulate. Though soil pores normally connect to interface with the atmosphere, soil air is different from atmospheric air. It differs in composition from place to place. Soil air also normally has a higher moisture content than the atmosphere. The content of carbon dioxide (CO_2) is usually higher as well, and that of oxygen (O_2) lower than accumulations of these gases found in the atmosphere.

Earlier we stated that only when soil pores are occupied by water or other substances does air fail to circulate in the soil. For proper plant growth, this is of importance, because in soil pore spaces that are water-dominated, air oxygen content is low and carbon dioxide levels are high, which restricts plant growth.

The presence of water in soil (often reflective of climatic factors) is essential for the survival and growth of plants and other soil organisms. Soil moisture is a major determinant of the productivity of terrestrial ecosystems and agricultural systems. Water moving through soil materials is a major force behind soil formation. Along with air, water, and dissolved nutrients, soil moisture is critical to the quality and quantity of local and regional water resources.

Mineral matter varies in size and is a major constituent of non-organic soils. Mineral matter consists of large particles (rock fragments) including stones, gravel, and coarse sand. Many of the smaller mineral matter components are made of a single mineral. Minerals in the soil (for plant life) are the primary source of most of the chemical elements essential for plant growth.

Soil organic matter consists primarily of living organisms and the remains of plants, animals, and microorganisms that are continuously broken down (biodegraded) in the soil into new substances that are synthesized by other microorganisms. These other microorganisms continually use this organic matter and reduce it to carbon dioxide (via respiration) until it is depleted, making repeated additions of new plant and animal residues necessary to maintain soil organic matter (Brady and Weil, 1996).

Now that we have defined soil, let's take a closer look at a few of the basics pertaining to soil, and some of the common terms used in any discussion related to soil basics.

Soil is the layer of bonded particles of sand, silt, and clay that covers the land surface of the earth. Most soils develop in multiple layers. The topmost layer (topsoil) is the layer of soil moved in cultivation, and in which plants grow. This topmost layer is an ecosystem composed of both biotic and abiotic components—inorganic chemicals, air, water, decaying organic material that provides vital nutrients for plant photosynthesis, and living organisms. Below the topmost layer is the subsoil (the part of the soil below the plow level, usually no more than a meter in thickness). Subsoil is much less productive, partly because it contains much less organic matter. Below is the parent material, the unconsolidated (and chemically weathered) bedrock or other geologic material from which the soil is ultimately formed. The general rule of thumb is that it takes about 30 years to form one inch of topsoil from subsoil; it takes much longer than that for subsoil to be formed from parent material—the length of time depending on the nature of the underlying matter (Franck and Brownstone, 1992).

Soil: Physical Properties

From the soil pollution technologist's point of view (regarding land conservation and methodologies for contaminated soil remediation through reuse and recycling), five major physical properties of soil are of interest. They are soil texture, slope, structure, organic matter, and soil color. Soil texture (the relative proportion of the various soil separates in a soil is a given and cannot be easily or practically changed significantly). It is determined by the size of the rock particles (sand, silt, and clay particles) or the soil separates within the soil. The largest soil particles are gravel, which consist of fragments larger than 2.0 mm in diameter.

Particles between 0.05 and 2.0 mm are classified as sand. Silt particles range from 0.002 to 0.05 mm in diameter, and the smallest particles (clay particles) are less than

0.002 mm in diameter. Though clays are composed of the smallest particles, those particles have stronger bonds than silt or sand, though once broken apart, they erode more readily. Particle size has a direct impact on erodibility. Rarely does a soil consist of only one single size of particle—most are a mixture of assorted sizes.

The slope (or steepness of the soil layer) is another given, important because the erosive power of runoff increases with the steepness of the slope. Slope also allows runoff to exert increased force on soil particles, which breaks them apart more readily and carries them farther away.

Soil structure (tilth) should not be confused with soil texture—they are different. In fact, in the field, the properties determined by soil texture may be considerably modified by soil structure. Soil structure refers to the combination or arrangement of primary soil particles into secondary particles (units or peds). Simply stated, soil structure refers to the way various soil particles clump together. The size, shape, and arrangement of clusters of soil particles called aggregates form naturally-formed larger clumps called peds. Sand particles do not clump because sandy soils lack structure. Clay soils tend to stick together in large clumps. Good soil develops small friable (easily crumbled) clumps. Soil develops a unique, stable structure in undisturbed landscapes, but agricultural practices break down the aggregates and peds, lessening erosion resistance.

The presence of decomposed or decomposing remains of plants and animals (organic matter) in soil helps not only fertility, but also soil structure—and especially the soil's ability to store water. Live organisms—protozoa, nematodes, earthworms, insects, fungi, and bacteria—are typical inhabitants of soil. These organisms work to either control the population of organisms in the soil or to aid in the recycling of dead organic matter. All soil organisms, in one way or another, work to release nutrients from organic matter, changing complex organic materials into products that can be used by plants.

Just about anyone who has looked at soil has probably noticed that soil color is often different from one location to another. Soil colors range from very bright to dull grays, to a wide range of reds, browns, blacks, whites, yellows, and even greens. Soil color is dependent primarily on the quantity of humus and the chemical form of iron oxides present.

Soil scientists use a set of standardized color charts (the Munsell Color Book) to describe soil colors. They consider three properties of color—hue, value, and chroma—in combination to produce many color chips to which soil scientists can compare the color of the soil being investigated.

Soil Separates

As pointed out earlier, soil particles have been divided into groups based on their size termed soil separates—sand, silt, and clay—by the International Soil Science Society System, the United States Public Roads Administration, and the United States Department of Agriculture. In this text, we use the classification established by the U.S. Department of Agriculture (USDA). The size ranges in these separates reflect major changes in how the particles behave, and in the physical properties they impart to soils.

In Table 11.2, the names of the separates are given, together with their diameters, and the number of particles in one gram of soil (according to USDA, 1975).

TABLE 11.2
Characteristics of Soil Separates (USDA)

Separate	Diameter (mm)	Number of Particles/Gram
Very coarse sand	2.00–1.00	90
Coarse sand	1.00–0.50	720
Medium sand	0.50–0.25	5,700
Fine sand	0.25–0.10	46,000
Very fine sand	0.10–0.05	722,000
Silt	0.05–0.002	5,776,000
Clay	Below 0.002	90,260,853,000

Sand, one of the individual-sized groups of mineral soil particles, ranges in diameter from 2 to 0.05 mm, and is divided into five classes (see Table 11.2). Sand grains are spherical (rounded) in shape, with variable angularity, depending on the extent to which they have been worn down by abrasive processes such as rolling around by flowing water during soil formation.

Sand forms the framework of soil and gives it stability when in a mixture of finer particles. Sand particles are relatively large, which allows voids that form between each grain to also be relatively large. This promotes free drainage of water and the entry of air into the soil. Sand is usually composed of a high percentage of quartz, because it is most resistant to weathering, and its breakdown is extremely slow. Many other minerals are found in sand, depending upon the rocks from which the sand was derived. In the short term (on an annual basis), sand contributes little to plant nutrition in the soil. However, in the long term (thousands of years of soil formation) soil with a lot of weatherable minerals in their sand fraction develop a higher state of fertility.

Another individual-sized group of mineral soil particles is silt (essentially microsand). Like sand (spherical and mineralogically similar), silt is smaller (it weathers faster and releases soluble nutrients for plant growth faster than sand)—too small to be seen with the naked eye. Too fine to be gritty, silt imparts a smooth feel (like flour) without stickiness. The pores between silt particles are much smaller than those in sand (sand and silt are just progressively finer and finer pieces of the original crystals in the parent rocks). In flowing water, silt is suspended until it drops out when flow is reduced. On the land surface, silt, if disturbed by intense winds, can be carried great distances, and is deposited as loess.

The clay soil separate is (for the most part) much different from sand and silt. Clay is composed of secondary minerals that were formed by the drastic alteration of the original forms, or by the recrystallization of the products of their weathering. Because clay crystals are plate like (sheeted) in shape they have a tremendous surface area-to-volume ratio, giving clay a tremendous capacity to adsorb water and other substances on its surfaces. Clay acts as a storage reservoir for both water and nutrients. There are many kinds of clay, each with different internal arrangements of chemical elements, which give them individual characteristics.

Soil Formation

Everywhere on Earth's land surface is either rock formation or exposed soil. When rocks formed deep in the Earth are thrust upward and exposed to the Earth's atmosphere, the rocks adjust to the unique environment, and soil formation begins. Soil is formed because of physical, chemical, and biological interactions in specific locations. Just as vegetation varies among biomes, so do the soil types that support that vegetation. The vegetation of the tundra and rain forest differ vastly from each other, and from vegetation of the prairie and coniferous forest—soils differ in similar ways.

In the soil-forming process, two related—but fundamentally different—processes are occurring simultaneously. The first is the formation of soil parent materials by weathering of rocks, rock fragments and sediments. This set of processes is carried out in the zone of weathering. The end point is to produce parent material for the soil to develop in and is referred to as C horizon material (see Figure 11.1). It applies in the same way for glacial deposits as for rocks. The second set of processes is the formation of the soil profile by soil-forming processes, which gradually changes the C horizon material into A, E and B horizons. Figure 11.1 illustrates two soil profiles, one on a hard granite, and one on a glacial deposit.

Soil development takes time and is the result of two major processes: weathering and morphogenesis. Weathering (the breaking down of bedrock and other sediments

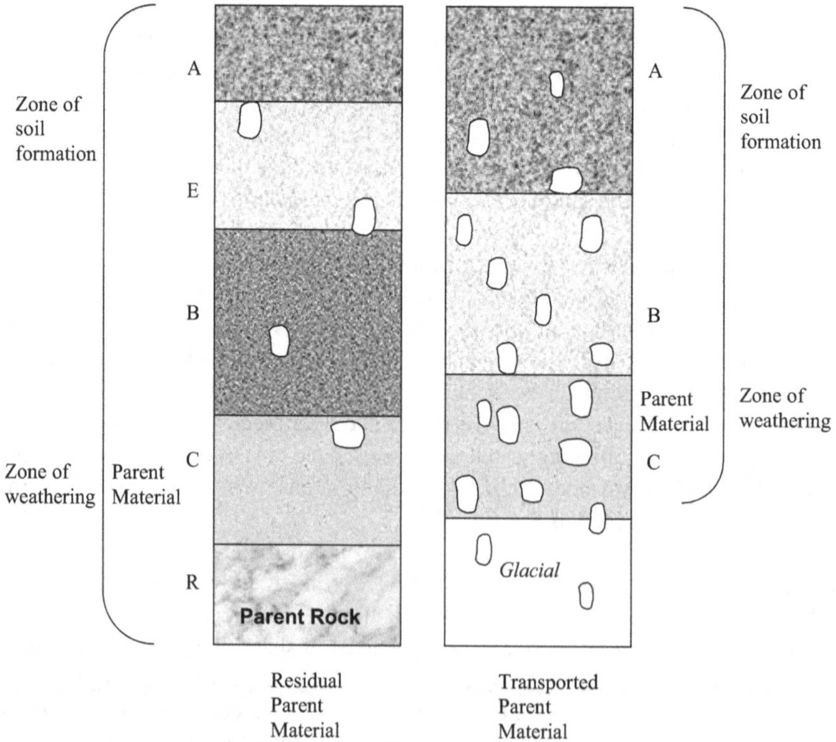

FIGURE 11.1 Soil profiles on residual and transported parent materials.

that have been deposited on the bedrock by wind, water, volcanic eruptions or melting glaciers) happens physically, chemically, or a combination of both.

Physical weathering involves the breaking down of rock primarily by temperature changes and the physical action of water, ice, and wind. When a geographical location is characterized as having an arid desert biome, the repeated exposure to very high temperatures during the day followed by low temperatures at night causes rocks to expand and contract, and eventually to crack and shatter. At the other extreme, in cold climates rock can crack and break because of repeated cycles of expansion of water in rock cracks and pores during freezing, and contraction during thawing. Physical weathering occurs where various vegetation types can not only spread their roots and grow, but the roots can exert enough pressure to enlarge cracks in solid rock, eventually splitting and breaking the rock. Plants such as mosses and lichens also penetrate rock and loosen particles. Bare rocks are also subjected to chemical weathering, which involves chemical attack and dissolution of rock. Accomplished primarily through oxidation via exposure to oxygen gas in the atmosphere, acidic precipitation (after having dissolved lesser amounts of carbon dioxide gas from the atmosphere) and acidic secretions of microorganisms (bacteria, fungi, and lichens), chemical weathering speeds up in warm climates and slows down in cold ones.

Physical and chemical weathering do not always (if ever) occur independently of each other. Instead, they normally work in combination. A classic example of the effect—the power of their simultaneous actions—can be seen in an ecological process known as bare rock succession, described in the chapter opening.

The final stages of soil formation consist of the processes of morphogenesis, or the production of a distinctive soil profile with its constituent layers or horizons. The soil profile (the vertical section of the soil from the surface through all its horizons, including C horizons) gives the environmental scientist critical information. When properly interpreted, soil horizons can provide warning on potential problems in using the land and talk much about the environment and history of a region. The soil profile allow us to describe, sample, and map soils.

Soil horizons are distinct layers, roughly parallel to the surface, which differ in color, texture, structure, and content of organic matter. The clarity with which horizons can be recognized depends upon the relative balance of the migration, stratification, aggregation, and mixing processes that take place in the soil during morphogenesis. In podzol-type soils (formed mainly in cool, humid climates), striking horizonation is quite apparent; in Vertisol-type soils (soils high in swelling clays), the horizons are less distinct. When horizons are studied, they are given a letter symbol to reflect the genesis of the horizon.

Certain processes work to create or destroy clear soil horizons. Formation of soil horizons that tend to create clear horizons by vertical redistribution of soil materials include the leaching of ions in soil solutions, movement of clay-sized particles, upward movement of water by capillary action, and surface deposition of dust and aerosols. Clear soil horizons are destroyed by mixing processes that occur because of organisms, cultivation practices, creep processes on slopes, frost heave, and by swelling and shrinkage of clays—all part of the natural soil formation process.

Soil Characterization

> Classification schemes of natural objects seek to organize knowledge so that the properties and relationships of the objects may be most easily remembered and understood for some specific purpose. The ultimate purpose of soil classification is maximum satisfaction of human wants that depend on use of the soil. This requires grouping soils with comparable properties so that lands can be efficiently managed for crop production. Furthermore, soils that are suitable or unsuitable for pipelines, roads, recreation, forestry, agriculture, wildlife, building sites, and so forth can be identified.
>
> —*H. D. Foth, p. 255, 1978*

When people become ill, they may go to a doctor to seek a determination (a diagnosis) of what is causing the illness, then hopefully, a forecast or outlook (prognosis) of how long they feel well again.

What do diagnosis and prognosis have to do with soil? Quite a lot. The diagnostic techniques used by a physician to identify the causative factors leading to an illness are analogous to the soil practitioner using diagnostic techniques to identify a particular soil. Sound far-fetched? It shouldn't because it isn't. Soil scientists must be able to determine the type of soil they study or work with.

Determining the type of soil makes sense, but what does prognosis have to do with all this? Consider that soil practitioners not only need to be able to identify or classify a soil type, but also that this information allows them to correctly predict how a pollutant will react or respond when spilled in that type of soil. The fate of the pollutant is important in determining the possible damage incurred to the environment—soil, groundwater, and air, because ultimately, a spill could easily affect all three. Thus, the soil practitioner must not only use diagnostic tools in determining soil type, but must also be familiar with the soil type, to judge how a pollutant or contaminant will respond when spilled in the soil type.

Let's take a closer look at the genesis of soil classification and where we are at present with it.

From the time humans first advanced from hunter-gatherer status to cultivators of crops they noticed differences in "good" (productive) soils and "bad" (unproductive) soils. This probably was the first effort in classifying soils—as good or bad. The ancient Chinese, Egyptians, Romans, and Greeks all recognized and acknowledged the differences in soils as media for plant growth. These early soil classification practices were based primarily upon texture, color, and wetness.

Soil classification as a scientific practice did not gain a foothold until the later eighteenth and early nineteenth centuries when the science of geology was born. Such terms (with an obvious geological connotation) as "limestone" soils, and "lake-laid" soils, as well as "clayey" and "sandy" soils came into being. The Russian scientist Dokuchaev was the first to suggest a generic classification of soils—that soils were natural bodies. Dokuchaev's classification work was then further developed by Europeans and Americans. The system is based on the theory that each soil has a definite form and structure (morphology), related to a combination of soil-forming factors. This system was used until 1960, when the U.S. Department of Agriculture published Soil Classification, A Comprehensive System. This classification system places major emphasis on soil morphology and gives less emphasis to genesis or the soil-forming factors as compared to previous systems. In 1975, Soil Classification, A Comprehensive System was replaced by soil taxonomy, which classifies objects according to their natural relationships. Soils are classified based on measurable properties of soil profiles.

Note that no clear delineation or line of demarcation can be drawn between the properties of one soil and those of another. Instead, a gradation (sometimes quite subtle—like from one shade of white to another) occurs in soil properties as one moves from one soil to another. Brady and Weil (1996) point out that "the gradation in soil properties can be compared to the gradation in the wavelengths of light as you move from one color to another. The changing is gradual, and yet we identify a boundary that differentiates what we call 'green' from what we call 'blue'" (p. 58). To properly characterize the primary characteristics of a soil, a soil must be identified down to the smallest three-dimensional characteristic sample possible (which makes sense when you consider the impossibility of studying an entire soil area at one time).

However, to accurately perform a soil sample characterization, a sampling unit must be large enough so that the nature of its horizons can be studied, and the range of its properties identified. The pedon (rhymes with head-on) is this unit. The pedon is roughly polygonal in shape and designates the smallest characteristic unit that can still be called a soil.

Because pedons occupy a very small space (from approximately 1 to 10m), they cannot, obviously, be used as the basic unit for a workable field soil classification system. To solve this problem, a group of pedons, termed a polypedon, is of sufficient size to serve as a basic classification unit (or commonly called a soil individual). In the United States, these groupings have been called a soil series.

There is a difference between a soil and the soil. This difference is important in the soil classification scheme. A soil is characterized by a sampling unit (pedon), which as a group (polyhedons) form a soil individual. The soil, on the other hand, is a collection of all these natural ingredients, and is distinguishable from other bodies such as water, air, solid rock, and other parts of the Earth's crust. By incorporating the difference between a soil and the soil, a classification system has been developed that is effective and widely used.

Diagnostic Horizons and Temperature and Moisture Regimes

Soil taxonomy uses a strict definition of soil horizons called diagnostic horizons, which are used to define most of the orders. Two kinds of diagnostic horizons are recognized: surface and subsurface. The surface diagnostic horizons are called epipedons (Greek epi over; pedon soil). The epipedon includes the dark (organic-rich) upper part of the soil, of the upper eluvial horizons, and/or sometimes both. Those soils beneath the epipedons are called subsurface diagnostic horizons. Each of these layers is used to characterize different soils in soil taxonomy.

In addition to using diagnostic horizons to strictly define soil horizons, soil moisture regime classes can also be used. A soil moisture regime refers to the presence of plant-available water or groundwater at a sufficiently elevated level. The control section of the soil (ranging from 10 to 30cm for clay and from 30 to 90cm for sandy soils) designates that section of the soil where water is present or absent during given periods in a year. The control section is divided into sections: upper and lower portions. The upper portion is defined as the depth to which 2.5 cm of water will penetrate within 24 hours. The lower portion is the depth that 7.5 cm of water will penetrate. Six soil moisture regimes are identified:

Aridic: characteristic of soils in arid regions

Xeric: characteristic of having extended periods of drought in the summer

Ustic: soil moisture is generally high enough to meet plant needs during growing season

Udic: common soil in humid climatic regions

Perudic: an extremely wet moisture regime annually

Aquic: soil saturated with water and free of gaseous oxygen

Table 11.3 lists the moisture regime classes and the percentage distribution of areas with different soil moisture regimes.

In soil taxonomy, several soil temperature regimes are also used to define classes of soils. Based on mean annual soil temperature, mean summer temperature, and the difference between mean summer and winter temperatures, soil temperature regimes are shown in Table 11.4.

The diagnostic horizons and moisture/temperature regimes just discussed are the main criteria used to define the various categories in soil taxonomy.

Soil Taxonomy

The U.S. Soil Conservation Service's soil classification system, soil taxonomy (which is based on measurable properties of soil profiles), places soils in categories (see Table 11.5).

Let's take a closer look at each one of these categories.

Order: soils not too dissimilar in their genesis. There are 11 soil orders in soil taxonomy. The names and major characteristics of each soil order are shown in Table 11.6.

Suborder: 55 subdivisions of order that emphasize properties that suggest some common features of soil genesis.

TABLE 11.3
Soil Moisture Regimes (Percent of Global Area Occupied by Soils With)

Moisture Regime	Percent of Soils
Aridic	35.9
Xeric	3.5
Ustic	18.0
Udic	33.1
Perudic	1.0
Aquic	8.3

Source: Adaptation from Eswaran, H. (1993). Assessment of global resources: Current status and future needs. *Pedologie*, 43, pp. 19–39.

TABLE 11.4
Soil Temperature Regimes (Percent of Global Areas Occupied by Soils With)

Soil Temperature Regimes/ Mean Annual Temperature	Percent of Soils
Pergelic (0°C)	10.9
Cryic (0°C–8°C)	13.5
Frigid (0°C–8°C)	1.2
Mesic (8°C–15°C)	12.5
Thermic (15°C–22°C)	11.4
Hyperthermic (>22°C)	18.5
Isofrigid (0°C–8°C)	0.1
Isomesic (8°C–15°C)	0.3
Isothermic (15°C–22°C)	2.4
Isohyperthermic (>22°C)	26.0
Water (NA)	1.2
Ice (NA)	1.4

Source: Adaptation from Eswaran, H. (1993). Assessment of global resources: Current status and future needs. *Pedologie*, 43, pp. 19–39.

TABLE 11.5
Subdivision of Soil Taxonomy Classification System (Shown in Hierarchical Order)

Category	Number of Taxa
Order	11
Suborder	55
Great group	Approximately 230
Subgroup	Approximately 1,200
Family	Approximately 7,500
Series	Approximately 18,500 in U.S.

Great group: diagnostic horizons are the major bases for differentiating approximately 230 great groups.

Subgroup: approximately 1,200 subdivisions of the great groups.

Family: approximately 7,500 soils with subgroups having similar physical and chemical properties.

Series: a subdivision of the family, and the most specific unit of the classification system. More than 18,000 soil series are recognized in the U.S.

Soil Orders

As stated earlier, 11 soil orders are recognized; they constitute the first category of the classification. The classification of the orders is illustrated in Table 11.6.

TABLE 11.6
Soil Orders (With Simplified Definitions)

Alfisol	Mild forest soil with gray to brown surface horizon, medium to high base supply (refers to number of interchangeable cations that remain in soil), and a subsurface horizon of clay accumulation
Andisol	Formed on volcanic ash and cinders and lightly weathered
Aridsol	Dry soil with Pedogenic (soil-forming) horizon, low in organic matter
Entisol	Recent soil without Pedogenic horizons
Histosol	Organic (peat or bog) soil
Inceptisol	Soil at the beginning of the weathering process with weakly differentiated horizons
Mollisol	Soft soil with a nearly black, organic-rich surface horizon and high base supply
Oxisol	Oxide-rich soil principally a mixture of kaolin, hydrated oxides, and quartz
Spodosol	Soil that has an accumulation of amorphous materials in the subsurface horizons
Ultisol	Soil with a horizon of silicate clay accumulation and low base supply
Vertisol	Soil with high activity clays (cracking clay soil)

Source: Soil Survey Staff. (1975). Soil Classification, A Comprehensive System. Washington, D.C: USDA Natural Resources Conservation Service; Soil Survey Staff. (1994). Keys to Soil Taxonomy. Washington, D.C.: USDA Natural Resources Conservation Service.

Soil Suborders

Soil orders are further divided into 55 suborders, based primarily on the chemical and physical properties that reflect either the presence or absence of water logging or genetic differences caused by climate and vegetation—to give the class the greatest genetic homogeneity. Thus, the Equals (formed under wet conditions) are "wet" (aqu for aqua); Alfisols become saturated with water sometime during the year. The suborder names all have two syllables, with the first syllable indicating the order, such as alf Alfisol and oll for Mollisol.

Soil Great Groups and Subgroups

Suborders are divided into great groups. They are defined largely by the presence or absence of diagnostic horizons, and the arrangements of those horizons. Great group names are coined by prefixing one or more additional formative elements to the appropriate suborder name. More than 230 great groups have been identified.

Subgroups are subdivisions of great groups. Subgroup names indicate to what extent the central concept of the great group is expressed. A Typic Fragiaqualf is a soil that is typical for the Fragiaqualf great group.

Soil Families and Series

The family category of classification is based on features that are important to plant growth such as texture, particle size, mineralogical class, and depth. Terms such as clayey, sandy, loamy, and others are used to identify textural classes. Terms used to describe mineralogical classes include mixed, oxidic, carbonatic and others. For temperature classes, terms such hypothermic, frigid, cryic, and others are used.

The soil series (subdivided from soil family) gets down to the individual soil, and the name is that of a natural feature or place near where the soil was first recognized. Familiar series names include Amarillo (Texas), Carlsbad (New Mexico), and Fresno (California). In the United States, there are more than 18,000 soil series.

SOIL MECHANICS/PHYSICS

Why does the Leaning Tower of Pisa lean? The tower leans because it was built on a nonuniform consolidation of a clay layer beneath the structure. This process is ongoing (by about 1/25 of inch per year)—and may eventually lead to failure of the tower.

The factors that caused the Leaning Tower of Pisa to lean (and affect using soil as foundational and building materials) are what this chapter is all about. The mechanics and physics of soil are important factors in making the determination as to whether a building site is viable for building. Simply put, these two factors can help to answer the question: will the soils present support buildings? This concerns us because wherever humans build, the opportunity for anthropogenic pollutants follows—and to clean up those pollutants, we must excavate below the surface of the soil again.

Soil Mechanics

The mechanics of soil are physical factors important to engineers because their focus is on soil's suitability as a construction material. Simply put, the engineer must determine the response of a volume of soil to internal and external mechanical forces. Obviously, this is important in determining the soil's suitability to withstand the load applied by structures of various types.

By studying soil survey maps and reports, checking with soil scientists and other engineers familiar with the region and the soil types of that region, an engineer can determine the suitability of a soil for whatever purpose. Conducting field sampling, to ensure that the soil product possesses the soil characteristics for its intended purpose, is also essential.

The soil characteristics important for engineering purposes include soil texture, kinds of clay present, depth to bedrock, soil density, erodibility, corrosivity, surface geology, plasticity, content of organic matter, salinity, and depth to seasonal water table. Engineers will also want to know the soil's space and volume (weight-volume relationships), stress-strain, slope stability, and compaction. Because these concepts are also of paramount importance to determining the fate of materials that are carried through soil, we present these concepts in this section.

Weight-Volume or Space and Volume Relationships

As we mentioned earlier, all-natural soil consists of at least three primary components (or phases), solid (mineral) particles, water, and/or air (within void spaces between the solid particles). Examining the physical relationships (for soils in particular) between these phases is essential.

Remember that all water that is not chemically attached acts as a void filler. The relationship between free water and void spaces depends on available water (moisture).

The volume of the soil mass is the sum of the volumes of the three components, or

$$VT = Va + Vw + Vs \tag{11.1}$$

The volume of the voids is the sum of *Va* and *Vw*. However, since weighing air in the soil voids must be done within the earth's atmosphere (as with other weighing) the weight of the solids is determined to compensate for that. The weight of air in the soil is factored in at zero. The total weight is expressed as the sum of the weights of the soil solids and water:

$$WT = Ws + Ww \tag{11.2}$$

The relationship between weight and volume can be expressed as

$$Wm = VmGm\gamma w \tag{11.3}$$

where

Wm = weight of the material (solid, liquid, or gas)
Vm = volume of the material
Gm = specific gravity of the material (dimensionless)
γw = unit weight of water

We can solve a few useful problems with the relationships described above. More importantly, this information about a location's soil allows engineers to mechanically adjust the proportions of the three major components, by reorienting the mineral grains by compaction or tilling. In remediation, a decision to blend soil types to alter the proportions (such as increasing or decreasing the percentage of void space) may be part of a site cleanup process.

DID YOU KNOW?

Void ratio or *soil porosity* is a measure of the empty (i.e., void) spaces in a soil, and is a fraction of the volume of voids over the total volume, between 0 and 1, or as a percentage between 0% and 100%.

Relationships between volumes of soil and voids are described by the void ratio, *e*, and porosity, ø. We must first determine the void ratio, which is the ratio of the void volume to the volume of solids:

$$e = Vv/Vs \tag{11.4}$$

The first step is to determine the ratio of the volume of void spaces to the total volume. We do this by determining the porosity (ø) of the soil, which is the ratio of void volume to total volume. Porosity is usually expressed as a percentage.

$$\emptyset = \text{Vv/Vt} \times 100\% \tag{11.5}$$

Two additional relationships are the terms: moisture content, (w) and degree of saturation, (S) relate the water content of the soil and the volume of the water in the void space to the total void volume

$$w = \text{Ww/Ws} \times 100\% \tag{11.6}$$

and

$$S = \text{Vw/Vv} \times 100\% \tag{11.7}$$

SOIL PARTICLE CHARACTERISTICS

Size and shape of particles in the soil, along with density and other characteristics, provide information to the engineer on shear strength, compressibility, and other aspects of soil behavior. These index properties are used to create engineering classifications of soil. Simple classification tests are used to measure index properties (see Table 11.7) in the lab or the field.

From the engineering point of view, the separation of the cohesive (fine-grained) from the incohesive (coarse-grained) soils is an important distinction. Let's take a closer look at these two terms.

A soil's level of cohesion describes soil particles' tendency to stick together. Cohesive soils contain silt and clay, which, along with water content, make these soils hold together through the attractive forces between individual clay and water particles. Because the clay particles so strongly influence cohesion, the index properties of

TABLE 11.7
Index Property of Soils

Soil Type	Index Property
Cohesive (fine-grained)	Water content
	Sensitivity
	Type and amount of clay
	Consistency
	Atterburg limits
Incohesive (coarse-grained)	Relative density
	In-place density
	Particle-size distribution
	Clay content
	Shape of particles

Source: Adaptation from Kehew, A.E. (2006). *Geology for Engineers and Environmental Scientists*, 3rd ed. UK: Pearson, p. 284.

cohesive soils are more complicated than the index properties of cohesionless soils. The soil's consistency—the arrangement of clay particles—describes the resistance of soil at various moisture contents to mechanical stresses or manipulations and is the most important characteristic of cohesive soils.

Sensitivity (the ratio of unconfined compressive strength in the undisturbed state to strength in the remolded state (see Equation 11.8)) is another important index property of cohesive soils. Soils with high sensitivity are highly unstable.

$$St = \frac{\text{strength in undisturbed condition}}{\text{strength in remolded condition}} \tag{11.8}$$

As we described earlier, soil water content also influences soil behavior. Water content values of soil (the Atterburg limits, a collective designation of the so-called limits of consistency of fine-grained soils determined with simple laboratory tests) are usually presented as the liquid limit (LL), plastic limit (PL) [Note: Plasticity is exhibited over a range of moisture contents referred to as plasticity limits], and the plasticity index (PI). The plastic limit is the lower water level at which soil begins to be malleable in the semi-solid state, but while the molded pieces still crumble easily with a little applied pressure. When the volume of the soil becomes nearly constant with further decreases in water content, the soil reaches the shrinkage state.

The upper plasticity limit (or liquid limit) is reached when the water content in a soil-water mixture changes from a liquid to a semi-fluid or plastic state and tends to flow when jolted. Obviously, soil that tends to flow when wet presents special problems for both engineering purposes and remediation of contamination.

The range of water content over which the soil is plastic, called the plasticity index, provides the difference between the liquid limit and the plastic limit. Soils with the highest plasticity indices are unstable in bearing loads—a key point to remember.

The best known and probably the most useful system of the several systems designed for classifying the stability of soil materials, is called the Unified System of Classification. This classification gives each soil type (14 classes) a two-letter designation, based on particle-size distribution, liquid limit, and plasticity index.

Cohesionless coarse-grained soils are classified by index properties including the size and distribution of particles in the soil. Other index properties (including particle shape, in-place density, and relative density) are important in describing cohesionless soils because they relate how closely particles can be packed together.

SOIL STRESS

Just as water pressure increases as you go deep into water, the pressure within soil increases as the depth increases. A soil with a unit weight of 75 pounds per cubic feet exerts a pressure of 75 psi at one-foot depth—and 225 psi at three feet. Of course, as the pressure on a soil unit increases, soil particles reorient themselves to support the cumulative load. This is critically important information to remember, because a soil sample retrieved from beneath the load may not be truly representative, once delivered to the surface. Representative samples are essential.

The response of a soil to pressure (stress), as when a load is applied to a solid object, transmits throughout the material. The load puts the material under pressure, which equals the amount of load divided by the surface area of the external face of the object over which it is applied. The response to this pressure or stress is called displacement or strain. Stress (like pressure) at any point within the object can be defined as force per unit area.

SOIL COMPRESSIBILITY

Compressibility (the tendency of soil to decrease in volume under load) is most significant in clay soils because of inherently high porosity. The actual evaluation process for these properties is accomplished in the consolidation test. This test subjects a soil sample to an increasing load. The change in thickness is measured after the application of each load increment.

SOIL COMPACTION

Compaction reduces void ratio and increases the soil density, which affects how materials (including pollutants) travel through soil. Compaction is accomplished by working on the soil to reorient the soil grains into a more compact state. Water content sufficient to lubricate particle movement is critical to obtaining efficient compaction.

SOIL FAILURE

Soil structural implications are involved with natural processes such as frost heave (which could damage a septic system and disturb improperly set footings, or shift soils under an improperly seated UST and its piping) as well as changes applied to soils during remediation efforts (e.g., when excavating to mitigate a hazardous materials spill in soil). When soil cannot support a load, soil failure occurs, which can include events as diverse as foundation overload, collapse of the sides of an excavation, or slope failure on the sides of a dike or hill. Because of the safety factors involved, a soil's structural stability is critically important.

Classifying the type of soil to be excavated before an excavation can be accomplished includes determining the soil type. Stable Rock, Type A, Type B, or Type C soil are the classifications soils will be categorized in but finding a combination of soil types at an excavation site is common.

Soil Types

Stable rock Generally stable (may lose stability when excavated) natural solid mineral material that can be excavated with vertical sides and will remain intact while exposed

Type A soil Most stable soil includes clay, silty clay, sandy clay, clay loam, and sometimes silty clay loam and sandy clay loam.

Type B soil Moderately stable soil Includes silt, silt loam, sandy loam and sometimes silty clay loam and sandy clay loam.

Type C soil Least stable soil includes granular soils like gravel, sand, loamy sand, submerged soil, as well as soil from which water is freely seeping, and unstable submerged rock.

Both visual and manual tests are used to classify soil for excavation. *Visual soil testing* concerns soil particle size and type. In a mixture of soils, if the soil clumps when dug, it could be clay or silt. The presence of cracks in walls and spalling (breaks up in chips or fragments) may indicate Type B or C soil. Standing water or water seeping through trench walls classifies the soil as Type C, automatically. *Manual soil testing* includes the sedimentation test, wet shaking test, thread test, and ribbon test. A sample taken from soil should be tested as soon as possible to preserve its natural moisture. Soil can be tested either on-site or off-site.

A sedimentation test determines how much silt and clay are in sandy soil. Saturated sandy soil is placed in a straight-sided jar with about five inches of water. After the sample is thoroughly mixed (by shaking it) and allowed to settle, the percentage of sand is visible. A sample containing 80% sand, for example, will be classified as Type C.

The wet shaking test is another way to determine the amount of sand versus clay and silt in a soil sample. This test is accomplished by shaking a saturated sample by hand to gauge soil permeability based on the following facts: (1) shaken clay resists water movement through it, and (2) water flows freely through sand and less freely through silt.

The thread test is used to determine cohesion (remember, cohesion relates to stability—how well the grains hold together). A representative soil sample is rolled between the palms of the hands to about 1/8″ diameter and several inches in length. The rolled piece is placed on a flat surface, then picked up. If the sample holds together for two inches, its considered cohesive.

The ribbon test is used as a backup for the thread test. It also determines cohesion. A representative soil sample is rolled out (using the palms of your hands) to 3/4″ in diameter, and several inches in length. The sample is then squeezed between thumb and forefinger into a flat unbroken ribbon 1/8″–1/4″ thick, which can fall freely over the fingers. If the ribbon does not break off before several inches are squeezed out, the soil is considered cohesive.

Once soil has been properly classified, the necessary measures for safe excavation can be chosen, based on both soil classification and site restrictions. The two standard protective systems include sloping or benching and shoring or shielding.

SOIL PHYSICS

A dynamic, heterogeneous body, soil is non-isotropic—it does not have the same properties in all directions. Because of soil properties vary directionally, various physical processes are always active in soil, as Windgardner (1996) makes clear: "all of the factors acting on a particular soil, in an established environment, at a specified time, are working from some state of imbalance to achieve a balance" (p. 63).

Soil practitioners must understand the factors involved in the physical processes that are active in soil. These include physical interactions related to soil water, soil grains, organic matter, soil gases, and soil temperature.

WATER AND SOIL

Water is not only a vital component of every living being, but also essential to plant growth, is essential to the microorganisms that live in the soil, and is important in the weathering process, which involves the breakdown of rocks and minerals to form soil and release plant nutrients. In this section we focus on soil water and its importance in soil. But first, we need to take a closer look at water—what it is and its physical properties.

Water exists as a liquid between 0° and 100°C (32°–212°F); exists as a solid at or below 0°C (32°F); and exists as a gas at or above 100°C (212°F). One gallon of water weighs 8.33 pounds (3.778 kg). One gallon of water equals 3.785 L. One cubic foot of water equals 7.5 gallons (28.35 L). One ton of water equals 240 gallons. One-acre foot water equals 43,560 cubic feet (325,900 gallons). Earth's rate of rainfall equals 340 cubic miles per day (16 million tons per second). Finally, water is dynamic (constantly in motion), evaporating from sea, lakes, and the soil; being transported through the atmosphere; falling to earth; running across the land; and filtering downward into and through the soil to flow along rock strata.

WATER: WHAT IS IT?

Water is often assumed to be one of the simplest compounds known on Earth. But water is not simple—nowhere in nature is simple (pure) water to be found. Here on Earth, with a geologic origin dating back over three to five billion years, water found in even its purest form is composed of many constituents. Along with H_2O molecules, hydrogen (H+), hydroxyl (OH–), sodium, potassium and magnesium, other ions and elements are present.

Water contains additional dissolved compounds, including various carbonates, sulfates, silicates, and chlorides. Rainwater (often assumed to be the equivalent of distilled water) is not immune to contamination, which it collects as it descends through the atmosphere. The movement of water across the face of land contributes to its contamination, taking up dissolved gases such as carbon dioxide and oxygen, and a multitude of organic substances and minerals leached from the soil.

SOIL PHYSICAL PROPERTIES

In soil physics, the physical properties (which are also a function of water's chemical structure) of water that concern us are density, viscosity, surface tension and capillary action. Let's take a closer look at each of these physical properties.

Density is a measure of the mass per unit volume. The number of water molecules occupying the space of a unit volume determines the magnitude of the density. As temperature (which measures internal energy) increases or decreases, the molecules vibrate strongly and frequently (which changes the distance between them), expanding or diminishing the volume occupied by the molecules.

As discussed previously, liquid water reaches its maximum density at 4°C, and its minimum at 100°C. In soil science work, the density may be a unit weight (62.4 pounds per cubic foot or 1 g/cm^3).

Viscosity is the measure of the internal flow resistance of a liquid or gas. Stated differently, viscosity is ease of flow of a liquid, or the capacity of a fluid to convert energy of motion (kinetic energy) into heat energy. Viscosity is the result of the cohesion between fluid particles and interchange of molecules between layers of different viscosities. High viscosity fluids flow slowly, while low viscosity fluids flow freely. Viscosity decreases as temperature rises for liquids.

Have you ever wondered why a needle can float on water? Or why can some insects stand on water? The reason: surface tension. Surface tension (or cohesion) is the property that causes the surface of a liquid to behave as if it were covered with a weak elastic skin. It is caused by the exposed surface's tendency to contract to the smallest possible area because of unequal cohesive forces between molecules at the surface.

What does surface tension have to do with soil? The surface tension property of water markedly influences the behavior of water in soils. Consider an example you may be familiar with, one that will help you understand surface tension and the other important physical properties of the water-soil interface. Water commonly rises in clays, fine silts, and other soils (you may already know this or, if not, think about it a moment—it is true). But why? How does this happen? Surface tension plays a major role—here's how: The rise of water through clays, silts and other soils is termed capillarity or capillary action (the property of the interaction of the water with a solid), and the two primary factors of capillary rise are surface tension (cohesion) and adhesion (the attraction of water for the solid walls of channels through which it moves).

Capillarity, as it relates to capillary rise in soil pores, can be better understood by referring to a fine glass tube immersed in water. The water rises in the tube (and soil pores), and the smaller the tube bore, the higher the water rises (Note: the height of rise in a capillary tube is inversely proportional to the tube diameter and directly proportional to the surface tension).

Why does the water rise? Because the water molecules are attracted to the sides of the tube (or soil pores) and start moving up the tube in response to this attraction. The cohesive force between individual water molecules ensures that water not directly in contact with the side walls is also pulled up the tube (or soil pores). This action continues until the weight of water in the tube counterbalances the cohesive and adhesive forces.

Keep in mind that for water in soil, the rate of movement and the rise in height of soil water are less than one might expect based on soil pore size, because soil pores are not straight like glass tubes, nor are the openings uniform. Also, many soil pores are filled with air, which may prevent or slow down the movement of water by capillarity.

A final word on capillarity—keep in mind that capillarity means movement in any direction, not just upward. Since the attractions between soil pores and water are as effective with horizontal pores as with vertical ones, water movement in any direction occurs.

The Water Cycle (Hydrologic Cycle)

Water is never stationary; it is constantly in motion. This phenomena occurs because of the water or hydrologic cycle. In simple terms, the water cycle can be explained as follows: The sun helps transfer water from lakes and oceans to the land. As the sun shines on the earth, the surface water is heated and evaporates, forming an invisible gas that mixes with the air. This gas is water vapor; it is pure water without any minerals or bacteria in it. Water vapor rises in the air, then cools, and condenses into tiny drops of water that form clouds. Further cooling may form drops large enough to fall as rain. In this way, water is brought from the oceans to the land, where it reappears in springs and wells, soaks into the ground, or runs off again through streams and rivers back to the ocean. Of course, the actual movement of water on earth is much more complex. Three different methods of transport are involved in this water movement: evaporation, precipitation, and runoff.

Evaporation of water is a major factor in hydrologic systems. Evaporation is a function of temperature, wind velocity, and relative humidity. Evaporation (or vaporization) is, as the name implies, the formation of vapor. Dissolved constituents (such as salts) remain behind when water evaporates. Evaporation of the surface water of oceans provides most water vapor, though water can also vaporize through plants, especially from leaf surfaces. This process is evapotranspiration. Ice can also vaporize without melting first. However, this sublimation process is slower than vaporization of liquid water.

Precipitation includes all forms in which atmospheric moisture descends to earth—rain, snow, sleet, and hail. Before precipitation can occur, the water that enters the atmosphere by vaporization must first condense into liquid (clouds and rain) or solid (snow, sleet, and hail) before it can fall. This vaporization process absorbs energy, which is released in the form of heat when the water vapor condenses. You can best understand this phenomena when you compare it to what occurs when water evaporates from your skin; this absorbs heat, making you feel cold. Note: The annual evaporation from ocean and land areas is the same as the annual precipitation.

Runoff is the flow back to the oceans of the precipitation that falls on land. This journey to the oceans is not always unobstructed—flow back may be intercepted by vegetation (from which it later evaporates), a portion is held in depressions, and some infiltrates into the ground. A part of the infiltrated water is taken up by plant life and returned to the atmosphere through evapotranspiration, while the remainder either moves through the ground or is held by capillary action. Eventually, water drips, seeps, and flows its way back into lakes, ponds, rivers, streams, and the oceans.

Soil Water

Have you ever wondered what happens to water after it enters the soil? For the average person, probably not, but if you are to work in the soil science field, the answer to this question is one that you need not only to know but must also have a full and complete understanding of. Water that enters the soil has (in simple terms) four ways it may go:

1. It may move on through the soil and percolate out of the root zone, where it may eventually reach the water table.
2. It may be drawn back to the surface and evaporate.
3. It may be taken up (transpired—used) by plants.
4. Finally, it may be "saved" in storage in the water profile.

What determines how much water ends up in each of these categories? It depends. Climate and the properties of the soil and the requirements of the plants growing in that soil all have impact on how much water ends up in each of the categories. But don't forget the influence of anthropogenic actions (what we like to call the heavy hand of man)—people alter the movement of water, not only by irrigation and stream diversion practices and by building, but also by choosing which crops to plant and the types of tillage practices employed.

SOIL CHEMISTRY

Almost every environmental and pollution problem we face today (and probably tomorrow) has a chemical basis. In short, in environmental science, studying such problems as greenhouse effect, ozone depletion, groundwater contamination, toxic wastes, air pollution, stream pollution and acid rain, would be difficult (if not impossible) without some fundamental understanding of basic chemical concepts. Of course, an environmental practitioner who must solve environmental problems and understand environmental remediation clean up processes (such as emission control systems or waste treatment facilities) must be well grounded in chemical principles and the techniques of chemistry in general, because many of them are being used to solve these problems.

This text assumes the reader who uses this text may or may not have some fundamental knowledge of chemistry. Thus, the chemistry topics have been selected with the purpose of review, providing only the essential chemical principles required to understand the nature of environmental problems we face, and the chemistry involved with scientific and technological approaches to their solutions.

WHAT IS CHEMISTRY?

Chemistry is the science concerned with the composition of matter (gas, liquid, or solid), and of the changes that take place in it under certain conditions.

Everything in the environment—every material, substance, and object—is a chemical substance or mixture of chemical substances. Your body is made up of literally thousands of chemicals. The chemical reactions and interactions we are a part of or cause daily—the food we eat, the clothes we wear, the fuel we burn, the vitamins we take in from natural or synthetic sources—are all products of chemistry created by nature or our own technology. Chemistry is about matter, and matter's constituents and consistency—and about measuring and quantifying matter.

What is matter? All matter can exist in three states: gas, liquid, or solid. Matter, composed of constantly moving, minute particles termed molecules, may be further divided into smaller units called atoms.

Molecules containing atoms of only one kind are called elements; those containing atoms of more than one kind are known as compounds.

Chemical compounds, produced by chemical actions that alter the arrangements of the atoms in the reacting molecules, alter when they experience an environmental change. Changes that could produce a chemical change can include heat, light, vibration, catalytic action, radiation, and pressure, as well as moisture (for ionization). Determining a compound's components by examination and possible breakdown is analysis and building up compounds from the components is synthesis. Bringing substances together without changing their molecular structures creates mixtures.

Virtually all compounds that contain carbon are organic substances. All other substances are inorganic substances.

Elements and Compounds

Material that has been separated from all other materials, pure substances are indistinguishable from each other, no matter what procedures were used to purify them or their origin. Copper metal, aluminum metal, distilled water, table sugar and oxygen are all pure substances. No matter what the original source is, all samples of table sugar are alike and are chemically indistinguishable from all other samples.

A substance is characterized as a material having a fixed composition, usually expressed in terms of percentage by mass. Distilled water is a pure substance consisting of approximately 89% oxygen and 11% hydrogen by mass. A substance that is not pure will have varying constituents—a lump of coal's carbon content may vary from 35% to 90% by mass, for example. Materials that are not pure substances are mixtures. Substances that can be broken down into two or more simpler substances are called compounds.

Substances that cannot be broken down or decomposed into simpler forms of matter are called elements, which are the basic substances of which all matter is composed. Presently, we know only 100+ elements, but well over a million compounds. Only 88 of the 100+ elements are present in detectable amounts on Earth, and many of these 88 are rare. Only 10 elements make up approximately 99% (by mass) of the Earth's crust, including the surface layer, the atmosphere, and bodies of water (see Table 11.8). As you see from this table, the most abundant element on Earth is oxygen, found in the free state in the atmosphere, as well as in combined form with other elements in numerous minerals and ores.

Table 11.8 includes symbols and atomic numbers for the ten listed elements. The symbols consist of either one or two letters, with the first letter capitalized. The atomic number of an element is the number of protons in the nucleus.

Classification of Elements

Each element may be classified as a metal, nonmetal, or metalloid. Metals are elements that are typically lustrous solids, good conductors of heat and electricity, melt and boil at high temperatures, and possess relatively high densities; they are normally malleable (can be hammered into sheets) and ductile (can be drawn into a wire). Examples of metals are copper, iron, silver, and platinum. Almost all metals are solids (none is gaseous) at room temperature (mercury being the only exception).

Elements that do not possess the general physical properties just mentioned (i.e., they are poor conductors of heat and electricity, boil at relatively low temperatures, do not possess a luster, and are less dense than metals) are called nonmetals. Most nonmetals (exception is bromine—a liquid) are either solids or gases at room

TABLE 11.8
Elements Making up 99% of Earth's Crust, Oceans, and Atmosphere

Element	Symbol	% of Composition	Atomic Number
Oxygen	O	49.5%	8
Silicon	Si	25.7%	14
Aluminum	Al	7.5%	13
Iron	Fe	4.7%	26
Calcium	Ca	3.4%	20
Sodium	Na	2.6%	11
Potassium	K	2.4%	19
Magnesium	Mg	1.9%	12
Hydrogen	H	1.9%	1
Titanium	Ti	0.58%	22

temperature. Nitrogen, oxygen, and fluorine are examples of gaseous nonmetals, while sulfur, carbon, and phosphorus are examples of solid nonmetals.

Several elements have properties resembling both metals and nonmetals. They are called metalloids (semimetal). The metalloids are boron, silicon, germanium, arsenic, tellurium, antimony, and polonium.

Physical and Chemical Changes

To maintain constant composition, a substance needs internal linkages among its units—between one atom or another. Called chemical bonds, these linkages control the processes that involve the making and breaking of these bonds—when the linkages change or break, a chemical change or chemical reaction has occurred. Combustion and corrosion are common examples of chemical changes that impact on our environment.

Consider some examples of chemical change.

A flame, brought into contact with a mixture of hydrogen and oxygen gases, creates a violent reaction. The covalent bonds in the hydrogen (H_2) molecules and of the oxygen (O_2) moles break, and new bonds form to produce molecules of water, H_2O. Remember that whenever chemical bonds are broken, formed, or both, a chemical change takes place. The chemical change that hydrogen and oxygen undergo produces water, a substance with new properties.

Heating mercuric oxide (a red powder) forms small globules of mercury and releases oxygen gas. The heat chemically changes mercuric oxide to form molecules of mercury and molecules of water.

A physical change (nonmolecular change), though, does not alter the molecular structure of a substance. Freezing, melting, or changing to vapor does not change the composition of the molecules. Ice, steam, and liquid water are comprised of molecules containing two atoms of hydrogen and one atom of oxygen—water in its normal three states. You can rip a substance, or saw it into small pieces, grind it into powder, or mold it into a different shape, without changing the molecules at all. Chemically, it is still the same substance.

How a substance behaves through chemical changes are its chemical properties. Characteristics that do not involve changes in a substance's chemical identity are its physical properties. These properties distinguish one substance from another, as DNA, for example, distinguishes one human from another.

Structure of the Atom

Hypothetically divide and subdivide and sub-subdivide a small piece of an element (say carbon) into the smallest piece possible. The result would be one particle of carbon. The smallest particle of an element that is still representative of the element is called an atom.

The atom, composed of particles (principally electrons, protons, and neutrons) is infinitesimally small. The simplest possible atom consists of a nucleus with a single proton (positively charged particle) that has a single electron (negatively charged particle) traveling around it—an atom of hydrogen. We say the hydrogen atom has an atomic weight of one, because of its single proton. An element's atomic weight is equal to the total number of protons and neutrons (neutral particles) in the nucleus of the element's atom. Electrons and protons bear the same magnitude of charge but are of opposite polarity.

An element's atomic number equals the number of protons in its nucleus. A neutral atom has the same number of protons and electrons; thus a neutral atom's atomic number is also equal to the number of electrons in the atom. The number of neutrons an atom has is always equal to or greater than the number of protons—except in the hydrogen atom.

An atom's protons and neutrons reside in the nucleus. Electrons reside primarily in atomic orbitals or electron shells, designated regions of space surrounding the nucleus. A prescribed number of electrons reside in any given type of electron shell. Other than hydrogen (which only has one electron), in an atom's innermost electron shell, two electrons are always close to the nucleus. In most atoms, other electrons are found in electron shells some distance from the nucleus.

Neutral atoms of the same element have an identical number of electrons and protons but may differ in the number of neutrons in their nuclei. Atoms of the same element with different numbers of neutrons are that element's isotopes.

Periodic Classification of the Elements

We discussed the periodic table (see Figure 3.1) in Chapter 3, but because of its importance in this discussion, a brief overview is provided in the following.

Scientists discovered through experience that chemical properties of the elements repeat themselves and summarized all such observations in the periodic law: The properties of the elements vary periodically with their atomic numbers.

Dimitri Mendeleev, using relative atomic masses, developed in 1869 the original form of what we know today as the periodic table, a chart of elements arranged in order of increasing proton number, showing the similarities of chemical elements with related electronic configurations. The elements fall into vertical columns known as groups. The atoms of the elements in a group all have the same outer shell structure, and an increasing number of inner shells. The alkali metals were traditionally shown on the left of the table (the groups numbered IA to VIIA, IB to VIIB) with 0 (for noble gases). Now, we usually classify all the elements in the middle of the table as transition elements, regarding the nontransition elements as main-group elements (numbered from I to VII, with the noble gases in group 0). The table's horizontal rows are periods. The first three are short periods; the next four (which include transition elements) are called long periods. The atoms of all the elements within a period have the same number of shells, and a steadily increasing number of electrons in the outer shell.

The periodic table tabulates a variety of information in one spot, making it an important tool for learning chemistry. Immediate determination of the atomic number of the elements is simple, because they are tabulated on the periodic table (see Figure 3.1), as is ready identification of which elements are metals, nonmetals, and metalloids. Usually, a graphically distinct line separates metals from nonmetals, with those elements lying to each immediate side of the line being metalloids. Metals fall to the left of the line, and nonmetals fall to the right.

Molecules and Ions

When elements (other than noble gases which exist as single atoms) exist in gaseous or liquid states at normal room temperatures and conditions, they consist of units that contain pairs of like atoms—molecules. Oxygen, hydrogen, chlorine, and nitrogen are elements we normally encounter as gases. Each exists as a molecule with two atoms. These molecules are designated by the notations O_2, H_2, Cl_2 and N_2.

For many compounds, the smallest particle is also the molecule. Molecular compounds contain atoms of two or more elements. Water molecules consist of two atoms of hydrogen and one atom of oxygen (H_2O), for example. Methane molecules have one carbon atom and four hydrogen atoms (CH_4).

Compounds do not always occur naturally as molecules. Many compounds occur as ions, aggregates of oppositely charged atoms or groups of atoms. Atoms become charged by gaining or losing some of their electrons. For example, atoms of metals that lose their electrons become positively charged. Atoms of nonmetals that gain electrons become negatively charged.

Chemical Bonding

In forming compounds, one element's atoms become attached to (or associated with) atoms of other elements by forces called chemical bonds. Chemical bonding is a strong force of attraction, and it holds atoms together in a molecule. Chemical bonds exist in various types. The transfer of electrons can form ionic bonds. For example, the electron configuration of the calcium atom has two electrons in its outer shell. An atom of chlorine has seven outer electrons. If the calcium atom transfers two electrons (one to each chlorine atom) the calcium atom changes to a calcium ion with the stable configuration of an inert gas. At the same time, having gained one electron, each chlorine atom becomes a chlorine ion, also with an inert-gas configuration. Bonding in calcium chloride is the electrostatic attraction between the ions.

Sharing of valence (the number of electrons an atom can give up or acquire to achieve a filled outer shell) forms covalent bonds in electrons. For example, hydrogen atoms have one outer electron. If in the hydrogen molecule (H_2), each atom contributes one electron to the bond, consequently each hydrogen atom has control of two electrons—one of its own and the second from the other atom. This gives it the electron configuration of an inert gas. In the water molecule (H_2O) the oxygen atom (with six outer electrons) gains control of an extra two electrons supplied by the two hydrogen atoms, and each hydrogen atom gains control of an extra electron from the oxygen.

We often classify chemical compounds into either of two groups, based on the bonding between their atoms. Chemical compounds that consist of atoms bonded together by ionic bonds are called ionic compounds, and compounds whose atoms are bonded together by covalent bonds are called covalent compounds.

Ionic and covalent compounds present interesting contrasts. For example, ionic compounds melt, boil, and become water soluble at higher temperatures than covalent compounds. Ionic compounds are nonflammable. Covalent compounds are flammable. Ionic compounds that are molten in water solutions conduct electricity, while molten covalent compounds do not. While ionic compounds generally exist as solids at room temperature, covalent compounds exist as gases, liquids, and solids at room temperature.

Chemical Formulas and Equations

Chemists developed a short method for writing chemical formulas, groups of symbols that represent elements. Sulfuric acid's formula is H_2SO_4. While it doesn't provide a recipe for making the acid, the formula does indicate that the acid is composed of two atoms of hydrogen, one atom of sulfur, and four atoms of oxygen. The formula does not tell you how to prepare the element, only what is in it.

A chemical equation describes what elements and compounds are present before—and after—a chemical reaction. Sulfuric acid poured over zinc causes the release of hydrogen and the formation of zinc sulfate in the following equation:

Zn + H2SO4 _ ZnSO4 + H2

(Zinc) (sulfuric acid) (zinc sulfate) (hydrogen)

One atom (also one molecule) of zinc unites with one molecule of sulfuric acid, which gives one molecule of zinc sulfate and one molecule (two atoms) of hydrogen. Notice that the same number of each element's atoms exist on each side of the arrow—but the atoms are combined differently.

Molecular Weights, Formulas, and the Mole

The molecular weight (the relative weight of a compound that occurs as molecules) is the sum of the atomic weights of each atom in the molecule. The water molecule's molecular weight is determined as follows:

$$\begin{aligned} &\text{2 hydrogen atoms} = 2 \times 1.008 = 2.016 \\ &\text{1 oxygen atom} = 1 \times 15.99 \quad = \underline{15.999} \\ &\text{Molecular weight of } H_2O \quad = 18.015 \end{aligned} \tag{11.9}$$

In other words, the molecular weight of a molecule is simply the sum of the atomic weights of all the constituent atoms. By dividing the mass of a substance by its molecular weight, the mass can be expressed as moles (mol). When (as they usually are) expressed in grams, the moles are written as g-moles. If mass is expressed in pounds, the result would be lb-moles. One g-mole contains $6.022 \times 1{,}023$ molecules (Avogadro's number, in honor of the scientist who first suggested its existence), and one lb-mole about $2.7 \times 1{,}026$ molecules.

$$\text{Moles} = \frac{\text{Mass}}{\text{Molecular Weight}} \tag{11.10}$$

The formula weight (the relative weight of a compound that occurs as formula units) is the sum of the atomic weights of all atoms that comprise one formula unit. Sodium fluoride's formula weight is determined as follows:

$$\begin{aligned} \text{1 sodium ion} &= 22.990 \\ \text{1 fluoride ion} &= \underline{18.998} \\ \text{Formula weight of NaF} &= 41.988 \end{aligned} \tag{11.11}$$

Physical and Chemical Properties of Matter

Two basic types of properties (characteristics) of matter exist: physical and chemical. Physical properties of matter (those that do not involve a change in the chemical composition of the substance) include hardness, color, boiling point, electrical conductivity, thermal conductivity, specific heat, density, solubility, and melting point. Changes in temperature or pressure may change these properties. Physical changes do not alter chemical composition of the substance. When you apply heat to solid ice, the ice converts to liquid (water), but no new substance is produced. Only the appearance has changed. Melting is a physical change. Dissolving sugar in water, heating a piece of metal, and evaporating water are other examples of physical changes.

The most used physical properties for describing and identifying kinds of matter are density, color, and solubility. Density (d) (mass per unit volume) is expressed by this equation:

$$d = \frac{\text{mass}}{\text{volume}} \tag{11.12}$$

The density, color, and solubility of a substance are important physical properties. In environmental science, they aid in determining of various pollutants, stages of pollution or treatment, the remedial actions required to clean up toxic/hazardous waste spills and other environmental problems.

Density depends on weight and space. All matter has weight and takes up space. While we may say that a certain material will not float because it is heavier than water, what we mean is that the material is denser than water. The element's density differs from any other element's density. Liquid and solid densities are normally given in units of grams per cubic centimeter (g/cm^3)—the same as grams per milliliter (g/mL). Using the physical property, color provides the advantage of not requiring chemical or physical tests. The degree to which a substance dissolves in a liquid (such as water) is its solubility.

Chemical properties are the properties involved in the transformation of one substance into another. When iron corrodes, oxygen combines with the iron and water, forming a new substance known as rust. When wood burns, oxygen in the air unites with the different substances in the wood, forming new substances. Changes that result in the formation of new substances are chemical changes.

States of Matter

The three states of matter are the solid state, the liquid state, and the gaseous state. The molecules or atoms are in a relatively fixed position in the solid state. While the molecules vibrate rapidly, they vibrate about a fixed point. This definite position of the molecules means that a solid holds its shape. A solid occupies a definite amount of space and has a fixed shape.

Lower the temperature of a gas and the gas's molecules slow down. If the gas cools sufficiently, the molecules slow down so much that they lose the energy they need to move rapidly throughout their container and may turn into liquid. A liquid is a material that occupies a definite amount of space, but which takes the shape of the container.

The atoms or molecules in some materials have no special arrangement—gases. A gas is a material that takes the exact volume and shape of its container.

How matter changes from one state to another is of primary interest to environmentalists, because often the change from one state to another has impact on environmental concerns. For example, changes in matter that aid in cleanup efforts include water vapor changing from the gaseous state to liquid precipitation, or a spilled liquid chemical changed to a semi-solid substance by addition of chemicals.

The Gas Laws

In the mixture of gases that composes the atmosphere, the most abundant elements are nitrogen, oxygen, argon, carbon dioxide, and water vapor (we address gases and the atmosphere in greater detail later). The pressure of a gas (the force that the moving gas molecules exert upon a unit area) is commonly expressed in unit of pressure (Newton per square meter, N/m^2) called a Pascal (Pa). The relationship that exists among the pressure, volume, and temperature of a gas (known as the ideal gas law) can be stated as

$$\frac{P1V1}{T1} = \frac{P2V2}{T2} \tag{11.13}$$

where P1, V1, T1 are pressure, volume, and absolute temperature at Time 1, and P2, V2, T2, are pressure, volume, and absolute temperature at Time 2. A gas is called perfect (or ideal) when it obeys this law.

A temperature of 0°C (273 K) and a pressure of 1 atmosphere (atm) have been chosen as standard temperature and pressure (STP). At STP the volume of 1 mole of ideal gas is 22.4 L.

Liquids and Solutions

While the most common solutions are liquids, solutions (homogenous mixtures) can be solid, gaseous, or liquid. A substance in excess in a solution is the solvent. The dissolved substance is the solute. When water is the solvent in a solution, it's called an aqueous solution. If the solute is present in only a small amount in a solution, it is called a dilute solution. When the solute is present in large amounts, the solution is a concentrated solution. By dissolving the maximum amount of solute possible in the solvent, a saturated solution is formed.

We frequently express the concentration (or amount of solute dissolved) in terms of the molar concentration (or molarity)—the number of moles of solute per liter of solution. A one molar solution, written 1.0M, has one-gram formula weight of solute dissolved in one liter of solution. In general

$$\text{Molarity} = \frac{\text{moles of solute}}{\text{number of liters of solution}} \tag{11.14}$$

Note that it is the number of liters of solution, not the number of liters of solvent that is used.

Example: Exactly 40 g of sodium chloride (NaCl), or table salt, were dissolved in water and the solution was made up to a volume 0.80 L of solution. What was the molar concentration, M, of sodium chloride in the resulting solution?

Answer: First find the number of moles of salt.

$$\text{Number of moles} = \frac{40\,\text{g}}{58.5\,\text{g/mole}} = 0.68 \text{ mole} \tag{11.15}$$

$$\text{Molarity} = \frac{0.68 \text{ mole}}{0.80 \text{ L}} = 0.85\text{M} \tag{11.16}$$

Thermal Properties

Environmental practitioners often need to know the thermal properties of chemicals and other substances, for use in hazardous materials spill mitigation and in solving other complex environmental problems. Whenever work is performed, friction usually produces a substantial amount of heat. Heat is a form of energy. The conservation of energy law tells us that the work done plus the heat energy produced must equal the original amount of energy available. That is,

$$\text{Total energy} = \text{work done} + \text{heat produced} \tag{11.17}$$

Environmental scientists, technicians and/or practitioners are concerned with several properties related to heat for substances.

A traditional unit for measuring heat energy is calories. A calorie (cal) is defined as the amount of heat necessary to raise one gram of pure liquid water by one degree Celsius at normal atmospheric pressure. Do not confuse the calories that we have defined with the one used when discussing diets and nutrition.

In SI units

$$1 \text{ cal} = 4.186 \text{ J (Joule)} \tag{11.18}$$

A kilocalorie (1,000 calories) is the amount of heat necessary to raise the temperature of one kilogram of water by 1°C.

The unit of heat in the British system of units is the British thermal unit (Btu). One Btu is the amount of heat required to raise one pound of water one degree Fahrenheit at normal atmospheric pressure (1 atm).

Specific Heat

While one kilocalorie of heat is necessary to raise the temperature of one kilogram of water one degree Celsius, different substances require different amounts of heat to raise one kilogram of the substance's temperature one degree. The amount of heat in kilocalories necessary to raise the temperature of one kilogram of the substance one degree Celsius is its specific heat.

The units of specific heat are kcal/kg°C or, in SI units, J/kg°C. The specific heat of pure water, for example, is 1.000 kcal/kg°C/4,186 J/kg°C.

The greater the specific heat of a material, the more heat required to elevate its temperature. The greater the mass of the material, or the greater the temperature change desired, the more heat needed.

The latent heat of fusion of a substance is the amount of heat necessary to change one kilogram of a solid into a liquid at the same temperature. The melting point is the temperature of the substance at which this change from solid to liquid takes place. The latent heat of vaporization is amount of heat necessary to change one kilogram of a liquid into a gas. The temperature of the substance at which this change from liquid to gas occurs is known as the boiling point.

ACID + BASES AND SALTS

By combining acids and bases in the proper proportions, they neutralize each other. Each loses its characteristic properties, and they form salt and water.

$$NaOH + HCl \rightarrow NaCl + H_2O \tag{11.19}$$

which is

sodium hydroxide + hydrochloric acid → sodium chloride + water

The acid-base-salt concept is a foundational chemical concept, originating at the very roots of chemistry. It is critical to the environment, the life processes, and to industrial chemicals.

The word acid is derived from the Latin acidus, meaning sour. One of the most identifiable properties of acids is sour taste (of course, though, never actually taste an acid in the laboratory or anywhere else). An acid (a substance that in water, produces hydrogen ions, H+) has the following properties:

1. conducts electricity
2. tastes sour
3. changes the color of blue litmus paper to red
4. reacts with a base to neutralize its properties
5. reacts with metals to liberate hydrogen gas

A base (a substance that produces hydroxide ions, OH-, and/or accepts H+), when dissolved in water, has the following properties:

1. conducts electricity
2. changes the color of red litmus paper to blue
3. tastes bitter and feels slippery
4. reacts with an acid to neutralize the acid's properties.

PH SCALE

By measuring the concentration of hydrogen ions (H+) in the solution we can determine whether a solution is an acid or a base. While the concentration can be expressed in powers of 10, it is more conveniently expressed as pH. Pure water, for example, has 1×10^{-7} grams of hydrogen ions per liter. The negative exponent of the hydrogen ion concentration is called the pH of the solution. The pH of water is 7—a neutral solution. A concentration of 1×10^{-12} has a pH of 12. A pH less than 7 indicates an acid solution and a pH greater than 7 indicates a basic solution (see Table 11.9).

Table 11.9 shows standard pH scale. The "p" is for the German word "poentz," and the H stands for hydrogen.

Substances found in our environment vary in pH value. Among the most important in environmental science are acid-based reactions. Various environmental problems

TABLE 11.9
pH Scale

pH	Concentration of H-Ions	Acidic/Basic
1	1.0×10^{-1} mol/L	Very acidic
2	1.0×10^{-2} mol/L	
3	1.0×10^{-3} mol/L	
4	1.0×10^{-4} mol/L	
5	1.0×10^{-5} mol/L	
6	1.0×10^{-6} mol/L	Acidic
7	1.0×10^{-7} mol/L	Neutral
8	1.0×10^{-8} mol/L	Basic
9	1.0×10^{-9} mol/L	
10	1.0×10^{-10} mol/L	
11	1.0×10^{-11} mol/L	
12	1.0×10^{-12} mol/L	
13	1.0×10^{-13} mol/L	
14	1.0×10^{-14} mol/L	Very basic

(for example, acid rain problems, hazardous materials spills into lakes or streams, and/or effect of point or nonpoint source pollution into our streams and rivers, lakes, and ponds) concern pH value. pH is also of concern in remediation or prevention. To protect local ecosystems, waste must often be neutralized before it can be released into the environment. In waste treatment processes (wastewater treatment, for example), if nitrogen is not removed, it can stimulate growth of algae in the receiving body of water—a condition affected by pH. Table 11.10 gives an approximate pH of some common substances.

Organic Chemistry

Organic chemistry, the branch of chemistry concerned with compounds of carbon, is incredibly varied and complex. Of the millions of different organic compounds known today, 100,000+ are not found in nature, but rather, are products of synthesis. In this text, we provide a very basic introduction to some of the organic substances most important to environmental science (important because of their toxicities as pollutants and other hazards).

All the familiar commodities that our technological world requires—motor and heating fuels, adhesives, cleaning solvents, paints, varnishes, plastics, refrigerants, aerosols, textiles, fibers, resins, among many others—have as their components organic compounds.

The principal concern from an environmental science perspective about organic compounds is that they are pollutants of water, air, and soil environments, as well as safety and health hazards. They are also (with few exceptions) combustible or flammable substances. They can cause a wide range of detrimental health effects,

TABLE 11.10
pH Examples

Substance	pH
Battery acid	0.0
Gastric juice	1.2
Lemons	2.3
Vinegar	2.8
Soft drinks	3.0
Apples	3.1
Grapefruit	3.1
Wines	3.2
Oranges	3.5
Tomatoes	4.2
Beer	4.5
Bananas	4.6
Carrots	5.0
Potatoes	5.8
Coffee	6.0
Milk (cow)	6.5
Pure water——————Neutral———	7.0
Blood (human)	7.4
Eggs	7.8
Sea water	8.5
Milk of magnesia	10.5
Oven cleaner	13.0

including the ability to cause damage to the kidneys, liver, and heart. Others depress the central nervous system. Several may cause some forms of cancer. Again, from the environmental science perspective, if human beings are subject to such health hazards from these compounds, how do these compounds impact delicate ecosystems?

Organic Compounds

Organic compound molecules have one common feature: one or more carbon atoms that covalently bond to other atoms—pairs of electrons the atoms share. Carbon atoms may share electrons with other nonmetallic atoms as well as with other carbon atoms. Methane, carbon tetrachloride, and carbon monoxide are compounds with moles in which the carbon atom is bonded to other nonmetallic atoms (see Figure 11.2).

Figure 11.2 shows carbon atoms sharing their electrons with the electrons of other nonmetallic atoms, like hydrogen, chlorine, and oxygen. The compounds that result from such electron sharing are methane, carbon tetrachloride, and carbon monoxide, respectively.

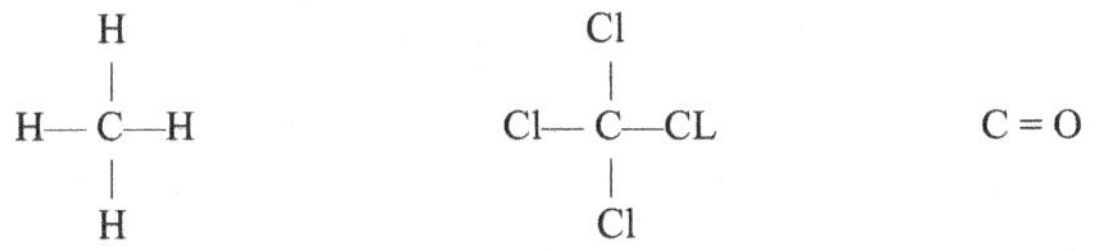

```
    H                 Cl
    |                 |
H—C—H           Cl—C—CL            C=O
    |                 |
    H                 Cl

Methane       Carbon tetrachloride    Carbon monoxide
```

FIGURE 11.2 Bonds to nonmetallic atoms

We find that two carbon atoms (when they share electrons with other carbon atoms) may form either of the following: carbon-carbon single bonds (C–C), carbon-carbon double bonds (C), and carbon-carbon triple bonds (C). Each bond (the bond is shown as a dash (–)) is a shared pair of electrons. Molecules of ethane, ethylene, and acetylene, compounds have molecules with only two carbon atoms. Molecules of ethane possess carbon-carbon single bonds; molecules of ethylene possess carbon-carbon double bonds; and molecules of acetylene possess carbon-carbon triple bonds.

Covalent bonds between carbon atoms in molecules of more complex organic compounds can link into chains, branched chains, or rings.

Hydrocarbons

hydrocarbons are the simplest organic compounds, compounds whose molecules are composed only of carbon and hydrogen atoms. Hydrocarbons are divided into two broad groups: aliphatic and aromatic hydrocarbons.

Aliphatic Hydrocarbons

Aliphatic hydrocarbons can be characterized by the chain arrangements of their constituent carbon atoms. They are divided into the alkanes, alkenes, and alkynes.

The alkanes (paraffins) are saturated hydrocarbons (the hydrogen content is at maximum) with the general formula CnH2n+2. In systematic chemical nomenclature, alkane names end in the suffix -ane. They form the alkane series methane (CH_4), ethane (C_2H_6), propane (C_3H_8), butane (C_4H_{10}), etc. The lower members of the series are gases; the high-molecular weight alkanes are waxy solids. Alkanes are present in natural gas and petroleum.

Alkenes (olefins) are unsaturated hydrocarbons (they may take on hydrogen atoms to form saturated hydrocarbons) that contain one or more double carbon-carbon bonds in their molecules. In systematic chemical nomenclature, alkene names end in the suffix -ene. Alkenes that have only one double bond form the alkene series starting with ethene—the gas that is liberated when food rots—(ethylene), CH2:CH2, propene, CH3CH:CH2, etc.

Alkynes (acetylenes) are unsaturated hydrocarbons that contain one or more triple carbon-carbon bonds in their molecules. In systematic chemical nomenclature, alkyne names end in the suffix -yne.

Aromatic Hydrocarbons

Aromatic hydrocarbons are unsaturated organic compounds containing a benzene ring in their molecules, or that have chemical properties like benzene, a clear, colorless, water-insoluble liquid that rapidly vaporizes at room temperature, whose molecular formula is C_6H_6. The molecular structure of benzene is represented by a hexagon with a circle inside.

ENVIRONMENTAL CHEMISTRY

Environmental chemistry is a blend of aquatic, atmospheric, soil chemistry, and the "chemistry" generated by human activities thereon. The three environmental media are what they are because of many scientific principles, including chemistry. We are interested in the chemistry that makes up these mediums as well as the chemical reactions that take place to preserve and to destroy these mediums. As we proceed through our treatment of these mediums, we will be concerned with the environmental impact of human activities, such as mining, acid rain, erosion from poor cultivation practices, disposal of hazardous wastes, photochemical reactions (smog), air pollutants such as particulate matter, greenhouse effect, ozone and water degradation problems related to organic, inorganic, and biological pollutants. All these activities and problems have something to do with chemistry. Remediation and/or mitigation processes are tied to chemistry as well.

THE BOTTOM LINE

The slow, inexorable progress of rock changing into soil gives us an ongoing demonstration of nature's unrelenting power. While we move around on the surface of the earth and push and shove soil into various stages and conditions, we forget one vital fact: we can do little to either slow or speed the natural process of soil creation.

Nature is a powerful and omnipresent force in our environments that we must both depend on and contend with. Human beings present another powerful force, and when you put their anthropogenic achievements into the equation (the heavy hand of man and what we can do (and do) to our environment), the repercussions of both the destruction that Nature can dish out and the impact to the environment that humans contribute can become overwhelming to comprehend. Fortunately for our peace of mind, we don't often think along this line of reasoning. Is this a good thing? Is it good to ignore Nature and what she can do, what she always does — around us? Is it important to think about the impact man has on his surroundings?

We answer yes to both questions. Many people simply don't care. They think, "We all know that man impacts his environment, but what the heck, the environment will take care of itself. And besides, why should we be concerned with what man does today to the environment—soon it will all be forgotten? No matter what we do we leave our mark on the land." We do impact the environment we use and inhabit more than we are sometimes aware. Have you ever walked through an old cemetery, one with crumbling and unreadable headstones? This can be a sobering reminder to all of us of our fragility—not only of human life, but of the shortness of memory.

Our point: We must not forget; we must not only increase our awareness of our impact on our environment, but we must also increase our memory spans—our sense of history. For long-range planning, we need long-range vision—and the ability to remember what has gone on before.

REFERENCES

ASTM (1969) *Manual on Water*. Philadelphia: American Society for Testing and Materials, 1969.

Brady, N.C. and Weil, R.R. (1996) *The Nature, and Properties of Soils*, 11th ed. Upper Saddle River, NJ: Prentice-Hall.

Carson, R. (1962). *Silent Spring, Boston*: Houghton Mifflin Company, 1962.

D'Amore, D.V. (2015). *Carbon Soil.* Washington, DC: NSAC.

Eswaran, H. (1993). Assessment of global resources: Current status and future needs. *Pedologie* 43(1):19–39.

Foth, H.D. (1978) *Fundamentals of Soil Science*, 6th ed. New York: John Wiley and Sons.

Franck, I. and Brownstone, D. (1992) *The Green Encyclopedia.* New York: Prentice-Hall.

Kemmer, F.N. (1979) *Water: The Universal Solvent.* Oak Ridge, Illinois: NALCO Chemical Company.

Konigsburg, E.M. (1996) *The View From Saturday*, New York: Scholastic,.

Macmillan Publishing Company (1992) *The Way Nature Works.* New York: Macmillan Publishing Company.

Soil Survey Staff (1975) *Soil Classification, A Comprehensive System.* Washington, D.C.: USDA Natural Resources Conservation Service.

Soil Survey Staff (1994) *Keys to Soil Taxonomy*. Washington, D.C.: USDA Natural Resources Conservation Service.

Spellman, F.R. (1999) *Environmental Science and Technology: Concepts and Applications.* Rockville, Maryland: Government Institutes.

United States National Park Service (2017). Lichens and Air Quality. Accessed 6/17/23 @ https://nps.gov/articles/lichens-and-air-quality.htm.

USDA (1975) *Soil Taxonomy: A Basic System of Soil Classification for Making and Interpreting Soil Surveys*. Washington, D.C.: USDA Natural Resources Conservation Service.

Windgardner, D.L. (1996). *An Introduction to Soils for Environmental Engineers.* Boca Raton, FL: Routledge.

12 Soil Organic Matter and Carbon

INTRODUCTION

Chapter 11 sets the stage, providing very basic coverage of Soil Science and Chemistry. The purpose is to provide foundational information for the material presented in this chapter. Before presenting important on soil organic matter (SOM) and soil organic carbon (SOC) pertinent key terms and definition (some of them presented earlier but deliberate repetition is warranted) are presented first.

KEY TERMS AND DEFINITIONS

Autotrophic respiration—metabolism of organic matter by plants.

Biomass—Total volume or biomass includes stem, bark, stump, branches, and foliage, especially if evergreen trees are being measured. When estimating biomass available for bioenergy, the foliage is not included, and the stump may or may not be appropriate to include depending on whether harvest occurs at ground level or higher. But regarding carbon content foliage and stumps and roots are carbon storage units—all important to this discussion. Both conversion and expansion factors can be used together to translate directly between merchantable volumes per unit area and total biomass per unit area.

Carbon sequestration—not to be confused with "carbon capture and storage," it is the process through which carbon dioxide is removed from the atmosphere, for example in forests through the process of photosynthesis. During this process, carbon dioxide is taken up through plants' leaves and incorporated into the plants' wood biomass.

Carbon cycle—all carbon sinks and exchanges of carbon from one sink to another by various chemical, physical, geological, and biological processes.

Carbon equivalent—The amount of carbon dioxide by weight emitted into the atmosphere would produce the same estimated radiative forcing as a given weight of another radiatively active gas. Carbon dioxide equivalents are computed by multiplying the weight of the gas being measured (for example, methane) by its estimated global warming potential (which is 21 for methane). "Carbon equivalent units" are defined as carbon dioxide equivalents multiplied by the carbon content of carbon dioxide (i.e., 12/44).

Carbon stock change—the change in carbon stocks over time, calculated by taking the difference between successive inventories and dividing by the number of years between these inventories for each national forest. When a positive change occurs this means that carbon is being removed from the atmosphere and sequestered

DOI: 10.1201/9781003432838-15

by the forests (i.e., carbon sink). However, a negative change means carbon is added to the atmosphere by forest-related emissions (i.e., carbon source).

Carbon dioxide equivalent—The amount of carbon dioxide by weight emitted into the atmosphere would produce the same estimated radiative forcing as a given weight of another radiatively active gas.

Carbon dioxide fertilization—is the phenomenon in which plant growth increases (and agricultural crop yields increase) due to the increased rates of photosynthesis of plant species in response to elevated concentrations of carbon dioxide in the atmosphere.

Ecosystem respiration—the total respiration of all organisms living in each ecosystem.

Flux—is the transfer of carbon from one carbon pool to another.

Gross primary production—is the sum of total canopy photosynthesis for a unit of area over a given unit of time.

Harvested wood products—This includes all wood materials (including bark) that leave harvest sites. Slash and other materials left at harvest sites are regarded as dead organic matter.

Heterotrophic respiration—is the metabolism of organic matter by bacteria, fungi, and animals.

Pool/reservoir—is any natural region or zone, or any artificial holding area, containing an accumulation of carbon or carbon-bearing compounds or having the potential to accumulate such substances.

Sequestration—is the process through which carbon dioxide is removed from the atmosphere, for example in forests through the process of photosynthesis. During this process, carbon dioxide is taken up through plants' leaves and incorporated into the plants' wood biomass.

Sinks—carbon reservoirs and conditions that take in and store more carbon (carbon sequestration) than they release. Carbon sinks can serve to partially offset greenhouse gas emissions. Forests and oceans are common carbon sinks.

Woody biomass—is the byproduct of management, restoration, and hazardous fuel-reduction treatments, as well as the product of natural disasters, including trees and woody plant (limbs, tops, needles, leaves, and other woody parts, grown in a forest, woodland, or rangeland environment).

SOIL ORGANIC CARBON (SOC)[1]

Soil organic carbon (SOC) is a very vigorous compound of the soil; each year, the amount of SOC processed by microorganisms within the soil is equal to the number of inputs from plant detritus. The pervasive dynamic nature of SOC is key to the ecosystem services, or "the benefits people obtain from ecosystems" (Millennium Ecosystem Assessment, 2003), that SOC provides.

Note that soil organic carbon is a vital sign of soil health. So, what is soil health? Well, soil health refers to the ability of soil to function effectively as a constituent

[1] Material presented in the following sections is Adapted from Berryman et al. (2020). *Soil Carbon*. Accessed 4/28/23 @ https://fs.usda.gov/km/pubs-journals/2020/rmrs-2020-berryman-eco/pdf.

of a healthy ecosystem (Schoenholtz et al., 2000, p. 335). Important soil functions such as nutrient mineralization, aggregate stability, trafficability, permeability in air, water retention, infiltration, and flood control are all linked to quantity and quality of SOC. In turn, these soil functions are linked with a broad range of ecosystem properties. For instance, high SOC in mineral soils is usually associated with high plant productivity (Oldfield et al., 2017), with subsequent positive implications for wildlife habitat, distribution, and abundance. Accordingly, ecosystem services can be degraded when SOC is altered or lost from forest or rangeland sites. A more complete understanding of the ecosystem and soil health at a particular site can be measured and monitored for SOC levels; indices of soil health incorporate measures of SOC and can be used to track changes in soil health over time and in response to management activities (Amacher et al., 2007; Chaer et al., 2009).

DID YOU KNOW?

SOM holds soil particles together through adhesion and entanglement, reducing erosion and allowing root movement and access to nutrients and water.

The largest pool of terrestrial C is contained in the globe's soils, with an estimated $2.27-2.77 \times 1{,}015$ kg of 2,270–2,770 petagrams (Pg) of C in the top 2–3 m of soil (Jackson et al., 2017). This represents a pool that is two to three times larger than the atmospheric and biotic C pools. All soil orders of North American and US soils store about 366 and 73.4 Pg C, respectively, in the top 1 m (Liu et al., 2013; Sundquist et al., 2009; USGCRP, 2018). In the United States it is in the non-intensively managed lands such as forests, wetlands, and rangelands where most of the SOC stock is located (Liu et al., 2012, 2014). In land uses, most SOC is concentrated near the surface, where it could be vulnerable to loss; 74.5% of North America's SOC occurs in the top 30 cm of mineral soils (Batjes, 2016; Scharlemann et al., 2014). Despite the importance of O horizons as a source of OM for building SOC, most assessments of SOC pools represent only mineral SOC and have omitted organic soil horizons that sit on top of the mineral soil.

A critical first step to understand how management activities can impact this national resource is to make an accurate assessment including ongoing monitoring of national SOC stocks. Because the SOC pool is large compared to other C pools (especially the atmosphere), a small change in SOC can produce a large change in atmospheric carbon dioxide levels. For instance, a global decrease in SOC of 5% in the upper 3 m would result in 117 Pg of C released into the atmosphere, causing an increase in the atmosphere C pool (829 Pg in 2013) of 14%, i.e., from 400 to 456 ppm carbon dioxide. Conversely, sequestering a small percentage in this large C pool translates into a substantial increase that is globally relevant. Nave et al. (2018) point out that site-level studies suggest that reforestation and other land-use and management changes increase SOC by 0.1–0.4 Mg C ha^{-1} $year^{-1}$, and a national-scale (conterminous United States) analysis suggests that reforesting topsoils are accumulating $13-21 \times 10^9$ kg or 13–21 teragrams (Tg)C $year^{-1}$ with the potential to sequester hundred more teragrams C within a century.

In the past twenty years, there have been several coordinated efforts to assess national-level SOC stocks in a statistically robust manner (Sidebar 12.1). Such assessments provide a baseline for detecting future change in United States SOC. However, there are some limitations to national-scale assessments. Varying depth of soil across the country and, consequently, SOC stocks. The reality is that most assessments do not consider soil deeper than 1 m, even though deeper soil can be an important reservoir of SOC (Harrison et al., 2010). Moreover, in any study of SOC on a specific site, it is difficult to detect SOC changes less than about 25% (e.g., Homann et al., 2001) due to soil heterogeneity and sampling and measurement error. Though laboratory incubations and other controlled experiments demonstrate that some perturbation will cause change (e.g., soil warming), it can be difficult to detect this change in the field. Errors in assessing SOC can be expected when the inclusion of rocks larger than 2 mm in samples, changes in texture, and sensitivity of SOC stocks to bulk density are considered (Page-Dumroese et al., 2003). Berryman et al. (2020) point out that improved measurement technology and statistical methods that account for different sources of uncertainty may help overcome these challenges and allow for detection of more subtle changes in SOC.

SIDEBAR 12.1—THE NATION'S SOIL ORGANIC CARBON STORES

Regarding a national scale for assessing the nation's soil organic carbon stores there have been several efforts to characterize soil organic carbon (SOC) but because of the high spatial variability of SOC content it is a major challenge to the accurate accounting of SOC. What this means is that large uncertainties may preclude the detection over time. Each effort has slightly different goals and objectives, but they all emphasize free and open data availability.

The USDA Forest Service's Forest Inventory and Analysis (FIA) program reports the status and trends of the nation's forest resource, across all ownerships. FIA collects information via their field campaigns; this information is on the area of forest land and its location; species, size, and health of trees; and total tree growth, mortality, and removals by harvest and land-use change. Note that in 2000, the US Forest Service enhanced the FIA program in several ways. For instance, it changed from a periodic survey to an annual survey (the field crew returns to each plot every 5–10 years), and it expanded the scope of data collection to include soil, understory vegetation, tree crown conditions, coarse woody debris, and lichen community composition on a subsample of plots. It also increased our capacity to analyze and publish the underlying data as well as information and knowledge derived from it. Regarding facilitating forest SOC estimation, FIA collects data on litter thickness and mass, C content, mineral soil bulk density, and rock fraction at repeatable depth intervals (0–10 cm, 10–20 cm) (O'Neill et al., 2005). Statistically sound estimates of forest floor and litter C and SOC can be obtained from these measurements when they are combined with other site attributes and ancillary information (Domke et al., 2016, 2017). An advantage of FIA assessments over others is the repeated nature of the survey; eventually, change in soil properties over time may be possible to detect.

DID YOU KNOW?

Ecosystem water storage increases through enhancing soil structure and increasing soils' effective surface area; SOM increases the amount of water that can be retained in the soil for plant and downstream use, reducing evaporative and runoff losses.

DID YOU KNOW?

The composition of an epiphytic lichen community is one of the best biological indicators of nitrogen and sulfur-based air pollution in forests. Their sensitivity results from their total reliance on atmospheric sources of nutrition. Because lichens are so sensitive to these pollutants, they are useful as an early indicator of improving or deteriorating air quality.

Soil C Accounting

Seems like when there is a good side to something (anything!) there is always a bad or downside too. This seems to be the case with FIA soil C accounting methodology. The problem is that nationwide assessments (surveys) like FIA may not be helpful for regional or local management challenges due to their coarse spatial resolution. Note that in 1976, the National Forest Management Act (is an amendment of the **Forest and Rangeland Renewable Resources Planning Act of 1974**) was enacted and set forth three points that would necessitate soil monitoring and analysis on national forests to inform planning. The first point was that land management could not produce substantial or permanent impairment of site productivity. Second, trees could be harvested only where soil, slope, or watershed conditions would not be irreversibly damaged. Last, harvesting had to protect soil, watershed, fish, wildlife, recreational, and aesthetic resources.

The bottom line: Efforts to quantify changes in SOC must rely on archived soil samples combined with new sampling and analysis to determine changes in surface and subsurface pool sizes (Berryman et al., 2020).

DID YOU KNOW?

Forests significantly impact the nations carbon footprint: forests and associated harvested wood products uptake the equivalent of 10%–15% of economy-wide carbon dioxide emissions annually.

MINERAL SOIL ORGANIC CARBON

Change is not static; it is dynamic—ever fluctuating, varying, and altering. In dealing with SOC and our understanding of SOC distribution and vulnerability have been limited due to slow to no change over time. But, again, change is inevitable and in SOC monitoring changes occurred (and still is) because of advancements in technology. These changes allow for fine-scale molecular and microbial investigations of SOC interactions with mineral soil; a new conceptual framework of SOC stabilization and destabilization has advanced that now improves our ability to SOC behavior. Our knowledge of the source and stabilization mechanisms of mineral-associated SOC has advanced.

Organic matter (OM) quality is used as a general descriptor for the combination of the chemical structure and elemental composition of OM that influences decomposition. In the past it was thought that the ability of an organism to effectively decay OM was directly related to the material's molecular composition (e.g. lignin content) and concentration of nutrients (such as N (nitrogen)). Note that these concepts are still useful when describing the decomposition dynamic of organic soils or organic soil horizons, but measures of OM quality have been obscure—elusive (Berryman et al., 2020). Moreover, these concepts break down in attempts to describe the dynamics of SOC associated with horizons dominated by mineral materials.

The relative importance of litterfall (aboveground sources) and root systems (belowground sources) of SOC is key to understanding impacts of disturbance and management on SOC. Today it is clear that across many ecosystems most SOC is derived from root inputs and not aboveground inputs. It is estimated that root inputs may account for five times as much SOC as aboveground sources (Jackson et al., 2017). The close association of mineral soil and roots may be the primary cause of the disproportionate importance of roots on SOC. In the past, the focus has been on forest floor liter or mass layer depth; however, there is a growing recognition of the role of fine root production and turnover as OM inputs. This recognition is having important implications for our ability to predict the response of SOC to disturbances that affect aboveground and belowground sources of OM (Berryman et al., 2020).

DID YOU KNOW?

It takes less energy to grow forest biomass and convert it to ethanol that it takes to grow corn and convert it to ethanol. In addition, the entire process emits fewer greenhouse gases when using forest biomass.

The old standard, concept, model, or paradigm suggested that OM entering the soil had three possible fates: (1) loss to the atmosphere as carbon dioxide, (2) incorporation into microbial biomass, and (3) stabilization as humic substances (Schnitzer and Kodama, 1977; Tate, 1987). Humic substances representing a major part of organic matter in soil, peat, coal, and sediments are described as dark-colored, recalcitrant, refractory, and heterogeneous organic compounds of high molecular weight which could be separated into fractions based on their solubility in acidic or alkaline solutions (Sutton and Sposito, 2005).

Now, the truth be told, recent advancements in technology have revealed that SOC is largely made up of identifiable biopolymers, and the perceived existence of humic substances was an artifact, a relic of the procedures used to extract the material (Kelleher et al., 2006; Kleber and Johnson, 2010; Lehman and Kleber, 2015; Marschner et al., 2008).

Using the old paradigm and concepts made it difficult to predict the long-term behavior of SOC pools. The thinking was that compounds were believed to be chemically recalcitrant and resistant to decomposition (e.g., lignin) sometimes turned over rapidly, whereas compounds thought to be labile—capable of changing state—(e.g., sugars) were demonstrated to persist for years (Grandy et al., 2007; Kleber and Johnson, 2010; Schmidt et al., 2011). It is these inconsistencies that uncovered key misconceptions of the old paradigm that prevented a predictive understanding of the vulnerability of SOC to change. Because of such shortcomings, the conceptual model that soil scientists use to describe SOC and stabilization is undergoing a change in basic assumptions (the model is changing) toward one that emphasizes the complex interactions between microorganisms and minerals in the soil.

This new model for understanding SOC stability postulates that SOC exists across a continuum of microbial accessibility, ranging from free, unprotected particulate materials and dissolved OM to organic substances that are stabilized against biodegradation through association with both. mineral surfaces or separately with occlusion within soil aggregates (Lehman and Kleber, 2015). Under this model, interfaces of the microbial community and soil minerals, instead of characteristics inherent in the SOC itself, are the main regulators of the pathways of OM stabilization and biodegradation. These factors may more accurately predict the behavior of the SOC pools and are also more easily measured than molecular properties of SOC, leading to new possibilities for management.

DID YOU KNOW?

Substantially more C is stored in the Earth's soils, about twice the amount of C that is present in the atmosphere.

Emerging under the new model are the basic mechanisms of SOC stabilization: sorption to mineral surfaces and occlusion within aggregates. Experience has led us to wonder and to suppose and to assume that whether OM is sorbed to the surface of mineral soil particles or occluded is dependent partly on the chemical characteristics of SOC and whether micrograms assimilate the SOC or use it as energy—an important concept. The development of the concept of substrate use efficiency (SUE; the proportion of substrate assimilated versus mineralized or respired) has found that the ability of an organism to effectively decompose and transform the structural components of OM into stabilized SOC is related not only to the chemical characteristics of SOC, but also to the composition of the soil microbial community (Cotrufo et al., 2013). Currently in this discussion, it is important to be redundant and to restate the following:

> Soil organic matter consists primarily of living organisms and the remains of plants, animals, and microorganisms that are continuously broken down (biodegraded) in the soil into new substances that are synthesized by other microorganisms. These other

> microorganisms continually use this organic matter and reduce it to carbon dioxide (via respiration) until it is depleted, making repeated additions of new plant and animal residues necessary to maintain soil organic matter.
>
> *(Brady and Weil, 1996)*

When microorganisms use the decomposition products of litter for energy (low SUE), that fraction remains as particulate OM stabilized within aggregates. High SUE from litter is used to stabilize on mineral surfaces, and eventually becomes dissolved OM because of microbial exudation and death lysis. Aggregate and mineral stabilized pools of C have limited capacity and are said to saturate at a certain level. By contrast, free particulate matter is thought to have no upper threshold or a very high threshold (Berryman et al., 2020).

SOC is thought to be an ecosystem property because it is a property that arises because of an exchange of material or energy among different pools and their physical environment. This is the case whether OM is stabilized within aggregates, through sorption on mineral surfaces, at depth, or in recalcitrant materials such as char. So, understanding the mechanisms that are important to the overall residence time of SOC as well as its response to a changing environment (Schmidt et al., 2011) will be valuable for monitoring and managing SOC. Further study and development of this model are likely to find numerous interacting pathways to stabilize C in soils that involve microbial accessibility and chemical recalcitrance.

DID YOU KNOW?

The processes of uptake, respiration, and decomposition of key nutrients such as nitrogen, calcium, and carbon enable the transferring from plant to soil to air and back to plant again. This is known as "nutrient cycling" which is a "biogeochemical" process because nutrients follow pathways that are mediated by biological, geological, and chemical players. Okay, so what are the players? The players include plants, soil microorganisms, weathered rock, and gases in the air and soil (Beldin and Perakis, 2009).

ASSESSING SOIL CARBON VULNERABILITY

Vulnerability, define vulnerability and what does vulnerability have to do with soil carbon content?

Okay, we can say that vulnerability is closely related to the terms susceptibility, weakness, helplessness, and so on and so forth. However, vulnerability—the term and definition—as used and applied to SOC (mineral soil and O horizons) refers to something else. Vulnerability of SOC to change refers to the susceptibility of SOC to change in the face of disturbance. Note that change could mean either increases or decreases, but usually the concern is with loss, the decrease of SOC.

Let's get back to vulnerability of soil organic carbon.

We could and can define vulnerability of SOC in terms of resistance and resilience. Soil C stores could resist losses as the result of a perturbation, or they could be resilient and recover SOC lost due to the perturbation. A system that is not affected by disturbance is thought to be resistant to change. Despite this, a system that loses SOC because of a disturbance, and regains lost SOC post-disturbance, is resilient but not resistant (Berryman et al., 2020).

Note that these concepts are important because SOC loss and recovery can affect the C storage of landscapes over long time-scales. Therefore, the implications of SOC management and response to change need to be considered on a larger scale in both geographic extent and time.

The vulnerability of SOC to change depends on the forcing (climate change, fire regime shift, invasive species, and disturbance).

So what is 'forcing' and how does it apply to forest soil storage of carbon? Sidebar 12.2 provides an example of forcing.

SIDEBAR 12.2—CLIMATE FORCING

The Intergovernmental Panel on Climate Change (IPCC) defines climate forcing as "An externally imposed perturbation in the radiative energy budget of the Earth climate system, for example, through changes in solar radiation, changes in the Earth's albedo or changes in atmospheric gases and aerosol particles." Thus climate forcing is a "change" in the status quo. IPCC takes the pre-industrial era (chase as the year 1750) as the baseline. The perturbation to direct climate forcing (also termed "radiative forcing") that has the largest magnitude and the least scientific uncertainty is the forcing related to changes in long-lived, well mixed greenhouse gases, in particular, carbon dioxide (CO_2), methane (CH_4), nitrous oxide (N_2O), and halogenated compounds (mainly CFCs).

This "radiative forcing" or heating effect is continually measured. Radiative forcing occurs when energy from the sun reaches the Earth. The planet absorbs some of this energy and radiates the rest back as heat. Note that a variety of physical and chemical factors—some natural and some influenced by humans—can shift the balance between incoming and outgoing energy, which forces changes in the Earth's climate. These changes are measured by the amount of warming or cooling they can produce. When the changes have a warming effect they are called "positive" forcing, whereas those that have a cooling effect are called "negative" forcing. When positive and negative forces are out of balance the result is a change in the Earth's average surface temperature. Greenhouse gases trap heat in the lower atmosphere and cause positive radiative forcing.

Atmospheric global greenhouse gas abundances are used to calculate changes in radiative forcing beginning in 1979 when NOAA's global air sampling network expanded significantly. NOAA's Annual Greenhouse Gas Index is defined as the change in annual average total radiative forcing by all the long-lived greenhouse gases since the pre-industrial era (1750).

Note that the carbon cycle and its response to multiple interacting drivers of global climate change are key aspects of the biospheric forcing of climate (Bonan, 2008)

THE BOTTOM LINE

We asked what is forest soil carbon vulnerability? Note that in this book soil carbon vulnerability is more accurately termed and defined as soil carbon disturbance. In the following chapter, forest soil carbon disturbance is discussed in detail.

REFERENCES AND RECOMMENDED READING

Amacher, M.C., O'Neill, K.P., Perry C.H. (2007). Soil Vital Signs: A New Soil Quality Index (SQI) for Assessing Forest Soil Health. Research Paper RMRS-RP-65. U.S. Department of Agriculture. Forest Service Rocky Mountain Research Station, Fort Collins, 12 p.

ASTM (1969). *Manual on Water*. Philadelphia: American Society for Testing and Materials.

Berryman, E. et al. (2020). Soil Carbon. Accessed 5/8/23 @ https://www.fs.usda.gov/rm/pubs-journals/2020/rmrs-2020-berrymn-eco/pdf.

Brady, N.C. and Weil, R.R. (1996). *The Nature and Properties of Soils*, 11th ed. Upper Saddle River, NJ: Prentice-Hall.

Batjes, M.I. (2016). Harmonized soil property values for broad-scale modelling (WISE30sec) with estimates of global soil carbon stocks. *Geoderma* 169:61–68.

Beldin, S. and Perakis, S. (2009). Unearthing Secrets of the Forest. Fact Sheet 2009-3078. United States Geological Survey, Washington, DC.

Bonan, G. (2008). *The Land Use Forcing of Climate: Models, Observations, and Research Needs*. Boulder, CO: Amerifix Science Teams.

Chaer, G.M., Myrold, D.D., Bottomley, P.I. (2009). A soil quality index based on the equilibrium between soil organic matter and biochemical properties of undisturbed coniferous forest soils of the Pacific Northwest. *Soil Biology and Biochemistry* 41(4):822–830.

Cotrufo, M.E., Wallenstein, M.D., Boot, C.M. et al. (2013). The Microbial Efficiency-Matrix Stabilization (MEMS) framework integrates plant litter decomposition with soil organic matter stabilization: Do labile plant inputs form stable soil organic matter? *Global Change Biology* 19(4): 988–995.

Domke, G.M., Perry, C.H., Walters, B.F., Woodall, C.W. Russell, M.B., Smith, J.F. (2016). *Estimating Litter Carbon Stocks on Forest Land in the United States*. Amsterdam, Netherlands: Elsevier Publishing

Domke, G.M., Perry, C.H., Walters, B.F., Nave, L.E., Woodall, C.W., Swanston, W. (2017). Toward inventory-based estimates of soil organic carbon in forests of the United States. *Ecological Applications* 27(4):1223–1235.

Grandy, A.S., Neff, J.C., Weintraub, M.N. (2007). Carbon structure and enzyme activities in alpine and forest ecosystems. *Soil Biology and Biochemistry* 39(1):2701–2711.

Harrison, R.B., Footen, P.W., Strahm, F.D. (2010). Deep soil horizons: Contribution and importance to soil carbo pools and in assessing whole-ecosystem response to management and global change. *Forest Science* 57(1):67–76.

Homann, P.S., Bormann, B.T., Boyle, J.R. (2001). Detecting treatment differences in sol carbon and nitrogen resulting from forest manipulation. *Soil Science Society of America Journal* 65(2):463–469.

Jackson, R.G., Lajtha, K., Crow, S.E. et al. (2017). The ecology of soil carbon pools, vulnerabilities, and biotic and abiotic controls. *Annual Review of Ecology, Evolution and Systematics* 48(1):419–445.

Kelleher, B.P., Simpson, M.J., Simpson, A.J. (2006). Assessing the fate and transformation of plant residues in the terrestrial environment using HR-MAS NMR spectroscopy. *Geochimica et Cosmochimica Acta* 70(6):4080–4094.

Kleber, M. and Johnson, M.G. (2010). Advances in understanding the molecular structure of soil organic matter: Implications for interactions in the environment. *Advances in Agronomy* 106:77–142.

Lehmann, J. and Kleber, M. (2015). The contentious nature of soil organic matter. *Nature* 528(7580):60–68.

Liu, S., Liu, J., Young, F.J. et al. (2012). Chapter 5: Baseline Carbon Storage, Carbon Sequestration, and Greenhouse-Gas Fluxes in Terrestrial Ecosystems in the Western United States. Professional Paper 1797. U.S. Department of the Interior, Geological Survey, Reston, Virginia.

Liu, S., Wei, Y., Post, W.M. et al. (2013). The Unified North American Soil Map and its implication on the soil organic carbon stock in North America. *Biogeosciences* 10:2915–2930.

Liu, S., Liu, J., Wu, Y. et al. (2014). Chapter 7: Baseline and Projected Future Carbon Storage, Carbon Sequestration, and Greenhouse-Gas Fluxes in Terrestrial Ecosystems of the Eastern United States. In: Zhu, Z., Reed, B.C. (eds) Baseline and Projected Future Carbon Storge and Greenhouse Gas Fluxes in Ecosystems of the Eastern United States. Professional Paper 1804. U.S. Department of the Interior, Geological Survey, Reston, pp. 115–156. Accessed 3/5/2023 @ https://pubs.usgs.gov/pp/1804.

Marschner, G., Brodowski, S., Dreves, A. et al. (2008). How relevant is recalcitrance for the stabilization of organic matter in soils? *Journal of Plant Nutrition and Soil Science* 171(1):91–110.

Millennium Ecosystem Assessment (2003). *Ecosystem and Human Well-Being: A Framework for Assessment*. Washington, DC: Island Press.

Nave, I.E., Domke, G.M., Hofmeister, K.L. et al. (2018). Reforestation can sequester two petagrams of carbon in U.S. topsoils in a century. *Proceedings of the National Academy of Sciences* 115(11):2776–2781.

O'Neill, K.P., Amacher, M.C., Perry, C.H. (2005). Soils of an Indicator of Forest Health, a Guide to the Collection, Analysis, and Interpretation of Soil Indicator Data in the Forest Inventory and Analysis Program. General Technical Report NC-GTR-258. U.S. Department of Agriculture, Forest Service, North Central research Station, St Paul, 53 p.

Oldfield, E.E., Wood, S.A., Bradford, M.A. (2017). Direct effects of soil organic matter on productivity mirror those observed with organic amendments. *Plant* and *Soil* 423(1–2):363–373.

Page-Dumroese, D.S., Brown, R.E., Jurgensen, M F., Mroz, G.D. (2003). Comparison of methods for determining bulk densities of rocky forest soils. *Soil Science Society of America Journal* 63(2):379–383.

Scharlemann, J.P.W., Tanner, E.V.U., Hiederer, R., Kapos, V. (2014). Global soil carbon: Understanding and managing the largest terrestrial carbon pool. *Carbon Management* 5:81–91.

Schmidt, M.W.I., Torn, M.S., Abiven, S. et al. (2011). Persistence of soil organic matter as an ecosystem property. *Nature* 478:49–56.

Schnitzer, M. and Kodama, H. (1977). Reactions of minerals with soil humic substances. In: Dixon, J.G., Weed, S.B. (eds) *Mineral and Their Roles in the Soil Environment*. Madison: Soil Science Society of America, pp. 741–770.

Schoenholtz, S.H., Miegroet, H.V., Burger, J.A. (2000). A review of chemical and physical properties as indicators of forest soil quality: Challenges and opportunities. *Forest Ecology and Management* 138(1):335–356.

Sundquist, E.T., Ackerman, K.V., Bliss, N.B. et al. (2009). Rapid Assessment of U.S. Forest and Soil Organic Carbon Storage and Forest Biomass Carbon-Sequestration Capacity. Open-File Report 2009-1283. U.S. Department of the Interior, Geological Survey, Reston, 15 p. https://pubs.er.usgs.gov/publication/ofr20091283.

Sutton, R. and Sposito, G. (2005). Molecular structure in soil humic substances: The new view. *Environmental Science* and *Technology* 39(23):9009–9015.

Tate, R.I. III (1987). *Soil Organic Matter: Biological and Ecological Effects*. New York: Wiley Interscience, 291 p.

U.S. Global Change Research Program (USGCRP) (2018). Second State of Carbon Cycle Report (SOCCR2): A Sustained Assessment Report. In: Cavallaro, N., Shrestha, G., Birdsey, R. et al. (eds). U.S. Global Change Research Program, Washington, DC, 878 p.

13 Forest Soil Carbon Disturbance

INTRODUCTION

Soil organic content (SOC) pools within forest soil are usually stable but there are natural and unusual vulnerabilities to climate change, environmental change, management, and disturbance that bring about change. First, the stability of any pool depends on the magnitude or, and controls on, its fluxes (inputs and outputs). The inputs are the quantity and quality of C fixed by the primary producers and altered by abiotic processes (e.g., fire); the outputs are regulated by microbial accessibility and microbial activity. The SOC pool magnitude and stability are affected by changes to OM inputs, quantity of OM inputs, microbial accessibility, or microbial activity. Examples of factors that could affect these inputs and outputs include the following:

- **Change affects quality of OM inputs**—the change in productivity can be from fire, harvesting, and removal or addition of biomass by mulching.
- **Changes affecting quality of OM inputs**—change in transformation of biomass by pyrolysis; change in species; changes in allocation of production (especially belowground versus aboveground production).
- **Changes affection microbial accessibility**—changes in the distribution of SOC with depth through erosion and deposition, leaching, bioturbation, and other influences; destruction of aggregates; destabilization of redox-active minerals; inputs of active minerals (e.g., ash deposition); changing the OC saturation state; changes in the quality and quantity of SOC inputs, which could affect priming (stimulation of decay of stabilized SOC).
- **Changes affecting microbial activity**—change in soil temperature and moisture, nutrient availability, freeze-thaw patterns, oxygen availability (i.e., redox), pH, or salinity; change in nutrient status from additions of substances such as herbicide or additions of N and sulfur from acid deposition.

DID YOU KNOW?

Bioturbation is defined as the reworking of soils and sediments by animals or plants; it includes ingestion, burrowing, and defecation of sediment grains. Bioturbation has a profound effect on forest soil environment.

DOI: 10.1201/9781003432838-16

VULNERABILITY VERSUS DISTURBANCE

Earlier, and often mentioned, forest soil SOC is vulnerable, susceptible, defenseless, exposed, at risk, and open to this that or whatever. And all this is true. However, in this book the emphasis is on a subset of vulnerability related to forest soil SOC issues and not on just vulnerability.

The subset?

Disturbance is a subset of vulnerability and can be further defined as a disruption, disorder, interruption, or simply as an annoyance. The truth be told, however, the disturbances to SOC in forest soil is not an annoyance but instead is a change whereby disturbances can affect the carbon cycle and can cause (accelerate) loss valuable storage of carbon.

So, what are the disturbances that can affect carbon storage and carbon dioxide release to the environment? Well, in this presentation seven disturbances are listed and described. The disturbances are as follows:

Disturbance one in this book is climate change.
Disturbance two is fire.
Disturbance three is harvesting and thinning.
Disturbance four is livestock grazing
Disturbance five is nutrient additions.
Disturbance six is tree mortality.
Disturbance seven is invasive species.

Disturbance one, "climate change" and increasing carbon dioxide is an ongoing occurrence. It is important to remember that a primary factor in soil formation is climate which has profound effects on SOC cycling (Jenny, 1941).

Let's get back to, for a moment at least, climate change. Although this is not a book focusing completely on climate change and all factors tied in directly or indirectly, so to speak, it is important to point out that climate change is a SOC disturbance in forest lands (and elsewhere).

Here's what we know about climate change (in the author's view):

"Through time and experience we have learned (yes, teachers learn, too) that whether we call it global warming, global climate change (humankind-induced global warming, under a broader label), or an inconvenient truth, the topic is a conundrum (riddle, the answer of which is a pun). As such, before diving into the many emotionally charged, heated class discussions about this "hot" topic (pun intended), we are reminded of two celebrated statements of just how complicated a conundrum can be. These celebrated conundrums are given as follows:

What is black and white and read all over? A newspaper.

"Why is a man in jail like a man out of jail?" there's no answer to it (Charles Dickens, 1843, *Martin Chuzzlewit*).

Consider this: Any damage we do to our atmosphere affects the other three environmental mediums: water, soil, and biota (us—all living things, including forests and soil included). Thus, the endangered atmosphere (if it is endangered) is a major concern (a life and death concern) to all of us (in F.R. Spellman's (2011) *Forest-Based Biomass Energy*. Boca Raton, FL: CRC Press).

Okay, let's get back to one of the major soil carbon disturbances: climate change.

When we have warming conditions, quantity and quality of OM inputs will be impacted as warming temperatures and shifting precipitation regimes lead to transitions in forest and rangeland communities (Clark et al., 2016). Changes in temperature and moisture levels alter microbial accessibility which, in turn, affects mineral complexation and leaching. And then there are the microbes who are also sensitive to changes in temperature and moisture availability. Moreover, changes directly tied to climate increase in carbon dioxide concentration will alter plant productivity, affect the quantity of C input in soil, as well as the relative contributions of roots and shoots to SOC, potentially increasing root-derived OM inputs (Phillips et al., 2012). Others have shown that litter quality will change or that species shifts could take place which changes the quality of C inputs to soil (MacKenzie et al., 2004).

DID YOU KNOW?

The units that are typically used when examining carbon densities—that is, the amount of carbon in a forest per unit area—include metric per hectare; metric tons per acre; and tons per acre.

SIDEBAR 13.1—SPECIFIES SHIFT AND CLIMATE CHANGE

Have you ever wondered about the species in your location? Have you wondered why you have this species of plants and animals in your location or close to where your habitat is and no other varieties, species and plant and animals? The question comes down to this: why are species located where they are? This is a question environmentalists, biologists, and ecologists have long been asking, and the classic answer is that vegetation and animals depend on climate, parent material, organism, topography, time, and most importantly (in the author's view) disturbance. The point is that organisms, plants, and animals can thrive in certain locations based on climate so long as they are not disturbed (Major, 1951).

It is also the author's view that climate change is occurring due to natural cycling and we humans are exacerbating the change in climate due to our pollution and disturbance of various ecological regions. Of course we also need to add Mother Nature to the mix. She can disrupt an environment in several ways and she certainly does.

Okay, prehistoric records show that the climate has changed through time, as have the roles or organisms (again, this includes humans) and other disturbances such as fire. Again, in present times, however, the climate is changing rapidly. There is no doubt about human-caused greenhouse gas emissions that have raised peak levels of atmospheric carbon dioxide about 400 ppm for the first time in at least 3 million years. Changing climate is having a profound effect on species distribution via changes in growth, reproduction, and mortality, with increasing likelihood of more marked changes in the coming decades. Climate changes can act to directly influence species shift (distribution) (e.g., wind, drought, floods) as well as indirectly (e.g., temperature and weather-related changes in patters of wildfire, insects, and disease outbreaks) (Iverson and McKenzie, 2014).

We know from years of study and tracking that some species shift or species range have shifted in recent decades, very probably in response to climate changes. For example, a meta-analysis (i.e., the statistical method to compare and combine effect sizes from a pool of relevant empirical studies; Mike et al., 2012) of more than 750 species (mostly arthropods, such as insects, spiders, or crustaceans) found an average rate of poleward migration of almost 17.0 km/decade (Chen et al., 2014). Parmesan and Yohe (2003) point out that an earlier analysis, using almost 100 birds, butterflies, and alpine herbs, reported an average poleward migration (i.e., a latitude shift) of 61 km/decade, and a significant mean advancement of spring events by 2.3 days per decade. For tree species, direct evidence of latitudinal shifts is more limited, somewhat scant. Well, note that there is some indirect evidence apparent in some studies in the eastern United States. Consider the report by Schuster et al. (2008), for example. In their report the researchers sought to quantify changes in tree species composition, forest structure, and aboveground forest biomass over the 1930–2006 timeframe in the deciduous Black Rock Forest in southeastern New York, USA. The results of the researchers' findings are consistent with other evidence indicating that North American temperate forests have function as significant C sinks for many decades. They also found that Black Rock Forest and other forests in the Highlands Province are still capable of sequestering carbon, especially in areas with canopies dominated by red oak and chestnut oak. However, the researchers noted recent changes in the stand structure, disturbance regimes, and influxes of new tree species combined with recent canopy mortality indicating the potential for substantial forest change in the future.

The bottom line: We know based on observation that climate change affects species shift or distribution or whatever. Anyway, after studying this and several other current issues concerning our environment I have come to realize that we do not know what we do not know about forest migration and all the factors involved and therefore we must have science, study, investigation, observation, and recordkeeping amended with a touch of common sense.

SIDEBAR 13.2—BLAST FROM THE PAST, BUT MAYBE FOR THE FUTURE, TOO.[1]

Note: The following is Another Conversation Between Mr. Snowshoe Hare and Mr. Deer Mouse—these two wise critters appear in a lot of my environmental books—it allows me to get back to Nature.

At the sunset hour, the forest is still; overhead, the sky is gray with clouds spitting rain, soaking the ground, and turning it into a muddy, sticky mess.

---

[1] Based on and Adapted from a conversation in Spellman, F.R. (2022). *Climate Migration*. Boca Raton: CRC Press.

What is now an open field and not that place of the past that normally had been scented with the smack of Douglas Fir and Ponderosa Pine trees separated and isolated by what was a natural grassland area. Within what had been the grassy area filled with Oregon grapes, snowberry, and chokeberry punctuated by shades of lime-green-stemmed yellow Monkey flowers bordered what was the peacock blue river, not at all blue now. A river that passed beneath what has been a wooden footbridge, but now is a scattered jack-straw-like pile of broken and charred timbers. And on the other side of what used to be the pristine river were the burnt remains of the outhouses and three-sided shelters that have been the cozy campsite refuges on the opposite shore of the river. The remote campsites have been sites to relax in, to get one with Nature, to rest after an arduous day of hiking. The entire place was—has been—accompanied by a steady release of pent-up stress accumulated by being associated with today's human race. Or in this case what was an association with the animal world … those by choice, of course. However, even before the climatic catastrophe altered this landscape into a battlefield like scene, a close association with Mr. Grizzly Bear, Mr. Wolverine, Mr. Cougar, and a few other Mr. and/or Ms. is not and was not recommended.

Yes, this is what this place is today. With its cold malevolent air hovering above the naked, muddy ground before its transformation to what could be described as a battlefield scene of death and destruction—no movement, no wind, no people. It's like they just vanished or were never there to begin with; that is, except for two stragglers.

"Mr. Deer Mouse, so good to see you again," says Mr. Snowshoe Hare while craning his neck to the left and to the right and back and forth repeatedly—this is a constant maneuver for Mr. Hare because unlike the rabbit, hares are not able to view front and back and all around with almost perfect 360° range of sight. Being alert for hungry foxes, wolves, coyotes, bobcats, and the deadly wolverine is just a normal and continuous and lifesaving and necessary habit. As far as Mr. Snowshoe Hare is concerned Mr. Fox, Mr. Wolf, Mr. Coyote, and Mr. Bobcat can look elsewhere for their culinary delight. (And the truth be told that is exactly what all those critters including human critters were doing; they were/are looking elsewhere). With the cover of darkness in the forest (that is, what was once a forest) comes the nocturnal predators, who seem to be more in need of nourishment than rest at night. However, Mr. Hare has noticed recently that it is difficult to spot any of those who would love to eat him—those types seemed to have moved on, to somewhere else.

They have moved on.

Mr. Mouse pauses a bit before saying, "Well, Mr. Hare it is good to see you too." Why the pause? Well, Mr. Mouse noticed that Mr. Hare appears to have the same problem he has: a loss of weight; an obvious change in appearance. Meanwhile, Mr. Mouse also is nervously craning his neck here and there and everywhere for the same reason as Mr. Snowshoe Hare; they are on predator watch, even though the predators are long gone; they are along with the other species, heading poleward with climate change. But for Mr. Mouse it is not just hungry foxes, wolves, coyotes, and bobcats that he must be on guard for, but also the Northern Hawk Owl. At this time of the evening the Owl is on the prowl and does not howl; moreover, the Owl is camouflaged in its plumage that makes it almost invisible to its victims. The truth be told the Northern Hawk Owl's beak might water just at the sight of Mr. Mouse, and any of his close relatives. Once spotted by the Owl it is time to break out the salt, pepper, and ketchup; well, that is, in the Owl's distinctive looking eyes. Even though this has always been the case—when he and his kind must be on constant watch for those who would consume them with or without salt, pepper, or ketchup there just doesn't seem to be anyone around—Mr. Mouse and Mr. Rabbit do not know, yet, that they are a couple of their kinds that have not gotten the message yet, but still, one never lets his guard down; no, never. Like humans, animals are creatures of habit. Routine and habit can sometimes keep you safe.

So, what is the message that Mr. Mouse and Mr. Rabbit have not gotten yet?

Well, nice of you to ask.

The truth be told, the climate in their indigenous area has changed, and is changing more each day so Mr. Mouse and Mr. Rabbit are only a couple of leftovers that have not moved on to those so-called "greener pastures."

Anyway, still alert as always, Mr. Snowshoe Hare says, "Have you noticed all the change south of here ... you know near what them humans call a city ... lately ... actually recently?"

Mr. Mouse, using one of his white feet to scratch his side says, "Yes, terrible ... I wondered at first what was going on around here ... I know all of us animals are prey for the other animals, but it is difficult if not impossible to sight any of our enemies or our old friends ... they seem to have vanished."

Mr. Hare nodded in agreement as he looked here, there, and everywhere, still on guard even though enemy and friend had departed weeks, days, and even hours ago seeking those green pastures to the north. And before Mr. Hare responded to Mr. Mouse's comments, he scanned around in the dimming light, at the shadows and the scorched Earth all around them. Then he looked at his friend and said, "Hard to believe what has happened to this place ... it is wasted ... pure waste ... all gone ... nothing to eat ... nothing to drink except for the raindrops and muddy water ... the smell of smoke is almost choking ... makes you wonder what happened? And why did it happen and where is everyone?"

Mr. Mouse listened to Mr. Hare while at the same time trying to look through the darkness at the wasteland surrounding them. Not much to see, he thought. Also, he felt afraid, scared not about predators lurking, but of the landscape itself. Maybe it was the silence. No wind blowing in the trees and boughs touching each other. Not a standing tree anywhere. Maybe it was that now there was too much. When the landscape is wide open without trees or foliage it seems so big, so scary, so terrifying, so daunting, and so forbidding. Maybe it was the lack of something to cling to, to seek cover under or within that was making him feel lost, disoriented. It scared him because he wondered what was next. Anyway, shuddering a bit but brushing himself off, he commented on Mr. Hare's comments by saying and adding: "You know, all that is true…we both are hungry and have lost weight and if it were not for the rain we would dry up, shrivel up, and shrink even more. We need to find them greener pastures, so to speak, up north where all our friends and others have gone … we need to migrate … yes, the forest has migrated up there toward what used to be great forest land for you, me, yours, and mine, don't you think?"

"What do I think? Well as those humans say, hells bells and time to ship out … for sure. Anyway, you and I seem to be the last of those lingering here in this new wasteland … that last forest fire, earthquake, and terrible storm has wiped this place dirty … it is disgusting and disturbing, for sure. We will have to move on, right now we are between a burned down forest and the promised land up north. And when we leave here and go there, we will have two worlds to enter…before all this destruction we had something to eat and to drink and places to hide … now nothing. But we will also be entering harm's way … making ourselves available to predators … as it was before this mess occurred. But if we are to have a chance to survive, we must move on … there simply is too little here for us and too many there also … that might be the crux of the problem; that is, there are too many for too little wherever we end up. Don't you agree?"

Without pausing, Mr. Mouse replies, "right on Mr. Rabbit … right on."

While, direct evidence of latitudinal shift is more limited. However, as was pointed out in Sidebar 13.1 indirect evidence is apparent in some studies in the eastern United States and in the far north at the treeline ecotones of black spruce and white spruce, or Siberian pine. A good example of indirect evidence of forest migration—migration that has a profound effect on SOC is an USDA Forest Service—Research and Development conducted by Woodall et al. (2009) and briefly described herein. In their excellent report they agree with the author's assumption that changes in tree species distributions are a potential impact of climate change on forest ecosystems. The goal of their study was to compare current geographic distributions of tree seedlings (trees with a diameter breast height ≤2.5 cm—0.98 inches) with biomass (trees with a diameter at breast height >2.5 cm) for sets of northern, southern, and general tree species in the eastern United State using a spatially balanced, region-wide forest inventory. Summarizing the researchers' findings, they used their tree seedlings' indicator to suggest that most northern study species are exhibiting a northward migration. Moreover, the researchers pointed out that the process of tree migration may continue or accelerate with a rate up to 100 km per century for numerous northern tree species.

Keep in mind that this study along with all others (including mine in the Cascades when I was studying Douglas fir stands) have and are hampered by a very important inflexible, unbending, and obstinate factor: Time with a capital T. Forests and their understory biomass, especially the trees themselves are slow growers, so to speak. Years are involved with forest migration. Under current conditions the movement of large seed plant dominated communities in geographic space takes time—not hours, days, or months but instead years. Therefore, it is accurate to state unequivocally that because of the time involved and the multitude of limitations that surround any result of this forest migration indicator of species shift is commonly labeled as indefinite, undecided. Much more research is called for. One aspect seems certain; however, forest trees, undergrowth and all associated with forest migration is a vulnerability including Mr. Hare and Mr. Mouse and friends is best described as a disturbance of SOC based on indirect evidence. Even in the face of suitable tree habitats appear to be changing; actual tree range shifts can take decades or more to detect.

DID YOU KNOW?

It takes less energy to grow forest biomass and convert it to ethanol that it takes to grow corn and convert it to ethanol. In addition, the entire process emits fewer greenhouse gases when using forest biomass.

THE BOTTOM LINE

Forest migration is real. Forest soil organic carbon is affected by forest migration. We know this and blame this on climate change but we also know that observing this is difficult simply because forest stands do not grow to any appreciable size for years, maybe decades.

REFERENCES AND RECOMMENDED READING

Batjes, H.H. (2016). Harmonized soil property values for broad-scale modelling (wise30sec) with estimates of global soil carbon stocks. *Geoderma* 269:6168.

Berryman, E. et al. (2020). Soil Carbon. Accessed 5/8/23 @ https://www.fs.usda.gov/rm/pubs-journals/2020/rmrs-2020-berrymn-eco/pdf.

Chaer, G.M., Myrold, D.D., Bottomley, P.J. (2009). A soil quality index based on the equilibrium between soil organic matter and biochemical properties of undisturbed coniferous forests soils of the Pacific Northwest. *Soil Biology and Biochemistry* 41(4):822–830.

Chen, I.C., Hill, J.K., Ohlemuller, R., Roy, D.B., Thomas, C.D. (2011). Rapid range shifts of species associated with high levels of climate warming. *Science* 333:1024–1026.

Chen, G. Lu., L., and Frierson, D.M. (2014). Phase speed spectra and the latitude of surface westerlies; interannual variability of global warm trend. *Journal of Climate* 21(22):5942–5959.

Clark, J.S., Iverson, L., Woodall, C.W. et al. (2016). The impacts of increasing drought on forest dynamics, structure, and biodiversity in the United States. *Global Change Biology* 22(70):2329–2352.

Dickens, C. (1843). *A Christmas Carol.* London: Sea Wolf Publishing.

Iverson, L. and McKenzie, D. (2014). Species Distribution and Climate Change. Climate Change Resource Center. Accessed 5/8/23 @ www.fs.usda.gov/ccrc/topics/species-distribution.

Jenny, H. (1941). *Factors of Soil Formation: A System Quantitative Pedology*. New York: McGraw-Hill.

Major, J. (1951). A functional, factorial approach to plant ecology. *Ecology* 32:392–412.

MacKenzie, M.D., DeLuca, T.H., Sala, A. (2004). Forest structure and organic horizon analysis along a fire chronosequence in the elevation forests of western Montana. *Forest Ecology and Management* 203:331–342.

Mike, M., Cheung, L., Roger, C.M., Ho, Y.L., Mak, A. (2012). Conducting a meta-analysis: Basics and good practices. *International Journal of Rheumatic Diseases* 15(2):129–135.

Parmesan, C. and Yohe, G. (2003). A globally coherent fingerprint of climate change impacts across natural systems. *Nature* 421:37–42

Phillips, R.P., Meier, I.C., Bernhardt, E.S. et al. (2012). Roots and fungi accelerate carbon and nitrogen cycling forests exposed to elevated CO_2. *Ecology Letters* 15(9):1042–1049.

Schuster, W.S.F., Griffin, K.L., Roth, H., Turnbull, M.H., Whitehead, D., Tissue, D.T. (2008). Changes in composition, structure, and aboveground biomass over seventy-six years (1930-2006) in the Black Forest, Hudson Highlands, southeastern New York State. *Tree Physiology* 28:537–549.

Woodall, C.W., Oswalt, C.M., Westfall, J.A., Perry, C.H., Nelson, M.D., Finley, A.O. (2009). Indicator of tree migration in forests of the eastern United States. *Forest Ecology and Management* 257:1434–1444.

14 Forest Carbon Status

INTRODUCTION

As we move on in this presentation recall and remember:

- **Carbon sequestration** is the removal of carbon dioxide from the air by plants.
- **Carbon storage** is the amount of carbon already bound up in the parts of woody vegetation.

These terms are redundant in this book on purpose—they are at the heart of this discussion. Truth be told, these terms are often misunderstood and are just plain confusing to many. Anyway, redundancy is good when it helps to define terms for understanding.

CARBON STATUS

Simply, and right to the point, forests are the largest terrestrial carbon sink in the world. Old-growth forests of the Pacific Northwest store more carbon per unit area than any other biome, anywhere on the globe. Much of this carbon is tied up in the huge trees like those shown in Figure 14.1.

FIGURE 14.1 Ponderosa Pine stand where the trees can grow to more than 200 in height and store large amounts of carbon. (U.S. Forest Service Public Domain Photo).

DOI: 10.1201/9781003432838-17

A Vital Army

When J.R.R. Tolkien, the author of *The Lord of the Rings*, once write, "I long to devise a setting in which the trees might really march to war." Well, the truth be told when Tolkien imagined trees marching to war, he probably couldn't have foreseen the truth of how applicable those words would one day be.

Applicable words?

Yes.

Applicable to what?

Well, those words are applicable in the sense that if we classify climate change as a battle for the globe's survival, then trees indeed will be a vital army holding the line (Androff, 2021). The trees hold carbon, while they grow and after they are harvested to build the structures of today and tomorrow. To grow a tree needs water, sunlight, minerals, and carbon dioxide. This is an important point because many ecologists and environmentalists and concerned individuals have a certain mantra whereby they envision planting trees—never stop planting trees—and this is the solution to not only the carbon storage function but also to saving the globe by increasing tree numbers to save us all.

Is this a realistic point of view? Not a bad idea but it is important to remember that, as stated earlier, trees and associated foliage need water and fertile soil to grow. Moreover, the trees also need sunlight and carbon dioxide. The point is that there are certain areas where it makes sense to plant trees and associated foliage to sequester carbon dioxide and to let the natural processes involved store carbon in their woody tissue.

Flux or Flow of Carbon Dioxide

According to Wuebbles et al. (2018), the flux or flow of carbon dioxide and other greenhouse gases into the atmosphere is the dominant contributor to the observed warming trend in global temperatures. The forest carbon cycle, however, sequesters (store) carbon dioxide from the atmosphere, accumulating substantial stores of carbon over time. Carbon uptake and storage of carbon are one of many ecosystems proceeded by forests and grasslands. As mentioned earlier, through the process of photosynthesis, growing plants remove carbon dioxide from the atmosphere stored in tin forest biomass—keep in mind that forest biomass includes not only trees but also plant stems, branches, foliage, and roots store carbon. A large portion of this carbon is eventually stored in soil. This uptake and storage of carbon from the atmosphere help modulate greenhouse gas concentrations in the atmosphere. Trees also release some carbon dioxide back into the atmosphere (e.g., emissions). This process is known as the *forest carbon cycle* (see Figure 14.2).

The forest carbon cycle begins with the sequestration and accumulation of atmospehre carbon dioxide due to tree growth. Again, to be clear and exact the accumulated carbon is stored in five different pools in the forest ecosystem: above-ground biomass (e.g., leaves, trunks, and limbs), belowground biomass (e.g., roots), deadwood, fallen leaves and stems, and soils. As trees or parts of tree

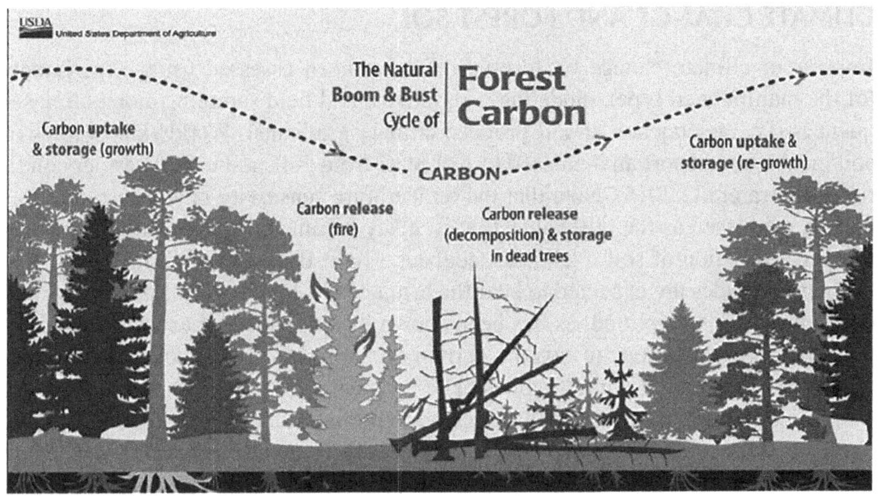

FIGURE 14.2 Forest carbon cycle. (From USDA (2019). USFS Washington, DC Office).

dies or are damaged, the carbon cycles though those different pools, specifically from the the living biomass pools to the deadwood, litter, and soil pools. The carbon cycle describes the way carbon moves between the Earth's atomsphere, hydrosphere, biosphere, and geosphere. Note that the length of time that carbon stays in each pool varies consideralby, ranging from months for litter to millennia in soil. The cycle continues as carbon flows out of the forest ecosystem and returns to the atmosphere through several process, including combustion (fire), respiration, and decomposition. Timber harvesting—where carbon is stored in harvested wood products (HWPs), while they are in use—is another way in which carbon leaves the forest ecosystem and enters the *product pool.* The carbon is stored in HWPs, while they are in use but eventually will return to the atmospehre upon the wood product's disposal and decompostion, which could take several years or more. In total, there are seven pools of forest carbon: five in the forest ecoysystem (mentioned above) and two in the HWP—one in the HWPs in use and the other HWPs in disposal sites.

DID YOU KNOW?

According to CRS (Congressional Research Service, 2022), carbon is not static; it is always moving through the pools of forested ecosystems. The size of the various pools and the rate at which carbon moves through them vary considerably over time. The amount of carbon sequestered in a forest relative to the amount of carbon released into the atmosphere is constantly changing with tree growth, death, and decomposition.

CLIMATE CHANGE AND FOREST SOC

Impacts of climate change on forest SOC have been assessed using experiments (of the manipulated type), modeling ecosystems, and field sampling along climosequences (i.e., a seuqence of soil profiels; climate gradients). Worldwide, studies of soil incubations report an increased loss of SOC from bulk soil under warmer conditions (Sierra et al., 2015). Note that the temperature sensitivity of soil respiration is usually expressed as the Q10 value, that is, a key parameter in depicting the magnitude and direction of soil C-climate feedback (Todd-Brown et al., 2013). Note that northern latitudes are expected to bear the brunt of this loss as permafrost thaws and decomposition is amplified, as has been shown by soil warming experiments. Also note that large quantities of organic carbon are stored in frozen soils (permafrost) within Arctic and sub-Arctic regions. The problem is that a warming climate can induce environmental changes that accelerate the microbial breakdown of organic carbon and the release of the greenhouse gases carbon dioxide and methane. Current evidence suggests a gradual and prolonged release of greenhouse gas emission in a warming climate and presents a research strategy to target poorly understood aspects of permafrost carbon dynamics (Schuur et al., 2015). Desipite indications of increased mineralization of SOC, coupled Earth system-climate models suggest a change in global SOC pools of −72 to +253 Pg (USGCRP, 2018). Predictions of growing SOC with warming are chiefly a result of a modeled upsurge in SOC of northern latitude soils, propelled by the effect of increased plant productivity on C inputs to soil (Genet et al., 2018). A recent study by Lynch et al. (2018) found that rapid climate warming in the Arctic threatens to undermine vast stocks of soil carbon (C) that have accrued over millennia, which could amplify the C-climate feedback. Moreover, field experiments have shown that shrub expansion in tundra, a phenomenon tied to climate change, may also promote stabilization of SOC. Note that this complex response of the soil-plant system to warming highlights the importance of a multifaceted approach to understnading climate change impacts on SOC stocks. A better approximation of a whole-ecosystem may be climosequences, giving longer-term response of SOC to climate change than shorter-term incubations. There is some indciation of reduced SOC in warmer areas compared to cooler areas (Lybrand et al., 2017; Wagai et al., 2008). Giardina et al. (2014) point out that along a tropical forest chronosequence, there was no trend in SOC storage or stabilization across a warming temperature gradient in spite of large inceases in plant-driven inputs. From these results we can reasonably assume that there is the need to support and maintain bulk density, aeration, and other soils properties that encourage aggregation and mineral-associated stabilization under a warming climate.

Plant growth may increase in productivity when carbon dioxide fertilization is effected or changed in some ecosystems (Hickler et al., 2008; Henry et al., 2005). In the past, higher productivity was thought to increase SOC stock (Harrision et al., 1993). Well, results of large-scale CO, enrichment experiments have not shown this to be the case, have not shown substantial increases in SOC as a result of carbon dioxide fertilization (Hungate et al., 1997; Schlesinger and Lichter, 2001). From these experiments it was found that increased C inputs were disproportionately partitioned into rapidly cycling, non-stabilized SOC pools. Moreover, Sulman et al. (2014)

pointed out that increases in root exudates have been shown to have a priming effect across many soils, stimulating the decay of SOC previously stabilized through mineral association.

DID YOU KNOW?

If the total amount of carbon released into the atmosphere by a given forest over a given period is greater than the amount of carbon sequestered in that forest, the forest is a *net source* of carbon emission to the atmosphere. If the forest sequesters more carbon than it releases into the atmosphere, the forest is a *net sink* of carbon.

To this point, we have discussed global warming and its trek northward, but it is also important to point out the important role of precipitation changes. Precipitation changes may impact SOC cycling in ways independent of simply increasing or decreasing average soil moisture, with effects especially pronounced in arid and semi-arid ecosystems such as deserts and grasslands (Berryman, et al., 2020). It is the shift in the timing of ranifall, rather than the amount alone, that has been shown to alter microbial activity by the way of enzyme and nutrient dynamics (Ladwig et al., 2015). A strong control over microbial mineralization of C in arid systems is based on precipiation event(s) size and time, which affect soil moisture between events (Cable et al., 2008). Moreover, Cable et al. (2008) point out that climate change predictions of the desert sourthwestern U.S. are for shifts in precipitation patterns. The impacts of climate change may be significant, because desert soil processes are strongly controlled by precipiation inputs ("pulsations") by the way of their effect on soil water availability. Additionally, moving from smaller, more frequent rainflall events to fewer, larger events can increase arid grassland productivity, and therefore increasing inputs of organic matter to soil (Thomey et al., 2011). The key point being made here is that shifts in microbial activity and plant production due to changing rainfall patterns are likely to affect SOC storage in water-limited ecosystems. The fly in the ointment here is that there is great uncertainty in projections of precipitations changes (IPCC, 2014).

THE BOTTOM LINE

The rate of carbon removal by forests from the atmosphere is influenced by several factors, including natural disturbances, forest age, management, successional pathways, climate and environmental factors, and availability of nutrients and water.

REFERENCES

Androff, A. (2021). *Trees are Climate Change, Carbon Storage Heroes*. Washington, DC: United States Forest Service-U.S. Department of Agriculture.

Berryman, E. et al. (2020). Soil Carbon. Accessed 5/8/23 @ https://www.fs.usda.gov/rm/pubs-journals/2020/rmrs-2020-berrymn-eco/pdf.

Cable, J.M., Ogle, K., Williams, D.G. et al. (2008). Soil texture drives responses of soil respiration to precipitation pulses in the Sonoran Desert: Implications for climate change.

Ecosystems 11:961–979.

CRS (Congressional Research Service) (2022). U.S. Forest Carbon Data: In Brief. Accessed 5/10/23 @ https://crsreports.congress.govR46313.

Genet, H., He, Y., Lyu, Z. et al. (2018). The role of driving factors in historical and projected carbon dynamics of upland ecosystems in Alaska. *Ecological Applications* 28(1):5–27.

Giardina, C.P., Litton, C.M., Crow, S.E, Asner, G.P. (2014). Warming-related increases in soil CO_2 efflux are explained by increased below-ground carbon flux. *Nature Climate Change* 4:822–827.

Harrison, K., Broecker, W., Bonani, G. (1993). A strategy for estimating the impact of CO_2 fertilization on soil carbon storage. *Global Biogeochemical Cycles* 7(1):69–80.

Henry, H.A., Cleland, E.E., Field, C.B., Vitousek, P.M. (2005). Interactive effects of elevated CO_2, N deposition and climate change on plant litter quality in a California annual grassland. *Occologia* 142(3):465–473.

Hickler, T., Smith, B., Prentice, I.C. et al. (2008). CO_2 fertilization in temperate FACE experiments is not representative of boreal and tropical forests. *Global Change Biology* 14(7):1531–1542.

Hungate, B.A., Holland, E.A., Jackson, R.B., Chapin III, E.S. Mooney, H.A., Filed, C.B. (1997). The fate of carbon in grasslands under carbon dioxide enrichment. *Nature* 388:576–579.

IPCC (Intergovernmental Panel on Climate Change) (2014). Climate Change 2014: Synthesis Report: Contribution of Working Groups I, II and III to the Fifth Assessment Report of the Intergovernmental Panel on Climate Change. In: Pachauri, R.K., Meyer, I.A. (eds). Core Writing Team Geneva, Switzerland, 151 p.

Ladwig, I.M., Sensibaugh, R.L., Collins, S.L., Thomey, M.L. (2015) Soil enzyme responses to varying rainfall regimes in Chihuahuan Desert soil. *Ecosphere* 6(3):1–10.

Lybrand, R.A., Heckman, K., Rasmussen, C. (2017). Soil organic carbon portioning and Δ14C variation in desert conifer ecosystems of southern Arizona. *Biogeochemistry* 134(3):261–277.

Lynch, L.M., Macmuller, M., Cotrufo, M.F. et al. (2018). Tracking the fate of fresh carbon in the Arctic tundra: Will shrub expansion alter response of soil organic matter to warming? *Soil Biology and Biochemistry* 120:134–144.

Schlesinger, W.H. and Lichter, J. (2001). Limited carbon storage in soil and litter of experimental forest plots under increased atmospheric CO_2. *Nature* 411:466–469.

Schuur, E.A.G., McGuire, A.D., Schadel, C. et al. (2015). Climate change and the permafrost carbon feedback. *Nature* 520(7546):171–179.

Sierra, C.A., Trumbore, S.E., Davidson, E.A. (2015). Sensitivity of decomposition rates of soil organic matter with respect to simultaneous changes in temperature and moisture. *Journal of Advances in Modeling Earth Systems* 7:335–356.

Sulman, B.N., Phillips, R.P., Oishi, A.C. et al. (2014). Microbe-driven turnover offsets mineral-mediated storage of soil carbon under elevated CO_2. *Nature Climate Change* 4(12):1099–1102.

Thomey, M.L., Collins, S.L., Vargas, R. et al. (2011). Effect of precipitation variability on net primary production and soil respiration in a Chihuahuan desert grassland. *Global Change Biology* 17:1505–1515.

Todd-Brown, K.E.O., Randerson, J.T., Post, W.M., Hoffman, F.M., Tarnocai, C., Schuur, E.A.G. et al. (2013). Causes of variation in soil carbon simulations from CMIPS Earth system models and comparison with observations. *Biogeosciences* 10:1717–1736.

Tolkien, J.R.R. (2021). *The Lord of the Rings Illustrated*. New York: William Morrow.

U.S. Global Change Research Program (USGCRP) (2018). Second State of Carbon Cycle Report (SOCCR2): A Sustained Assessment Report. In: Cavallaro, N., Shrestha, G., Birdsey, R. et al. (eds). U.S. Global Change Research Program, Washington, DC, 878 p.

USDA (2019). *The Forest Carbon Cycle*. Washington, DC: United States Department of Agriculture.

Wagai, R., Mayer, L.M., Kitayama, K., Knicker, H. (2008). Climate and parent material controls organic matter storage in surface soils: A three pool, density-separation approach. *Geoderma* 147(1–2):23–33.

Wuebbles, D.J. et al. (2018). *Carbon on National Forests and Grassland*. Washington, DC: United States Department of Agriculture.

15 Forest SOC Disturbance
Fire Carbon is Measured and Reported at Different Scales and Units

1 Metric Ton = 1 Ton = 1.102 Short Tons = 1000 kg = 1, 204.62 lbs

INTRODUCTION

Forest litter—to remove or not to remove? This is the pertinent question facing those who manage or live in forest areas. Forest litter is a fire hazard—it's the kindling for the blaze, so to speak. While this is true and an important factor to consider on the other side of the forest litter, issue is the knowledge that forest litter is benefical in that it is a C storage material that also serves as a habitat and species benefit while reducing wildfire risk and improving wildlife habitat. Aside from the habitat and species benefits, we also get the general benefit of removing leaf litter, woody debris, and other potential fuels from the forest floor.

DID YOU KNOW?

The potential of forest soils to sequester C depends on many biotic and abiotic variables such as: forest type, stand age and structure, root activity and turnover, temperature and moisture conditions, and soil physical, chemical, and biological properties (Page-Dumroese, 2003).

FOREST CARBON DYNAMICS

Forest carbon dynamics are driven by different anthropogenic (human-caused) and ecological disturbances. Note that the *antropogenic distrubances* are planned activiites, such as timber harvest, whereas *ecological disturbances* unplanned, such as weather events (e.g., ice storms droughts, hurricanes), insect and disease infestations, and wildfires. Usually, disturbances result in tree mortality, causing the transfer of carbon from the living pools to the deadwood, litter, soil and product pools, and/or eventually to the atmosphere. However, over time, if the disturbed forest site regenerates, the carbon releases by distrubances will be offset. Note that if the site changes to a differt type of land use (e.g., agriculture, home, and highway buiding), the carbon releases may not be offset.

DOI: 10.1201/9781003432838-18

Fire Effects on SOC

Any fire alters the amount and distriubiton of C pools in forest soils (Wells et al., 1979). Fire effects on SOC have to do with what this author calls the quantity and quality factor (Q and Q factor) of C inputs to soil, such as the forest floor, and affecting conditons that control microbial activity and access to SOC. Whether a prescribed burn or wildfire, fire has important first-order effects on SOC that are tied to fire intensity and duration of heating (Neary et al., 1999). When fire mineralizes surface OM it reduces total C pools—SOC plus forest floor—and OM inputs to SOC. Fires in peatland and other organic soils are difficult to control, and can persist for long periods of time, combusting large amounts of SOC (Reddy et al., 2015). Partial combustion of the O horizon is common in mineral soils. This is espescially the case in shrubby or forested sites with high fuel-loading near the surface (Neary et al., 1999). Experience has shown that even controlled prescribed burns can generate enough heat to consume the forest floor and reduce O horzon and SOC storage (Boerner et al., 2008; Sackett and Haase, 1992; Sanchez Meador et al., 2017).

Soil bulk density and parent material have an effect on soil heating. Actual soil heating is strongly attenuated with a depth as little as 2.5 cm in the mineral soil profile and is well buffered during light surface fires (DeBano et al., 1977). Even so, there is the potential for high loss of SOC by the way of combustion during fire given high surface SOC concentrations. When soil temperatures exceed about 150°C, soils start to lose signifcnat SOC (Araya et al., 2017; Neary et al., 1999). Earlier research studies have quantified thresholds of soil heating for loss of different chemical fractions of SOC (Gonzalez-Perez et al., 2004). Well, because of a more modern and enlightened outlook (based on research and observation over time) the SOC current model (paradigm), it is now reasoned (by many) that the loss of aggregates could be a more valuable indicator for overall SOC storage postfire. Moreover, in a soil-heating laboratory experiment, soils exposed to temperatures that would be expected in high-intensity, high-severity fires had proportionately less SOC stored in macroaggregates than soils exposed to low tempertures (Araya et al., 2017). Evidence shows that even temperatures in low- and moderate-severity burns, if long enough in duration, can degrade soil aggregates (Albalasmeh et al., 2013). More reseach is needed in this area and at the present time this is not progressing as it should because there remains a lack of knowledge, understanding of how common of burn severity, such as partly decayed organic matter on the florest floor—aka duff and—and crown consumption, relate to measurable soil effects (Kolka et al., 2014).

Note that many current studies are underway with many focused on the possible effect of higher atmospheric CO_2 levels on forest-soil C accumulated and also of increasing interest being placed, focused on management practices, such as prescribed burns, harvesting, site preparation, reforestation, drainage, and fertilization, have on soil C pools and cycling (Page-Dumrose et al., 2003).

The first-order loss of SOC because of combustion may be partially offset by creation of heat-altered C such as soot, charcoal, or biochar, collectively known as pyrogenic C (pyC). PyC is produced by incomplete combustion of organic matter and is persuasive throughout the atmosphere as well as in soils and sediments. Note, however, that it is difficult, in most cases, to detect significant increases in pyC from only one fire. The main problem is complexity; factors controlling pyC formation and accumulations are complicated, intricate and multifarious, to say the least. Also, pyC formation varies by soil type, climate, and ecosystem (Czimczik and Masiello, 2007). Boot et al. (2015) report that in a fireprone ponderosa pine forest in Colorado, there was no difference in pyC content in the soil between recently burned and unburned areas, implying a role for legacy of past burns or erosion, or both, in the present day content of pyC. In regard to the same wildfire, postfire erosion and sedimentation were found to be an important control over spatial distribution of pyC (Cotrufo et al., 2016). DeLuca and Aplet (2008) making an educated guess-estimate that charcoal may account for 15%–20% of the total C in temperate, coniferous forest mineral soils and that some forest management practices (e.g., salvage logging, thinning) may reduce soil pyC content and long-term storage.

DID YOU KNOW?

A recent study of the possible effect of higher atmospheric CO_2 levels on forest-soil C accumulations demonstrated that increased tree growth, changing internal C allocations, and alteration of climate temperature, and precipitation patterns were likely involved (e.g., Caspersen et al., 2000).

Second-order impacts of fire on the O horizon and SOC may take longer to become apparent. These effects result from fires's direct impact on soil microbial biomass, soils chemical characteristics, and plant productivity. After severe, high-intensity fire, the soil microbial community can shift in composition and size, which can impact microbial SOC transformations (Knelman et al., 2017; Prieto-Fernandez et al., 1998).

Note that many wildfires only cause minimal damage to the land and pose few threats to the land or people downstream; however some fires require special effort to prevent probems afterwards. Loss of vegetation exposes soil to erosion; water runoff may increase and cause flooding; sediments may move downstream and damage houses of fill reservoirs putting endangered species and community water supplies at risk.

Fire-induced increases at soil pH and initial increase in N availability can also affect microbial activity and mineralizaton of the O horizon and SOC (Gonzalez-Perez et al., 2004; Hanan et al., 2016; Kurth et al., 2014; Raison, 1979).

After a forest fire, the first priority is emergency stabilization in order to prevent further damage to life, property, or natural resouces. This stablization period may continue for up to a year or longer. The stablization effort focuses on those lands unlikly to recover naturally, on their own, from wildland fire damage. Stablization efforts are made more difficutl due to projected climate change which may increase the incidence and severity of widlfires in some forest regions—this interference could have a major impact on the soil C pools (Berryman, 2020).

Note: The 411 on Meta-Analysis

Before moving on with this discussion it is important to explain what a meta-analysis is and what *it is all about. Simply, a meta-analysis is a statisitical analysis thte combines the results of multiple* scientific studies. It is all about addressing the same question with multiple participants, various studies, and results (findings). What happens in meta-analysis is that multiple participants report their individual findings or measurements all of which are normally expected to have some sort of error or errors. The aim then is to use approaches from statistics to derive a pooled (key word is "pooled") estimate closest (or thought to be) to the unkown common truth based on how any errors are perceived. Note that meta-analytic results are considered the most trustworthy source of evidence by the evidence-based medicine literature (Herrera et al., 2022; CEBM, 2009; Nice, 2014).

Formal guidance for the conducting and reporting on of meta-anayses is provided by the Cochrane Handbook. *The Cochrane Handbook* contains essential information (guidance) for preparing and maintaining Cochrane Reviews of the effects of health interventions. Cochrane Reviews are systematic reveiws carried out by an international network of researchers with almost 30,000 contributors from several countries.

Forest fire can cause a darkening of the soil surface and therefore reduce surface albedo which, in turn, can raise soil temperatures, increasing SOC mineralization rates. Loss of vegetation reduces the forest floor. The bare floor surface is left vulnerable to erosion, exposing deeper SOC for decomposition and loss of CO_2 (Berryman et al., 2020). It is still unclear, however, whether postfire erosion could increase SOC sedimentation enough to counter CO_2 losses (Cortrufo et al., 2016; Doetterl et al., 2016). During a meta-analysis it was found that 10 years postfire, SOC increased across various forested sites, which could be attributed to a combination of ancillary effects and pyC reaction (Johnson and Curtis, 2001). In those locations exposed to repeated burning, these ancillary or secondary effects collect over decades, with net effects on SOC that may vary by ecosystem type. Pellegrini et al. (2017) point out that a recent meta-analysis found an overall increase in SOC in frequently burned grasslands. The thinking here is that these changes are largely tied to effects of fire on nutrient availability and plant productivity. Exceptionally large amounts of C may be lost due to indirect effects of wildfire. Analysis has shown that in permafrost soils, wildfire increases the active layer depth, ultimatly leading to increased C loss as carbon dioxide in the long term (Zhang et al., 2015a, b).

DID YOU KNOW?

Observation and experience have shown that only a portion of the burned area receives emergency stabilization measures. In the severely burned locations, areas where water runoff will be excessive, or steep slopes above valuable facilities are the focus areas. Note that whatever treatment is employed must be put into action as soon as possible, to hedge off possible storm damage.

Pyrogenic C (i.e., char, biochar, black carbon) may not exacly follow the emerging model of SOC stabilization. Pyrogenic C is the product of incomplete combustion of OM and fossil fuels along a range of increasing alteration relative to its original OM from char to soot. It has been found that pyC can persist in soils and sediments for centuries to millenia and so thought to be resistant to ruin. Note that pyC affects many factors important for SOC stabilization. Liang et al. (2006) point out, for example, that it increases cation exchange capacity, promotes water and nutrient retention, and reduces soil bulk density, encouraging microbial activity. A heavy utilizer of pyC is ectomycorrhizal fungi (Tedersoo et al., 2010). It's probably pyC's complex molecular structure with condensed aromatic rings that enables it to resist degradation.

How does pyC resist degradation?

Brodowski et al. (2006) and Wagai et al. (2009) suggest that because pyC is often found in association with mineral surfaces and within soil aggregates it may promote these stabilization processes, which in turn allow it to resist decomposition. One thing is certain (at the present time); we are lacking nationwide estimates or pyC in soil, although a recent global analysis estimated that pyC srepresents about 14% of total SOC (Reisser et al., 2016).

DID YOU KNOW?

The jury is still out about climate change effects on SOC; they remain highly uncertain. To the point, we need research to better understand how expected increases in precipitation and temperature variability—rather than changes in the mean—impact SOC vulnerability (Berryman et al., 2020).

Management and SOC

To understand the role that management can play in fire-SOC dynamics, we can simply look at the role or reported effects linked to fire behavior, which depends on fuel-loading, weather, and topography. For instance, slash piles can generate extremely high temperatures when burned. Note that the U.S. Forest service burens piles of slash (woody debris) in an effort of reducing hazaroud fuel—part of prescribed burning techniques. These piles are made from the slash left after mechanical thinnng or cutting trees in the forsest.

With regard to temperatures in slash burns they are higher than typical broadcast burns and lead to chemcial and microbial communtity transformations with potential feedbacks to the SOC process (Esquilin et al., 2007; Massman, 2012). Note that broadcast burns are controlled applications of fire to fuels, under specified conditions that allow fire to be confined to the predetermined area, and produce the fire behavior and characteristics required to meet forest health objectives. Korb et al. (2004) reported that slash piles in ponderosa pine forest in Arizon were found to have lower SOC in the top 15 cm of mineral soil 7 months after burning. However, Frandsen and Ryan (1986) report that high soil moisture and reduced bulk density near the surface can decrease surface heating.

The creation of heat-altered C such as soot, charcoal, or biochar—together known as pyC—this works to offset the first-order loss of SOC. Only one fire, however, is usually not enough to register and to detect significant inceases in pyC. It's all about complexity (likely to vary by soil type, climate, and ecosystem); that is, factors controlling pyC formation and accumlation (Czimczik and Masiello, 2007).

THE BOTTOM LINE

Any forest fire alters the amount and distribution of C pools in forest soil (Wells et al., 1979). Non-prescribed forest fires, wildfires, are a serious problem. Many of our ecosystems, especially those in the western United States, are overloaded with surface fuels that have accumualted over the years from fire suppression. This type of stand condition is conducive to wildfire and can trigger catastrophic changes in soil productivity if fire severity is high (Sands, 1983; Harvey et al., 1999). Positive effects of pyC are carbon sequestration, carbon dioxide capture, improving aggregate stability and easier and water capacity, stabilization of contaminants, sorption or removal of pollutants, and catalysts for microbes. Well, with the good there usually comes some bad or negative effects of just about everything and this is the case with pyC including: toxicity to plants, microbes, soil animals, and invertebrates; they are carriers of contaminants; carbon fluxes; heavy metals; alter soil pH; and, release organic chemicals (Berryman et al., 2020). Keep in mind that a majority of a forest stand's C storage likely occurs above the ground or in the deeper mineral soil horizons (Grigal and Ohmann, 1992; Tilman et al., 2000). The point is that changes in mineral soil C (or lack of change) *may not* be an indicator of total-size C losses, because mostly C loss from fire occurs in the forest floor material.

REFERENCES AND RECOMMENDED READING

Albalasmeh, A.A, Berli, M., Shafer, D.S., Ghezzehei, T. (2013). Degradation of moist soil aggregates by rapid temperature rise under low intensity fire. *Plant Soil* 362(1–2):335–344.

Araya, S.N., Fogel, M.L., Berhe, A.A. (2017). Thermal alteration of soil organic matter properties: A systematic study to infer response of Sierra Nevada climosequence soils to forest fires. *Soil* 3:31–44

Berryman, E. et al. (2020). Soil Carbon. Accessed 5/8/23 @ https://www.fs.usda.gov/rm/pubs-journals/2020/rmrs-2020-berrymn-eco/pdf.

Boerner, R.E.J., Huang, J., Hart, S.C. (2008). Fire, thinning, and the carbon economy: Effects of fire surrogate treatments on estimated carbon storage and sequestration rate. *Forest Ecology and Management* 255:3081–3097.

Boot, C.M., Haddix, M., Paustian, K., Cotrufo, M.F. (2015). Distribution of black carbon in ponderosa pine forest floors and soils following the High Park wildfire. *Biogeosciences* 12(10):3029–3039.

Brodowski, S., John, B., Flessa, H., Amelung, W. (2006). Aggregate-occluded black carbon in soil. *European Journal of Soil Science* 57:539–546.

Caspersen, J.P. et al. (2000). Contributions of land-use history to carbon accumulation in U.S. forests. *Science* 290:1148–1151.

CEBM (2009). Levels of Evidence. Center for Evidence-Based Medicine. University of Oxford. Accessed 5/29/23 @ https://www.cebm.ox.ac.uk/resources/levles-of-evidnece/oxford-centre-for-evidnece=based=medican-levels-of evidence-march-2009).

Cotrufo, M.F., Boot, C.M., Kampf, S. et al. (2016). Redistribution of pyrogenic carbon from hillslopes to steam corridoes following a large montane wildfire. *Global Biogeochemical Cycles* 30(9):1348–1355.

Czimczik, C.I. and Masiello, C.A. (2007). Controls on black carbon storage in soils. *Global Biogeochemical Cycles* 21(3). doi 10.1029/2006GB002798.

DeBano, L.F., Dunn, P.H., Conrad, C.E. (1977). Fire's Effect on Physical and Chemical Properties of Chaparral Soils. Research Paper PSW: RP-145. U.S. Department of Agriculture Forest Service, Pacific Southwest Forest and Range Experiment Station, Berkeley, 21 p.

DeLuca, T.H. and Aplet, G.H. (2008). Charcoal and carbon storage in forest soils of the Rocky Mountain west. *Frontiers in Ecology and the Environment* 6:18–24.

Doetterl, S., Bethe, A.A., Nadeu, E. et al. (2016). Erosion, deposition, and soil carbon: A review of process-level controls, experiment tools and models to address C cycling in dynamic landscapes. *Earth-Science Reviews* 154:102–122.

Esquilin, A.E.J., Stromberger, M.E., Massman, W.J. et al. (2007). Microbial community structure and activity in Colorado Rocky Mountain forest soil scarred by slash pile burning. *Soil Biology and Biochemistry* 39:1111–1120.

Frandsen, W.H. and Ryan, K.C. (1986). Soil moisture reduces belowground heat flux and soil temperatures under a burning fuel pile. *Canadian Journal of Forest Research* 16:244–248.

Gonzalez-Perez, J.A., Gonzalez-Vila, F.J., Almendros, G., Knicker, H. (2004). The effect of fire on soil organic matter: A review. *Environment International* 30(6):855–870.

Grigal, D.F. and Ohmann, L.F. (1992). Carbon storage in upland forests of the Lake States. *Soil Science Society of America Journal* 56:273–296.

Hanan, E.J., Schimel, J.P., Dowdy, K., D'Antonio, C.M. (2016). Effects of substrate supply, pH, and carbon net nitrogen mineralization and nitrification along a wildfire-structured age gradient in chaparral. *Soil Biology and Biochemistry* 95:87–99.

Harvey, A.E., Graham, R.R., McDonald, G.I. (1999). Tree Species Composition Changes-Soil Organism Interaction: Potential Effect on Nutrient Cycling and Conservation in Interior Forest. In Proceedings: Pacific Northwest Forest and Rangeland Soil Organism Symposium: Organism Functions and Processes, Management Effects on Organisms and Processes, and Role of Soil Organisms in Restoration, General Technical Report PNW-GTR-461, USDA Forest Service, Pacific Northwest Research Station, Portland, OR, pp. 137–145.

Herrera Ortiz, A.F., Camacho, E.C., Rojas, J.C., Camacho, T.C., Guevara, S.Z., Cuenca, N.T.R., Perdomo, A.V., Herazo, V.D.C., Malo, R.G. (2022). Practical guide to perform a systematic literature review and meta-analysis. *Principles and Practices of Clinical Research* 7(4):47–57.

Johnson, D.W. and Curtis, P.S. (2001). Effects of forest management on soil C and N storage: Meta-analysis. *Forest Ecology and Management* 140(2–3):227–238.

Knelman, F.E., Graham, E.B., Ferrenberg, S. et al. (2017). Rapid shifts in soil nutrients and decomposition enzyme activity in early succession following forest fire. *Forests* 8(9):347.

Kolka, R., Sturtevant, B., Townsend, T. et al. (2014). Post-fire comparisons of forest floor and soil carbon, nitrogen, and mercury pools with fire severity indices. *Soil Science Society of America Journal* 78:558–565.

Korb, J.E., Johnson, N.C. Covington, W.W. (2004). Slash pile burning effects on soil biotic and chemical properties and plant establishment. *Restoration Ecology* 12(1):52–62.

Kurth, V.J., Hart, S.C., Ross, C.S. et al. (2014). Stand replacing wildfires increase nitrification for decades in southwestern ponderosa pine forests. *Oecologia* 175:395–407.

Liang, B., Lehmann, J., Solomon, D., Kinyangi, J., Grossman, J., O'Neil, Skjemstad, B.J., Thies, J., Luizao, F.J., Peterson, J., Neves, E.G. (2006). Black carbon increases cation exchange capacity in soils. *Soil Science Society of American Journal* 70(5): 1719–1730

Massman, W.J. (2012). Modeling soil heating and moisture transport under extreme conditions: Forest fires and slash pile burns. *Water Resources Research*. 48:10548.

Neary, D.G., Klopatek, C.C., DeBano, L.F., Follion, P.F. (1999). Fire effects on belowground sustainability: A review and synthesis. *Forest Ecology and Management* 122:51–71.

Nice (2014). Developing NICE Guidelines: The Manual. Accessed 5/29/23 @ https://www.nice.org.uk/media/default/aboutwhat-we-do/our-programmes/devleoping-nice-guidelines-the-manual.pdf.

Page-Dumroese, D.S., Brown, R.E., Jurgensen, M.F., Mroz, G.D. (2003). Comparison of methods for determining bulk densities of rocky forest soils. *Soil Science Society of America Journal* 63(2):379–383.

Pellegrini, A.F.A., Ahlstrom, A., Hobbie, S.E. et al. (2017). Fire frequency drives decadal changes in soil carbon and nitrogen and ecosystem productivity. *Nature* 553:194–198.

Prieto-Fernandez, A., Acea, M.J., Carballas, T. (1998). Soil microbial and extractable C and N after wildfire. *Biology and Fertility of Soils* 27(2):132–142.

Raison, K.J. (1979). Modification of the soil environment by vegetation fires, with reference to nitrogen transformations: A review. *Plant* and *Soil* 51:73–108.

Reddy, A.D., Hawbaker, T.I., Wurster, F. et al. (2015). Quantifying soil carbon loss and uncertainty from a peatland wildfire using multitemporal LiDAR. *Remote Sensing of Environment* 170:306–316.

Reisser, M., Purves, R.S. Schmidt, M.W.I., Abiven, S. (2016). Pyrogenic carbon in soils: A literature-based inventory and a global estimation of its content in soil organic carbon and stocks. *Frontiers in Earth Science* 4:80.

Sackett, S.S. and Haase, S.M. (1992). Measuring Soil and Tree Temperatures During Prescribed Fires with Thermocouple Probes. General Technical Report PSW-GTR 131. U.S. Department of Agriculture. Forest Service, Pacific Southwest Research Station, Berkeley, 15 p.

Sanchez Meador, A., Springer, J.D., Huffman, D.W. et al. (2017). Soil functional responses to ecological restoration treatments in frequent-fire forests of the western United States: A systematic review. *Restoration Ecology* 25(4):497–508.

Sands, J. (1983). Modification of the soil environment in vegetative fire. *Plant and Soils* 51:73–108.

Smith, R. (1983). Physical Changes to Sandy Soils Planted in Radiata Pine. In: Ballard, R., Gessel, S.P. (eds) IUFRO Symposium on Forest site and Continuous Productivity, General Techncal Report. PNW-GTR-163, USDA Forest Service, Pacific Northwest Station Portland, OR.

Tedersoo, L., May, T.W., Smith, M.E. (2020). Ectomycorrhizal lifestyle in fungi: Global diversity, distribution, and evolution of phylogenetic lineages. *Mycorrhiza* 20:217–263.

Tilman, D. et al. (2000). Fire suppression and ecosystem carbon storage. *Ecology* 81:2680–2685.

Tolkien, J.R.R. (2021). The *Lord of the Rings Illustrated.* New York: William Morrow.

USDA (2019). *The Forest Carbon Cycle*. Washington, DC: United States Department of Agriculture.

USDA (2023). *After Fire*. Washington, DC: United States Department of Agriculture.

Wagai, R., Mayer, L.M., Kitayama, K. (2009). Nature of the "occluded" low-density fraction in soil organic matter studies: A critical review. *Soil Science and Plant Nutrition* 55(1):13–25.

Wuebbles, D.J. et al. (2018). *Carbon on National Forests and Grassland*. Washington, DC: United States Department of Agriculture.

Wells, C.G. et al. (1979). *Effects of Fire on Soil: A State-of review*. WO-GTR-7. Washington, D.C.: USDA Forest Service.

Zhang, Y., Wolfe, S.A., Morse, P.D. et al. (2015a). Spatiotemporal impacts of wildfire and climate warming on permafrost across a subarctic region, Canada. *Journal of Geophysical Research: Earth Surface* 120(11):2338–2356.

Zhang, B., Zhou, X., Zhou, L., Ju, R. (2015b). A global synthesis of belowground carbon responses to biotic disturbance: A meta-analysis. *Global Ecology and Biogeography* 24(2):126–138.

16 Forest Disturbance
Harvesting and Thinning

INTRODUCTION

To this point we have discussed how important forests are to sequestering and storing C. Earlier it was pointed out that forest carbon stocks are closely tied to forest biomass, so factors that increase tree growth substantially will subsequently increase rates of carbon storage within forests (McKinley et al., 2011; Ryan et al., 2010). Nitrogen deposition, for example, from industrial and agricultural activities has increased soil nitrogen availability, allowing trees to increase carbon capture and contribute to the capacity of existing forests to sequester carbon more effectively (Nadelhoffer et al., 1999). The point here is that forest operations, such as harvesting and thinning, alter SOC by reducing C input quality by means of forest flow and root OM inputs as the stand regenerates. Moreover, microbial accessibility and activity are altered through the disturbance of the soil surface, which changes temperature and moisture regimes.

Keep in mind that higher atmospheric carbon dioxide levels, changes to patterns of temperature and rainfall, and changing forest management strategies all contribute to higher rates of carbon dioxide storage in existing forest (Ainsworth and Rogers, 2007; Cole et al., 2010). However, there is some doubt surrounding the ability of forests to continue to incorporate additional carbon dioxide with additional changes in the climate (Bellassen and Luyssaert, 2014).

Now we know, based on research and experience, that forest management activities can be used to increase the amount of carbon that is sequestered in forests, as well as the amount of carbon stored in wood products (Ryan et al., 2010). We also know that the amount of additional carbon that can be sequestered depends greatly upon the condition of the forest (e.g., forest type, age, health) and the forest management practice in question, making it important to consider change to carbon stocks across the entire system to assess trade-offs between different pools.

FOREST HARVESTING

Note that because of a result of high heterogeneity in SOC, it can be exceedingly difficult to detect changes because of forest harvesting in any specific study. The results of many individual experiments are integrated in meta-analyses and can be used to recognize changes that are largely steadfast across studies, even when heterogenicity masks treatment effects within a single study. A few studies including meta-analyses and review articles conclude that the effect of harvest is a reduction in SOC, depending on soil and forest type determining the magnitude of C loss (Jandl et al., 2007; Johnson and Curtis, 2001; Nave et al., 2010). Nave et al. (2010) described an 8% average drop in SOC stocks after harvesting over all forest and soil types studied.

DOI: 10.1201/9781003432838-19

And thinking has been somewhat reinforced concerning the impact of whole-tree harvesting; it has been theorized by a few that with the roots remaining in the soil there would not be a significant or total loss of SOC stocks if the O horizons are left undisturbed (Jang et al., 2016; Powers et al., 2005)—to date, this appears to be the case. Experience indicates that postharvest reductions in SOC occur because of disturbance that occurs during harvesting (Achat et al., 2015a, b; James and Harrison, 2016). Note, however, advancements in the understanding of how harvesting impacts below-ground processes are difficult because most studies focus on the first 30 cm of the soil profile or even just the forest floor. Experts point out that valid estimation of changes in ecosystem C is not valid unless sampling is conducted at 20 cm (approximately 7.9 inches) or more (Goldstein et al., 2020).

The bottom line: The depth of soil sampling and testing influences SOC findings.

THINNING FOREST STANDS

Thinning forest stands is conducted to benefit natural resources and to protect communities. However, there are potential downsides to thinning forest stands that must be taken into consideration; thus thinning should be conducted by following guidance by law and policy and grounded in real science—and never grounded in feel-good science.

Anyway, thinning forest stands, as opposed to harvesting for timber, has become a common practice to achieve various silviculture objectives. Regarding SOC and thinning effects, depending on residue management, they are variable. Residues of the thinning process include the principal parts of the cut trees: bole, branches, leaves, and the related roots that have been detached. All these components have a path toward decomposition, which is, in general, hastened due to physical disturbance. Varying the timing according to Schaedel et al. (2017) or intensity according to D'Amore et al. (2015) of thinning may mitigate C losses. Experience has shown that thinning and competition control in forest stands have a much smaller impact on soil characteristics and therefore affect SOC stocks less than forest biomass harvesting operations (Berryman et al., 2020). Intensely managed forests may provide the benefit of C sequestration or could lead to a release of C to the atmosphere (Harmon et al., 1990). Moreover, herbicide application to improve seedling growth is double-edged sword: on positive edge is the actual improvement in seedling growth which has a positive, but on the negative edge is the damage to below-ground C (Markewitz, 2006).

Allowing woody biomass harvesting to help meet the Nation's demand for alternative energy sources is questionable because we do not know what we do not know about the short- and long-term effects of such removals.

What we do know (I hope) is that a readily available source of biomass material is the slash from thinning operations in the forest: woody debris such as deadfalls, leaf falls, branch falls, and assorted twigs and do not forget creatures and microbes (important ones) which take up a lot of space within a forest stand.

And when removed … what happens next?

Well, that is the question, for sure. There exists two-sides of opinion on this problem. On one side the point is made that forest fires can be bad news and failure to remove understory brush and the rest is a real mistake—it is simply allowing the buildup of very combustible tinder, brushwood, sticks, twigs, firewood, bare trunks, stumps, and ashen tombstones, kindling in the forest … all marking a once thriving forest stand. Then there is the other side, the ecologist and environmentalist point of view, and they might take the notion that forest understory brush and woody debris are a plus—a huge plus; a nature-made storage house of C—and to be let be.

Note that biomass harvesting and removal of woody residues by burning or for bioenergy are, according to Berryman et al. (2020), a major concern in many forest ecosystems because of the potential unfavorable effects on productivity (Janowiak and Webster, 2010), Ectomycorrhizae (Harvey et al., 1976), long-term nutrient cycling (Harmon et al., 1994), soil moisture content (Maser et al., 1988), N_2 fixation (Jurgensen et al., 1987), and rejuvenation success (Schreiner et al., 1996). Another important point overlooked to this point in this discussion is that woody debris is important to a critical element of biodiversity and nutrient cycling. Harvey et al. (1981) noted that harvesting has the potential to disturb soils and reduce the amount of woody residues, particularly in dry forest types. But there is another side or view to this detail because several studies have shown that coarse wood retention has very little effect on SOC or nutrients, possibly because all soils have been influenced by coarse wood at some time (Luo and Xu, 2022). Then again, coarse wood serves as a C storage pool, creates wildlife forage areas, boosts fungal diversity, supports erosion control, and increases moisture retention. In addition, if soils have very low buffering capacity due to soil parent chemistry and historical impacts of atmospheric deposition, biomass harvesting reduces the total amount of nutrients left on-site. On sites, that are extremely nutrient limited, are particularly marked by long-term anthropogenic acidification, overgrazing), wildfire, or excess OM piling and burning. Jang et al. (2016) point out that understanding inherent soil chemistry and composition, resilience to nutrient losses, and ecosystem dynamics dependent on nutrient cycling throughout a rotation or longer are necessary for assessing long-term sustainability. Note that nutrient cycling refers to the repeated pathway of nutrients or elements from the environment through one or more organisms back to the environment. These nutrient cycles include the nitrogen cycle, carbon cycle, phosphorous cycle, and so on.

THE BOTTOM LINE—THE FOREST PARADOX

This chapter has pointed out that there exists a Forest Paradox—two different views as to what to do with woody debris, leaves, branches, deadfalls, and so forth? So what we are talking about here is a puzzle, really an inconsistency between the two views. The first view is that we need to leave the understory woody debris and other forms of biomass as they are. They are storage sites of C and important to all of us. The second view is that all the forest understory and woody debris are kindling for the next forest fire.

The question is who is right? … and herein lies the puzzle … the paradox.

REFERENCES

Achat, D.L., Fortin, M., Landmann, G. et al. (2015a). Forest soil carbon is threatened by intensive bioharvesting. *Scientific Reports* 5(1):15991.

Achat, D.L., Deleuze, C., Landmann, G. et al. (2015b). Quantifying consequences of removing harvesting residues on forest floor soils and tree growth-a meta-analysis. *Forest Ecology and Management* 348:124–141.

Ainsworth, E.A. and Rogers, A. (2007). The response of photosynthesis and stomatal conductance to rising [CO_2]: Mechanism and environmental interactions. *Plant, Cell & Environment* 30(3):258–270.

Bellassen, V. and Luyssaert, S. (2014). Carbon sequestration: Managing forests in uncertain times. *Nature* 506(7487):153–166.

Berryman, E. et al. (2020). Soil Carbon. Accessed 5/8/23 @ https://www.fs.usda.gov/rm/pubs-journals/2020/rmrs-2020-berrymn-eco/pdf.

Cole, C.T., Anderson, J.E., Lindroth, R.L., Waller, D.M. (2010). Rising concentrations of atmospheric CO_2 have increased growth in natural stands of quaking aspen (Populus tremuloides). *Global Change Biology* 16(8):2186–2197.

D'Amore, D.V., Oken, K.L., Herendeen, P.A. et al. (2015). Carbon accretion in unthinned and thinned young-of the Alaskan prehumid coastal temperate rainforest. *Carbon Balance and Mangement* 10(1):25.

Goldstein, A., Turner, W.R., Spawn, S.A., Anderson-Teixeira, K.J., Cook-Patton, S., Fargione, J., Gibbs, H.K., Griscom, B., Hewson, J.H., Howard, J.F., Ledezma, J.C., Page, S., Koh, L.P., Rockstrom, J., Sanderman, J., Hole, D.G. (2020). Protecting irrecoverable caron in Earth's ecosystems. *Nature Climate Change* 10:287–295.

Harmon, M.F., Ferrell, W.K., Franklin, J.F. (1990). Effects of carbon storage on conversion of old-growth forests to young forests. *Science* 247:699–702.

Harmon, M.E., Sexton, J., Caldwell, B.A., Carpenter, S.E. (1994). Fungal sporocarp mediated losses of Ca, Fe, K, Mg, Mn, N, P, and Zn from conifer logs in the early stages of decomposition. *Canadian Journal of Forest Research* 24(9):1883–1893.

Harvey, A.E., Larsen, M.I., Jurgensen, M.F. (1976). Distribution of ectomycorrhizae in a mature Douglas-fir/larch forest soil in western Montana. *Forest Science* 22(4):393–398.

Harvey, A.E., Larsen, M.I., Jurgensen, M.F. (1981). Rate of Woody Residue Incorporation into Northern Rocky Montana Forest Soils. Research Paper INT-RP-282. U.S. Department of Agriculture, Forest Services Intermountain Forest and Range Experiment station, Ogden Utah, 5 p.

James, J. and Harrison, R. (2016). The effect of harvest on forest soil carbon a meta-analysis. Forests 7(12):308.

Jandl, R., Lindner, M., Vesterdal, L. (2007). How strongly can forest management influence soil carbon sequestration? *Geoderma* 137(3):253–268.

Jang, W., Page-Dumroese, D.S., Keyes, C.R. (2016). Long-term changes from forest harvesting and residue management in the northern Rocky Mountains. *Soil Science Society of America Journal* 80(3): 727–741.

Janowiak, M.K. and Webster, C.R. (2010). Promoting ecological sustainability in woody biomass harvesting. *Journal of Forestry* 108(1):16–23.

Johnson, D.W. and Curtis, P.S. (2001). Effects of forest management on soil C and N storage: Meta-analysis. *Forest Ecology and Management* 140(2–3):227–238.

Jurgensen, M.F., Larsen, M.J., Graham, R.T., Harvey, A.E. (1987). Nitrogen fixation in woody residue of northern Rocky Mountain conifer forest. *Canadian Journal of Forest Research* 17(10):1283–1288.

Luo, Y. and Xu, J. (2022). *Soil Organic Matter Dynamics*. Amsterdam, Netherlands: Elsevier.

Markewitz, D. (2006). Fossil fuel carbon emissions from silviculture impact on net carbon sequestration in forests. *Forest Ecology and Management* 236:153–161.

Maser, C., Cline, G.F., Cromack, K. Jr., et al. (1988). What We Know About Large Trees That Fall to the Forest Floor. In: Maser, C., Tarrant, R.F., Trappe, J.M., Franklin, J.F. (eds) From the Forest to the Sea: A Story of Fallen Trees. General Technical Report PNW-GTR-229, U.S. Department of Agriculture: Forest Service, Pacific Northwest Research Station, Portland Oregon, pp. 25–46.

McKinley, D.C., Ryan, M.G., Birdsey, R.A., Giardina, C.P., Harmon, M.E., Heath, L.S., Houghton, R.A., Jackson, R.B., Morrison, J.F., Murray, B.C. (2011). A synthesis of current knowledge on forest and carbon storage in the United States. *Ecological Applications* 21(6): 1902–1924.

Nadelhoffer, K.J., Emmett, B.A., Gunderson, P., Kjonaas, O.J., Koopmans, C.J., Schleppi, P., Tietema, A., Wright, R.F. (1999). Nitrogen deposition makes a minor contribution to carbon sequestration in temperate forests. *Nature* 398:145–148.

Nave, L.E., Vance, E.D., Swanston, C.W., Curtis, P.S. (2010). Harvest impacts on soil carbon storage in temperate forests. *Forest Ecology and Management* 259:857–866.

Powers, R.R., Scott, D.A., Sanchez, F.G. (2005). The North American long-term soil productivity experiment: Findings from the decade of research. *Forest Ecology and Management* 220(1):31–50.

Ryan, M.G., Harmon, M.E., Birdsey, R.A., Giardina, C.P., Heath, L.S., Houghton, A., Jackson, R.B., McKinley, D.C., Morrison, J.F., Murray, B.C. (2010). A synthesis of the science on forests and carbon for US forests. *Ecological Applications* 21(6):1902–1924.

Schaedel, M.S., Larson, A.J., Affleck, D.L. et al. (2017). Early forest mining changes aboveground carbon distribution among pools, but not total amount. *Forest Ecology and Management* 389:187–198.

Schreiner, F.G., Krueger, K.A., Houston, D.B., Happe, P.J. (1996). Understory patch dynamics and ungulate herbivory in old-growth forests of Olympic National Park, Washington. *Canadian Journal of Forest Research* 26(2):255–265.

17 Forest Disturbance
Ungulate Herbivory

INTRODUCTION

Oh, not so long ago …
It really was just yesterday
measured on the geologic
timeline when we used to
remove all dead material
from the forest
to sweep it clean
and reduce fire threat
But we got smart:
dead trees are
vital for the forest
and take many different forms.
They can free fall
and become part of the forest floor
or remain standing,
becoming what is known as a
snag … a standing dart and certain
critter's delight.
Tree snags are good … but there is more.
The trees that fall to the forest floor
are important … they help keep the soil moist
by soaking up precipitation
as well as serving as a shelter
for many forms of life … such as
insects, amphibians, mammals
and macroinvertebrates/
microinvertebrate species
in the forest. They continually replenish
the soil by slowly releasing nutrients.
One of the most important roles
for these fallen forest trees and their limbs
is serving as a nursey for young seedlings.
As the logs and limbs lie on the ground—
the herbaceous layer—they serve as
the forest's carpet; that is,

DOI: 10.1201/9781003432838-20

leaves and cones accumulate on top,
slowly decomposing and turning into soil.
Then it is the seeds that fall into the fertile soil,
growing into young seedlings.
And don't forget understory growth and
woody debris; they are storehouses of
carbon and that is what this book is about.
The bottom line: Trees (along with plants)
are the respiratory system of the Earth,
shaping the atmosphere which sustains us today.

F.R. Spellman (2011)

Seedlings, understory growth (aka the forests furnishings), woody debris, grubs, and seeds are all important to the forest and especially to those roaming or grazing the forest who might be hungry. Forest areas have always been attractive for wild species including wild turkey, deer, quail, and assorted songbirds. A common practice today and surely in the future is to integrate trees, livestock, and forage into a single system on one site. Agroforestry practice, or silvopasture, is the turning livestock in the woods. This practice provides ungulate (i.e., animals with hooves) herbivores (i.e., plant eaters like cattle (e.g., cows), goats, or sheep shade from the trees; this shady condition lengthens the forage growing season and improves forage quality and provides cattle with improved comfort and reduces stress. Other potential livestock choices include horses, turkeys, chickens, ostriches, emu, or game animals such as bison, deer, elk, and caribou. The forage and browse components can include shrubs, grass, legumes, and forbs. The final silvopasture component is the soil and its fertility, appropriate pH, and well-developed structure. Other important ingredients include proper drainage and erosion control.

Note that the practice of turning cattle out to the woods to forage sounds is like a simple enough practice and in some cases it is simple whereby cattle are just turned loose among the trees to basically fend for themselves. The truth is that the practice of silvopasture is a management activity that has been successfully implemented in many locations throughout North America. Silvopasture is more than just turning out cattle to feed in the forest. Silvopastures are inherently sustainable systems. They increase biological diversity, protect water quality, reduce soil erosion, and improve the water hold capacity of the soil; moreover, according to USDA National Agroforestry Center (2014) what is required is a high degree of management skill and some intense labor. If not managed properly the practice risks short-term and long-term environmental failure. Managers must determine suitable overstory woody species, compute forage availability, balance livestock numbers and grazing rotations accordingly, and must have knowledge of tree canopy management needs, blow down potentials, and control of sapling damage, and understanding herbivore/plant interactions generated by the higher stock density (of livestock) afforded by agronomic management. If controlled ungulate herbivory is properly implemented and managed numerous economic and environmental benefits can be realized.

TBCF and Others

Before moving on to a discussion of animal grazing in forests and grasslands and their impact on SOC it is important to define a few important elements/parameters/key words that are part and parcel to our discussion herein and to follow.

Key Terms/Acronyms

- TBCF—total below-ground carbon flux
- GPP—gross primary production
- BNPP—below-ground primary production
- MAT—mean annual temperature
- MAP—mean annual precipitation

Note: These key terms are foreign to those who are not soil scientists, ecologists, and other environmental and soil science practitioners. They are important, however, to our discussion here. You are probably wondering what these terms have to do with carbon storage and sequestration. Quite a lot. That is, if we are to go into detail about deep soil science, heavy agronomy, and environmental sciences/engineering disciplines then we could take this to a more technological discussion to an even higher level, so to speak. However, for our purposes in this text we will only introduce the basics—the key basics. The *mean annual temperature* (MAT) and mean annual precipitation (MAP) covaried strongly across the global forest data base. However, MAT was the most important variable explaining observed patterns in blow-ground C processes in a study by Litton and Giardina (2008).

The Basics

In the beginning of their professional training the basic terms for those future scientists who study forests, or for those skilled in planting, managing, or caring for trees—*foresters*—have long been interested in two related systems of measurements of global primary production. *Gross primary production* (GPP) is the total amount of carbon dioxide "fixed" by land plants per unit time through the photosynthetic reduction of carbon dioxide into organic compounds. GPP is important because a large fraction of GPP supports plant mass-autotrophic respiration (R_a)—a key parameter of the carbon cycle—with the remainder allocated to the (NPP) of plant structural biomass in stems, leaves, and fruits. Labile carbohydrates such as sugars and starch, and, to a much lesser extent, volatile organic compounds (VOCs) are used in plant defense and signaling (that is, in *plant signaling* it involves the conveying of information within the plant cells from receptor systems to effectors). Therefore, we can state NPP as

$$\text{NPP} = \text{GPP} - R_a \tag{17.1}$$

The fraction of gross primary product that is *total below-ground carbon flux* (TBCF) and the fraction of TBCF that is *below-ground net primary production* (BNPP) represent globally significant C fluxes that are vital in regulating ecosystem C balance.

Well, the fly in the ointment, so to speak, is that our global estimates of the partitioning of GPP to TBCF and of TBCF to BNPP, as well as the absolute size of these fluxes, remain tentative, vague, undefined, and highly uncertain.

LIVESTOCK GRAZING

Forests and rangelands can contain large amounts of SOC because grasses allocate a high percentage of biomass to roots. Forest and rangeland SOC stocks are related to plant productivity, but management activities have important effects on SOC stocks (Silver et al., 2010). Grazing by livestock can manipulate several factors that have control over SOC content with complex interactions that make it difficult to forecast the net effect on SOC. Firstly, grazing influences the quantity of OM that returns to the soil. Secondarily, grazing affects OM quality by changing plant physiology and ecological processes. Secondary feedback can occur if nutrient removal through grazing reduces grassland and forest productivity.

Research shows that grazing rate, duration, and intensity can interact with wind erosion, site properties, and restoration activities to cause both increases and decreases in SOC (Pineiro et al., 2010). Herbivores alter the quality of OM inputs by reducing C/N ratios of plant shoots and roots. Note that Frank and Groffman (1998) point out that lower C/N ratios in plant litter increase decomposition rates and net N mineralization by reducing microbial demand for N; to be precise, N stocks are high enough to promote mineralization despite immobilization through microbial assimilation. These changes in decomposition rates suggest that microbial activity and substrate use efficiency are changed and that while decomposition rates may increase a return of carbon dioxide to the atmosphere, a portion of the C will return to the soil in dissolved forms that may become stabile on mineral surfaces (Berryman et al., 2020).

Conant et al. (2001) point out that forest management and conversion into grassland influence soil carbon; the techniques designed to increase forage production may also increase the quantity of inputs to SOC, and in turn accumulating atmospheric C as a C sink. Note that it is nitrogen that is typically the nutrient limiting primary production in grasslands [and in forests] and thus SOC content (Pineiro et al., 2010; Spellman, 2011). Correct ungulate grazing management systems make it possible to control SOC when N content and grassland/forest productivity are maintained. Note that high stocking rates lead to decreased production (Conant et al., 2001), so systems for instance as slow rotation grazing with moderate stocking levels will increase vegetative heterogenicity and increase soil aggregate stability (Conant and Paustian, 2002; Fuhlendorf and Engle, 2001).

DID YOU KNOW?

Forests absorb a large amount of atmosphere carbon dioxide in photosynthesis, but this is considerable uncertainty around regional estimates of the balance between photosynthesis and ecosystem respiration, which determines the net storage rate of C (King et al., 2015).

Soil stability is important for aggregation and microbial accessibility of SOC. Livestock grazing in forests alters forest dynamics by reducing the biomass and density of understory grasses and wedges; moreover, livestock grazing reduces the abundance of fine fuels, which formerly carried low-intensity fires through forests (Belsky and Blumenthal, 1995). In addition, experience shows that grazing can increase the rates of erosion, which exports SOC from a forest area. The reality is that grazing in arid and semiarid systems can lead to a destabilization of soil surfaces that subsequently leads to losses of soil nutrients to wind and water erosion (Neff et al., 2005). Grazing by domestic livestock causes losses in soil nutrients which can lower fertility, which, in turn, reduces plant productivity. Note that at the present time it is unclear whether the process of post fire erosion increases sedimentation rates that outweigh other losses.

Pineiro et al. (2010) point out that local adaptations of grazing systems have been shown to increase net productivity, N storage, and, because of these pathways, SOC storage. Through the direct action of grazing abrupt changes in intensity in grazing systems ultimately reduce net C storage in soils by altering plant communities. Moreover, when intense grazing of cattle in areas of adapted low disturbance occurs its control of the soil microbial community, decomposition of plant litter, N availability, and SOC are lost through alteration of plants—this is especially the case where slow-growing native plants are prevalent (Klumpp et al., 2009).

DID YOU KNOW?

Grazing by domestic livestock has contributed to increasingly dense U.S. western forests and to changes in tree species composition.

Note that multiple factors are involved in the response of SOC to grazing including climate, soil properties, landscape position, plant community composition, and grazing management practices (Pineiro et al., 2010). One of the considerations that can help predict impacts of grazing on SOC stock is the sensitivity of rangelands to seasonal drought. In the U.S. Southwestern forests and rangelands, they are particularly sensitive to drought. In drought conditions the annual loss of C is a common occurrence due to low plant productivity (Svejcar et al., 2008). In these cases, managers may reduce grazing intensity during drought periods to ensure the recovery of forest and grassland productivity the following year. Brown et al. (2010) point out that the effects of the reduced stocking rate on SOC in forest and rangelands remain highly uncertain—the problem is there is a lack of data in arid southwestern systems.

THE BOTTOM LINE

Over the last century, the structure, composition, and dynamics of western interior forests have changed dramatically. While it is true that animal gazing within the forests has changed the ecological composition of grazed forests, it is also changed due to clearcutting. For those forest areas not extensively logged many have experienced great increases in tree density and changes in species composition, often forming dense stands of fire- and disease-sensitive trees (Belsky and Blumenthal, 1995).

REFERENCES

Belsky, A.J. and Blumenthal, D.M. (1995). Effects of livestock grazing on stand dynamics and soils in Upland forests of the interior west. *Conservation Biology* 11(2):315–327.

Berryman, E. et al. (2020). Soil Carbon. Accessed 5/8/23 @ https://www.fs.usda.gov/rm/pubs-journals/2020/rmrs-2020-berrymn-eco/pdf.

Brown, J., Angerer, J., Salley, S.W. et al. (2010). Improving estimates of rangeland carbon sequestration potential in the US Southwest. *Rangeland Ecology & Management* 63(1):147–154.

Conant, R.T. and Paustian, K. (2002). Spatial variability of soil organic carbon in grasslands: Implications for detecting change at different scales. *Environmental Pollution* 116(supplement):127–S135.

Conant, R.T., Paustian, K., Elliot, E.T. (2001). Grassland management and conversion into grassland: Effects on soil carbon. *Ecological Applications* 11(2):343–355.,

Frank, D.A. and Groffman, P.M. (1998). Ungulate vs. landscape control of soil C and N processes in grasslands of Yellowstone National Park. *Ecology* 79(7):2229–2241.

Fuhlendorf, S.D. and Engle, D.M. (2001). Restoring heterogeneity on rangelands: Ecosystem management based on evolutionary grazing patterns. *Environmental Science* 51(8): 625–632.

King, A.W., Andres, R.J., Davis, K.J., Hafer, M., Hayes, D.J., Huntzinger, de Jong, B., Kurz, W.A., McGuire, A.D., Vargas, R. (2015). North American net terrestrial CO_2 exchange with the atmosphere 1990-2009. *Biogeosciences* 12:399–414.

Klumpp, K., Fontaine, S., Attard, E. (2009). Grazing triggers soil carbon loss by altering plant roots and their control on soil microbial community. *Journal of Ecology* 97(5):876–885.

Litton, C.M. and Giardina, C.P. (2008). Below-ground carbon flux and partitioning: Global patterns and response to temperature. *Functional Ecology* 22:941–954.

Neff, J.C., Reynolds, R.L., Belnap, J., Lamothe, P. (2005). Multi-decadal impacts of grazing on soil physical biogeochemical properties in southeast Utah. *Ecological Applications* 15(1):87–95.

Pineiro, G., Paruelo, J.M., Oesterfield, M., Jobbagy E.G. (2010). Pathways of grazing effects on soil organic carbon and nitrogen. *Rangeland Ecology & Management* 63(1):109–119.

Schuman, G.E., Janzen, H.H., Herrick, J.F. (2002). Soil carbon dynamics and potential carbon sequestration by rangelands. *Environmental Pollution* 116(3):393–396.

Silver, W.L., Ryals, R., Eviner, V. (2010). Soil carbon pools in California's annual grassland ecosystems. *Rangeland Ecology & Management* 63(1):128–136.

Spellman, F.R. (2011). *Forest-Biomass Energy*. Boca Raton, FL: CRC Press.

Svejcar, T., Angell, R., Bradford, J.A. et al. (2008). Carbon fluxes on North American rangelands. *Rangeland Ecology & Management* 61:465–474.

USDA (2014). *Agroforestry Notes*. Washington, DC: United States Department of Agriculture-National Agroforestry Center.

18 Forest Disturbance
Nutrient Additions

FOREST FERTILIZATION

When fertilizing nutrients are added (applied) to forest soils, either through nutrient management protocols or acid rain deposition of N the result can be positive (gains), negative (losses), or neutral (no perceived or detectable change) in SOC stocks. However, in determining if fertilizing nutrients provide gains, losses, or neutral we return to the standard saying often used to describe scientific facts: we do not know what we do not know about—in this case—about all the numerous factors involved in gaging the effect of nutrient addition to forest lands (Blagodatskaya and Kuzyakov, 2008; Jandl et al., 2007; Janssens et al., 2010). The truth be told fertilization of forests and its ultimate affects/effects has not been widely studied up to now. However, we do know from observation and research that forest fertilization has been shown to increase or decrease SOC by increasing productivity, increasing SOC mineralization rates, shifting production to aboveground vegetation components, and depressing certain enzyme activity (Jandl et al., 2007; Van Miegroet and Jandl, 2007). One other thing, Johnson and Curtis (2001) point out that the effects of forest land fertilization on SOC have been found to be site specific, but most studies show an increase in SOC stocks.

DID YOU KNOW?

The storage and flow of C into and out of forest lands can differ under the influence of dominant tree species because of species-based variation in C production, decomposition, retention, and harvest-based export (Gahagan et al., 2015).

AGROFORESTRY

Agroforestry is the intentional integration of trees and shrubs into crop and animal farming systems to create economic, environmental, and social benefits. A wide range of public benefits is obtained by people who manage forests, farms, and other lands by supplying clean water, supporting wildlife habitat, and adapting to climate change, as well as sequestering carbon, basically, a multifunctional approach. This means that they will sequester carbon as an additional benefit while enhancing and/or protection crops or livestock production or providing other benefits. Like forest trees, trees planted in agroforestry practices also sequester carbon and play an important role in reducing greenhouse gas concentrations. As mentioned earlier,

DOI: 10.1201/9781003432838-21

carbon sequestration also can improve soil health and help with adapting to climate change. Note that the potential for carbon sequestration in the United States through the agroforestry system is significant, simply because there is a large amount of land under agricultural production.

DID YOU KNOW?

The key characteristics of agroforestry practices that help increase carbon sequestration are that they include perennial plants and they include a diversity of plants (USDA, 2014).

Note that perennial systems, like agroforestry, generally sequester more carbon than annual vegetation because they grow for more days of the year than annual crops. Moreover, perennial plants commonly store more carbon each year, while annual plants gradually release carbon after they die off each year and decompose. In general, trees sequester carbon in their trunks and branches, increasing the amount stored as they grow.

Fixing Nitrogen

We can view agroforestry as having the extra benefit from planting N_2-fixing shrubs and trees as an economical nitrogen source for crops (Danso et al., 1992). Biological nitrogen-fixing BNF tree species are associated with higher forest SOC; accumulation of SOC has been reported to be about 12–15 g of C for every gram of N fixed (Binkley, 2005). The point is that nitrogen (along with potassium and phosphorus) is essential for photosynthesis, and is the primary compound of plant protoplasm, which builds plant cells. Nitrogen enhances leafy growth, supports the creation of healthy flower buds, and aids fruit set. Moreover, it is a catalyst for other minerals. Note that plants can effortlessly reduce all the nitrogen available in the soil in each area. Also, depletion occurs when nitrogen is leached naturally from the soil through action of sun and water.

It is ironic that nitrogen forms most of the Earth's atmosphere but can't uptake the nitrogen from the air. Nitrogen (N_2), an extremely stable gas, again, is the primary component of the Earth's atmosphere (78%). The nitrogen cycle is composed of four processes. Three of the processes—fixation, ammonification, and nitrification—convert gaseous nitrogen into usable chemical forms. The fourth process—denitrification—converts fixed nitrogen back to the unusable gaseous nitrogen state. Although nitrogen is an essential ingredient for plant growth, it is chemically very inactive, and before most of the biomass can incorporate it, it must be *fixed*. Special nitrogen-fixing bacteria found in soil and water fix nitrogen. Thus, microorganisms play a major role in nitrogen cycling in the environment. These microorganisms (bacteria) can take nitrogen gas from the air and convert it to nitrate. This is called *nitrogen fixation*. Some of these bacteria occur as free-living organisms in the soil. Others live in a *symbiotic relationship* (a close relationship between two organisms of different species, and one where both partners benefit from the association) with plants. An example of a symbiotic relationship, related to nitrogen, can be seen, for example, in the roots of peas. These roots have small swellings along their length. These contain millions of symbiotic bacteria, which can take nitrogen gas from the atmosphere and convert it to nitrates that can be used by the plant. Then the plant is plowed into the soil after the growing season to improve the nitrogen content. Price (1984) describes the nitrogen cycle as an example "of a largely complete chemical cycle in ecosystems with little leaching out of the system." Simply, the nitrogen cycle provides various bridges between the atmospheric reservoirs and the biological communities.

Nitrogen-fixing plants can be incorporated in forest lands in a range of different ways. Having stated this note that it is always best to plant nitrogen-fixing plants in forest lands in regions where the bacterial content of the forest soils and soil type is characteristic of the temperate climate wherein the forestland is located. Moreover, while in some areas and situations non-native species may be worth consideration in certain circumstances; however, it is always smart to choose nitrogen-fixing plants that are native to the bioregion. But again, nitrogen-fixing plants and trees are not suited for all forest regions. Indeed, in cool temperature regions, there may only be a few plants that can grow successfully. For example, in some cool temperate regions only alder species can survive. In other cool temperate regions maybe the alder and laburnum have the best chance for survival.

In regions where space within the forestland is limited, it is best to plant shrubs and not trees. Nitrogen-fixing shrubs can fit into under-story areas and are grown to provide feedstock for mulches or fruit trees can be planted and harvested in under-story sections of the forest.

There are several varieties of nitrogen-fixing shrubs, all dependent on the local climate. Some of these nitrogen-fixing shrubs include the following:

- Autumn Olive
- Russian Olive
- Gorse
- Wax Myrle/American Bayberry
- Broom
- Buffaloberries
- Goumi
- Sea Buckthorn
- Elaeagnus x ebbingei
- Bitterbrush/Cliff-Rose
- Mountain Mahogany
- Mountain Misery
- California Lilac

Again, be advised these are not suitable for all regions and climate zones. Consider, for example, California Lilac (Ceanothus), it can provide some forest understories with a blast of blue color. This pollinator-friendly, flowering, and drought-tolerant shrub can be free-standing or used as forest under-story nitrogen-fixing shrub. The problem is that California Lilac requires full sun and some partial shade—full sun coverage under the canopy of a tall forest is rare or just totally non-existent—in the partial shade this shrub does well; however, it requires more sunlight than shade.

For farmers/harvesters who plant nitrogen-fixing herbaceous plants in forest lands in order to subsist on the food-producing-plants or profit economically from after harvest sales a few of the nitrogen-fixing herbaceous plants include the following:

- Soybeans
- Red Clover
- Peanuts/Groundnuts
- Lupins
- Green Beans/French Beans
- Fava Beans
- Field Peas
- Pidgeon Peas
- Everlasting Sweet Pea
- Alfalfa
- Red Clover
- American Vetch
- White Clover
- Tufted Vetch

Note that this is by no means a comprehensive list. Yet this list gives us a place to begin, to start when it comes to selecting nitrogen fixers that might be integrated into forest lands.

Also note that agroforestry systems with multiple types of plants tend to use resources efficiently. These nitrogen-fixing plants take up carbon dioxide, reduce production of other greenhouse gases, and store carbon in the soil at different depths.

It's All about Trees

Agroforestry is about trees and crops and/or nitrogen-fixing shrubs. Adding trees to an agricultural system tends to store more carbon than is stored in a similar agricultural system without trees. For example, silvopastures often sequester more carbon than pastures without trees.

Okay, so what is silvopasture?

Silvopasture is a management activity that can be successfully implemented in many parts of North America. Silvopastures can be created by either planting trees into pastures or thinning stands of trees and planting forages. In either case, silvopasture managers coordinate tree thinning and pruning practices to modify the canopy density in ways that complement sustained forage production through most of the rotation and meet the needs of canopy species (USDA, 2014).

When I state here that it is all about trees, it is. Simply, forests with nitrogen-fixing trees (N-fixers) typically accumulate more carbon (C) in soils than similar forests without N-fixing trees (Resh and Binkley, 2002).

THE BOTTOM LINE

On a worldwide scale, the largest source of nutrient additions to forest soils is atmospheric N deposition derived from both natural and anthropogenic sources. Kanakidou et al. (2016) reported that anthropogenic N, from fossil fuel combustion and recirculated cropland fertilizers, accounts for about 60% of the approximately 130 Tg of N deposited worldwide each year. Chronic N addition experiments consistently show the SOC increase under higher N availability (Frey et al., 2014). Note that this result has been attributed to greater productivity due to the fertilization effect, as well as reductions of OM decomposition. Frey et al. (2014) found that in addition to more C stored in tree biomass, there were significant shifts in SOC chemistry due to shifts in the microbial community (fewer fungi). Increases in SOC were attributed to a reduction in decomposition rate due to the lower abundance in fungi in the soil (Berryman et al., 2020). Regarding tree and shrub components of agroforestry practices that contribute to carbon sequestration by using carbon dioxide for photosynthesis and storing carbon aboveground in tree trunks and branches, as well as below-ground in roots and the soil it is important to recognize that this carbon can stay in the tree or soil for a long time.

REFERENCES

Berryman, E. et al. (2020). Soil Carbon. Accessed 5/8/23 @ https://www.fs.usda.gov/rm/pubs-journals/2020/rmrs-2020-berrymn-eco/pdf.

Blagodatskaya, F. and Kuzyakov, Y. (2008). Mechanisms of real and apparent priming effects and their dependence on soil microbial biomass and community structure: Critical review. *Biology and Fertility of Soils* 45(2):115–131.

Binkley, D. (2005). How nitrogen-fixing trees change soil carbon. In: Binkley, D., Menyailo, O., (eds) *Tree Species Effects on Soils: Implications for Global Change*. NATO Science Series IV. Earth and Environmental Sciences (NAIV. Vol, 55). Dordrecht: Springer, pp. 155–164.

Danso, S.K.A., Bowen, G.D., Sanginga, N. (1992). Biological nitrogen fixation in trees in argo-ecosystems. *Plant Soil* 141(2):177–196.

Frey, S.D., Ollinger, S., Nadelhoffer, K. et al. (2014). Chrome nitrogen additions suppress decomposition and sequester soil carbon in temperate forests. *Biogeochemistry* 121(2):305–316.

Gahagan, A., Giardina, C.R., King, J.S., Binkley, D., Pregitzer, K.S., Burton, A.J. (2015). Carbon fluxes, storage and harvest removals through 60 years of stand development in red pine plantations and missed hardwood stands in Norther Michigan, USA. *Forest Ecology and Management* 337:88–97.

Jandl, R., Lindner, M., Vesterdal, L. (2007). How strongly can forest management influence soil carbon sequestration? *Geoderma* 137(3):253–268.

Janssens, I.A., Dieleman, W., Luyssaert, S. et al. (2010). Reduction of forest for respiration in response to nitrogen deposition. *Nature Geoscience* 3(5):315–322.

Johnson, D.W. and Curtis, P.S. (2001). *Effects of Forest Management on Soil C and N Storage: Meta-Analysis*. Amsterdam, Netherlands: Elsevier.

Kanakidou, M., Myriokefalitakis, S., Daskalakis, N. et al. (2016). Past, present, and future atmospheric nitrogen deposition. *Journal of the Atmospheric Scinces* 73:2039–2047.

Price, P.W. 1984. *Insect Ecology*. New York: John Wiley & Sons.

Resh, S.C., Binkley, D., Parrotta, J.A. (2002). Greater soil carbon sequestration under trees nitrogen-fixing trees compared with eucalyptus species. *Ecosystems* 5:217–231.

USDA (2014). Forest Grazing, Silvopasture, and Turning Livestock into the Wood. AF-Note 46. Washington, DC: United States Department of Agriculture.

Van Miegroet, H. and Jandl, R. (2007). Are nitrogen-fertilized forest soils sinks or sources of carbon? *Environmental Monitoring and Assessment* 128(1–3):121–131.

19 Forest Disturbance
Tree Mortality

INTRODUCTION

It has been my experience from childhood to the present time to have witnessed many broken and otherwise dead trees in the forests of the U.S. Pacific Northwest, Talkeetna, Alaska, California Sequoias and Redwoods (see Figures 19.1 and 19.2), South America, and in the Appalachians along the Appalachian Trail (most often in the Virginia section of the Trail). When looking back at my youth and my associated ignorance, I recall looking at and sometimes kicking deadfalls to test their state of decay. I recall looking at the still standing tall dead trees. I also recall looking at ugly snags and wondering how "somethings" so large could have reached such a disgusting demise—remnants of forest giants of the past and now just left-over waste.

How time changes just about everything! Well, with education and many hours spent in several forests and always spotting standing dead trees, snags, and deadfall I now realize the actual importance of the dead trees is a paradox, basically a contradiction based on differing viewpoints.

FIGURE 19.1 The author standing in front of the remains of a Giant Sequoia in Sequoia National Park, California.

DOI: 10.1201/9781003432838-22

FIGURE 19.2 Author inside fallen Giant Sequoia. Sequoia National Park, California.

THE PARADOX: DEAD OR ALIVE

Dead trees are essential to the health of forests.

Dead standing trees, snags, and deadfalls are kindling for intense wildfires.

Two different points of view.

Are we better off leaving alone the dead trees in the forest or is it best to remove them? Herein lies the paradox and two different views—two strongly held opposite views.

Let's get back to why dead standing trees, snags, and deadfalls are essential to the health of forests. They help kick-start the new forest by offering habitat, food, and other resources to insects, birds and mammals. This is the mindset of those who think it important to let be, leave well enough alone. Included in this mindset is the attitude that even if wildfire occurs in a forest that has been basically left alone this is no big deal. Why? Well, because after a wildfire life returns quickly, within days. Dead standing trees, snags, and deadfalls are some the of the most valuable wildlife structures in the forest and help support numerous animals. The dead trees are decomposed by fungi, bacteria, and other life forms which aid new plant growth by retuning important nutrients in the forest ecosystem.

There is one obvious contradiction, the ol' fly in the ointment in the opinion of those who foster the attitude that wildfires are good, a tool of Mother Nature to make things better in forestlands. So, what is the contradiction, the ol' fly in the ointment? Simply, the contradiction, the ol' fly in the ointment is us; a former student of mine called us, all forms of life as "we'ums."

The point is not all forests are remote. Forests that are not remote are subject not only to we'ums settling within but also grazing cattle and the forests' other lifeforms.

Is it a good plan or action to remove dead woody debris from the forest or all the dead woody debris to be a significant carbon sink?

You be the judge.

WHEN DEAD IS NOT REALLY DEAD

In my years of teaching environmental engineering, science, and health college courses for lower- and upper-level students I had a lot of crossover lectures between these three major subject areas/professions. This crossover activity was not by chance but was planned deliberately. When lecturing on global pollution, climate change, and environmental health topics I quickly go to the forests, so to speak. I go to the forests to explain what will happen to life on Earth if we do not respect and protect the forests of the world. I harp on this important area because they are so essential to life on Earth, period.

To this point in the book it has been pointed out that dead woody debris whether still dead standing, snag, or litterfall all contain carbon and until completely incinerated contain carbon and can store the carbon for decades, and longer.

So, the point is when a tree dies, the carbon contained within lives on.

THE CARBON FRACTIONS

It is and has been widely used and professed for ages that elemental carbon is found in all living things—carbon is life.

Well, this statement is true, of course. But what is not commonly known or understood is that in the forest, for example, dead standing trees, snags, woody debris, and litterfall all contain carbon; C, the 4th element in the Periodic Table is vital to life and in dead forest pool storage of C, which comprises about 8% carbon storage globally, is important to sustaining life.

We know all this but what we do not know is important, maybe critical—probably is to sustaining life on Earth. So, what is it that we do not know about C storage in dead wood? Answer: we are uncertain in quantifying dead wood carbon (C) stocks because there presently exists a lack of accurate dead wood C fractions (CFs) that are employed to convert dead woody biomass into C. Most of the C estimation protocols or procedures employ a default dead wood CF of 50%, but live tree reports suggest that this value is an over-estimate (Martin et al., 2021).

Again, forests in the U.S. are an important carbon sink or pool. Carbon dioxide uptake by forests in the contiguous U.S. offsets about 12%–19% of our total carbon dioxide emission each year (Ryan et al., 2010). As noted, carbon sequestration is only one of many services provided by forests; others include clean water, clean air, biodiversity, wood products, wildlife habitat, food, and recreation.

The long-term capacity of forest ecosystems to capture and store carbon depends in large part on their health, productivity, resilience, and adaptive capacity (Franco, 2021). The forest ecosystem is dynamic and is affected by temporal and spatial variability in temperature and precipitation, insect and disease epidemics, wildfires, catastrophe storms, and human activity (Franco, 2021).

A paradigm shift has occurred in forest management involving their older procedures and thinking that focused on optimizing a given good or service (e.g., volume harvested per surface unit) to a new view on optimizing not maximizing. What this means is that carbon sequestration in forest ecosystems and effective substituting forest-based materials for fossil fuel-intensive materials in the context of providing other needed goods and services can be a sustainable, a long-term means of responding to events, happenings such as climate change.

Sustainable carbon fraction management involves effectively managing ecosystems by restoring, maintaining, and enhancing their health and productivity. Also, apply management practices that increase sequestration of offset emission, as well as take into account the carbon fractions in wood products and substitutions that result from managed forests in the long term (Franco, 2021).

DID YOU KNOW?

Carbon turnover refers to the amount of time carbon is cycling in a system in units of years. This is often referred to as mean residence time.

NET PRIMARY PRODUCTION AND GROSS PRIMARY PRODUCTION

$$\text{NPP} = \text{GPP} - \text{respiration (by plants)}$$

Gross Primary Production is the sum of leaf-level photosynthetic rates across a canopy—rates across the canopy determines Gross Primary Production and its components. As with just about all production processes there are forcings. These forcings related to the drivers of carbon process rates and storage (e.g., temperature, incident light, and phytotoxic gases such as O_3 (Ozone)) act on leaf-level photosynthesis and on gross primary production through effects on carbon uptake, allocation, and loss (Ryan, 1991; Amthor and Baldocchi, 2001; Baldocchi and Amthor, 2001; Litton et al., 2007, Litton and Giardina, 2008; Bonan, 2008, Giardina et al., 2014).

DID YOU KNOW?

The atmospheric concentration of carbon dioxide increased by more than 30% between 1850 and 2003, and the present concentration is higher than that at any time in the past 400,000 years.

DID YOU KNOW?

Most plant species on Earth use C3 photosynthesis, in which the first carbon compound produced contains three carbon atoms. In this process, carbon dioxide enters a plant though is microscopic pores on plant leaves, *stomata*, wherein a series of complex reactions occur and the enzyme Rubisco fixes carbon into sugar though the Calvin-Benson cycle. Certain plants use alternative forms of photosynthesis, called Crassulacean acid metabolism (CAM) and C4.

Observation and experience bring about a change in certain things, practices, hopes, and the rest. In this case what we have come to realize is that a foundational driver of leaf and canopy photosynthesis is sunlight. What is really going on here is that water is consumed in forest trees and plants by photosynthesis and transpiration. The latter process—responsible for around 90% of water use—is driven by the evaporation of water from the leaves of forests trees and plants. Transpiration allows plants to transport water and mineral nutrients from the soil. But now we know that there is more and in this case we now know that temperature and moisture are forcing variables that are increasingly receiving attention because they exert strong influences on plant and ecosystem productivity.

Note that these variables change for much of the globe—dramatically in some places (U.S. CCSP, 2007). Under optimal moisture and sunlight, the response of canopy photosynthesis to warming is parabolic, risking to as often very plastic (seasonally and across species is) maximum between 20°C and 30°C (Giardina, 2017). Canopy photosynthesis declines (is limited) when exposed below optimum temperature which limits enzyme reaction rates; when above the optimum temperature, warming causes cellular membranes to become less stable and alters the relative solubilities of CO_2 and O_2, and enzyme reaction rates limit canopy photosynthesis. In wet climates, drying can result in increases in gross primary production (Baldocchi and Amthor, 2001). The point is water is assimilated with carbon dioxide during photosynthesis to produce carbohydrates; thus, it is a strong determinant of gross primary production (Giardina, 2017).

DID YOU KNOW?

Longer-term patterns in water availability influence plant communities, made of species that are associated with traits adapted to water and temperature conditions of a location (Giardina, 2017).

Note that for a given species especially long-lived trees, reduction in water availability due to climate variability, episodic drought, or longer-term baseline shifts in precipitation can cause a series of diurnal to seasonal to inter-annual responses that affect gross primary production.

Regarding gross primary production (GPP) and net primary production (NPP) in forestlands the former is the amount of chemical energy, typically expressed as carbon biomass that primary producers create in each length of time. The latter is the rate at which all the autotrophs in an ecosystem produce net useful chemical energy. Note that NPP is available to be directed toward growth and reproduction of primary producers—this fact is important because it means it is also available for consumption by herbivores.

Both GPP and NPP are typically expressed in units of mass per unit area per unit times interval. In forestland ecosystems, mass carbon per unit area per year ($g C m^{-2} yr^{-1}$) is most often used as the unit or measurement.

Note: The terms "production" and "productivity" are used interchangeably but this does not mean that there are distinctions often drawn between the two terms.

Atmospheric Chemicals and Productivity

Another factor that affects plant and ecosystem productivity is atmospheric chemistry. Increasing levels of atmospheric carbon dioxide can stimulate gross primary production and carbon sequestration. Keep in mind, however, that carbon is allocated to aboveground and belowground carbon pools depend on species and site conditions (King et al., 2015; Luo et al., 2004; Norby et al., 2005; Liu et al., 2005). Present thinking is that productivity gains are most likely the result of both direct stimulation of photosynthesis (Norby et al., 2005), as week as indirect effects on growing season length (Taylor et al., 2008), and associated increases in resource-use efficiencies (Baldocchi and Amthor, 2001). On the other hand, gases such as ozone (O_3), which is also increasing across temperate and tropic regions, can have the opposite effect on plants (King et al., 2015). Cascading effects on soil carbon formation occur whenever ozone damages leaves, compromises growth, and reduces carbon storage (Loya et al., 2003).

Vegetation Species Composition (Genotype Effects) on Soil Productivity

Effects on ecosystem productivity can be significant depending on vegetation species composition or genotype effects. In the practice of forestry, the base of the pyramid, the foundational practice is tree selection and improvement, and gains in productivity have been substantial in the past century (Ryan et al., 2010). These species-level effects on productivity relate primarily to physiological trait differences across species or within species across genotypes. Moreover, species effects on productivity include biological plant invasions where the invaders increased the productivity of a site by increasing the supply of limiting resources (via nitrogen fixation), by increasing stand-level capacity to fix carbon at the leaf and/or canopy level, or by altering disturbance regimes (Vitousek, 1990; Binkley and Fisher, 2013; Peltzer et al., 2010). Gahagan et al. (2015) point out that in addition to leaf and canopy physiology, tissue chemistry (along with return of nutrients, detritus and associated compounds to soil), plant allocation patterns, and resource-use efficiencies can vary dramatically across species occupying the same sites.

Site Disturbance

Possibly the strongest constraint on carbon storage in forestland is disturbance (Pregitzer and Euskirchen, 2004; Ryan et al., 2010). The word "disturbance" is a big, wide-ranging term. In forestland disturbance can range from single tree mortality to tree falls to stand and landscape-scale even such as ice storms, blowdowns, or wildfire. The term also can mean annoyance, bother, irritation, and so on and so forth. In this text, the term disturbance used to describe forestland is meant to point out disruption of the status quo in the forest stand and thus a reduction of carbon storage. As disturbances escalate in degree from single trees to whole stands to entire landscapes, the effect on landscape-scale carbon stocks also increases (Ryan et al., 2010). Note if the forestland is undisturbed for several years and results in old-growth forest conditions then conditions store some of the largest quantities of aboveground biomass in the terrestrial biosphere.

THE BOTTOM LINE

Whenever more than 50% of the forest canopy is lost due to tree mortality it affects carbon stocks in the forest floor and soil. High tree mortality can be caused by drought and bark beetles (subfamily Scolytidae) in Western U.S. pine and spruce (*Abies* spp.) forests and by invasive pests (detailed in the next chapter). Berryman et al. (2020) point out that harvesting and mass tree mortality effects on SOC are similar; however, there are important differences: mortality occurs more slowly than most harvesting operations; mortality events do not usually kill as many trees as are harvested in a typical operation; and, dead trees are commonly in place through the mortality event, although some limited post mortality harvesting is conducted in high use areas (campgrounds and national forest hiking and recreation areas).

Edburg et al. (2012) point out that forest tree mortality events result in a reorganization of detritus over several years, impacting OM inputs to SOC formation. Moreover, Warnock et al. (2016) noted that root OM to the soil may increase as trees die but may later decline due to reduced live tree density; consequently, microbial activity in the rhizosphere (i.e., the soil region in vicinity of roots) is altered after tree mortality. In the first few years following mortality, litterfall is expected to increase as dead trees drop their needles and fine branches: subsequently, litterfall will decline, reducing forest floor mass in the longer term (Zhang et al., 2015a, b). Longer-term inputs to the soil are larger branches and boles, as wind topples standing dead trees throughout the next several years. Accordingly, mortality will change the rate and type of OM matter input to the soil. Changes in nutrient dynamics could alter SOC mineralization rates as soil-extractable N levels increase and SOC/N ratios decrease (Clow et al., 2011; Morehouse et al., 2008; Trahan et al., 2015). Ultimately, changes in microclimate post mortality, brought about by canopy loss and decreases in transpiration, could impact detrital C processing (Berryman et al., 2013). Regarding the duration of these effects it depends on how fast the remaining living trees spread out their canopies to counteract or offset for the loss of the overstory.

DID YOU KNOW?

Many of the temperate forests of the eastern United States have been cleared for agriculture (Bonan, 2008).

Observation over time has shown that following tree mortality events some changes in the SOC cycling occur due to the reorganization of detritus. Evidence from stable isotopes suggests shifts in C substrate type used for root and heterotrophic respiration starting in the first year after mortality (Maurer et al., 2016). Various studies have reported decreases in microbial biomass C (MBC) and increases in the aromaticity of dissolved organic C (DOC) in the soil, which may affect SOC balance (Brouillard et al., 2017; Kana et al., 2015; Trahan et al., 2015). Even with these changes in SOC substrates and cycling rates, changes in SOC stocks following tree mortality events are often undetectable and, on average, minor compared to impacts on soil respiration, DOC, and MBC (Morehouse et al., 2008; Zhang et al., 2015b). This implies that though individual process rates may be affected by tree mortality, the balance between inputs to and outputs from the SOC pool may be constant enough to lead to undetectable changes in SOC (Berryman et al., 2020).

DID YOU KNOW?

Aromaticity is an important indicator of the origin, stability, and chemical reactivity of soil humic materials (Schnitzer et al., 1991).

Note that Brouillard et al. (2017) report that changes in SOC may be difficult to detect because they could be highly dependent on the amount of tree mortality. In bark beetle-impacted lodgepole pine (*Pinus contorta*) forests of Colorado, soil respiration 8 years after mortality depended on the relative amount of living versus dead trees. There is a question about scale predictably, however, because plot-level impacts of biome disturbance may not scale predictably to the forest or watershed.

Edburg et al. (2012) point out that forest tree mortality events result in a reorganization of detritus over several years, impacting OM inputs to SOC formation. Moreover, Warnock et al. (2016) noted that root OM to the soil may increase as trees die but may later decline due to reduced live tree density; consequently, microbial activity in the rhizosphere is altered after tree mortality.

REFERENCES

Amthor, J.S. and Baldocchi, D.D. (2001). Terrestrial higher-plant respiration and net primary production. In: Roy, J., Saguier, B., Mooney, H. (eds) *Terrestrial Globe Productivity Past, Present and Future.* San Diego, CA: Academic Press.

Baldocchi, D.D. and Amthor, J.S. (2001). Canopy Photosynthesis: History, measurements and models. In: Roy, J., Saguier, B., Mooney, H. (eds) *Terrestrial Globe Productivity Past, Present and Future.* San Diego, CA: Academic Press.

Berryman, E.M., Marshall, J.D., Rahn, T. et al. (2013). Decreased carbon limitation of litter respiration in mortality-affected pinon-juniper woodland. *Biogeosciences* 10:1625–1634.

Berryman, E. et al. (2020). Soil Carbon. Accessed 5/8/23 @ https://www.fs.usda.gov/rm/pubs-journals/2020/rmrs-2020-berrymn-eco/pdf.

Binkley, D. and Fisher, R. (2013). *Ecology and Management of Forest Soils.* Chichester: Wiley, 347 p.

Bonan, G.B. (2008). Forests and climate change forcings, feedbacks, and the climate benefits of forests. *Science* 320:1444.

Brouillard, B.M., Mikkelson, K., Bokman, C.M. et al. (2017). Extent of localized tree mortality influences soil biogeochemical response in a beetle-infested coniferous forest. *Soil Biology and Biochemistry* 114:309–318.

Clow, D.W., Rhoades, C., Briggs, J., et al. (2011). Responses of soil and water chemistry to mountain pine beetle induced tree mortality in Grand County, Colorado, USA. *Applied Geochemistry* 26:S174–S178.

Edburg, S.L., Hick, J.A., Brooks, P.D., Pendall, E.G., Ewers, B.E., Norton, U., Gochis, D.J., Gutmann, E.D., Meddens, A.J. (2012). Cascading impacts of bark beetle-caused tree mortality on coupled biogeophysical and biogeochemical processes. *Frontiers in Ecology and the Environment* 10:416–424.

Franco, C.R. (2021). *Considering Carbon in Land Management*. Washington, DC: USDA.

Gahagan, A., Giardina, C.R., King, J.S., Binkley, D., Pregitzer, K.S., Burton, A.J. (2015). Carbon fluxes, storage and harvest removals through 60 years of stand development in red pine plantations and missed hardwood stands in Norther Michigan, USA. *Forest Ecology and Management* 337:88–97.

Giardina, C.P., Litton, C.M., Crow, S.E, Asner, G.P. (2014). Warming-related increases in soil CO_2 efflux are explained by increased below-ground carbon flux. *Nature Climate Change* 4:822–827.

Giardina, C.P. (2017) Ecosystem Carbon Storage and Fluxes. USDA General Technical Report WO-95. Forest Service. USDA Accessed 6/5/23 @ https://data.fs.usda.gov/wwwbeta/sites/default/files/s_document/wo-95.

Kana, J., Tabovska, J., Santruckova, H. (2015). Excess organic carbon in mountain spruce forest soils after bark beetles outbreak altered microbial N transformations and mitigated N-saturation. *PLos One* 10(7):e0134165.

King, A.W., Andres, R.H., Davis, K.J., Hafer, M., Hayes, D.J., Huntzinger, D.N., de Jong, B., Kurz, W.A., McGuire, A.D., Vargas, R. (2015). North American net terrestrial CO_2 exchange with the atmosphere 1990-2009. *Biogeosciences* 12:399–414.

Litton, C.M. and Giardina, C.P. (2008). Below-ground carbon flux and partitioning: Global patterns and response to temperature. *Functional Ecology* 22:941–954.

Litton, C.M., Raich, J.W., Ryan, M.G. (2007). Review: Carbon allocation in forest ecosystems. *Global Change Biology* 13:2089–2109.

Liu, L., King, J.S., Giardina, C.P. (2005). Effects of elevated concentrations of atmospheric CO_2 and tropospheric O_3 on leaf litter production and chemistry in trembling aspen and paper birch communities. *Tree Physiology* 25:1511–1522.

Loya, W.M., Pregitzer, K.S., Karberg, N.J., King, J.S., Giardina, C.P. (2003). Reduction of soil carbon formation by tropospheric ozone under elevated caron dioxide. *Nature* 425:705–707.

Luo, Y., Su, B.O., Currie, W.S., Dukes, J.S., Finzi, A., Hartwig, U., Hungate, B., McMurtrie, R.E., Oren, R., Paton, W.J., Pataki, D.E., Shaw, R.M., Zak, D.R., Field, C.B. (2004). Progressive nitrogen limitation of ecosystem responses to risking atmospheric carbon dioxide. *Bioscience* 54:731–739.

Martin, A.R., Domke, G.M., Doraisami, M., Thomas, S.C. (2021). Carbon fractions in the world's dead wood. *Nature Communications*. Accessed 6/4/23 @ https://doi.org/10.1038/s41467-021-21149-9|www.nature.com/naturecommunicaitons.

Maurer, G.E., Chan, A.M., Trahan, N.A. et al. (2016). Carbon isotopic composition of forest soil respiration in the decade following bark beetle and stem girdling disturbances in the Rocky Mountains. *Plant Cell & Environment* 19:1513–1523.

Morehouse, K., Johns, T., Kaye, M. (2008). Carbon and nitrogen cycling immediate following bark beetle outbreaks in the southwestern ponderosa pine forests. *Forest Ecology and Management* 255:2698–2708.

Norby, R.J., DeLucia, E.H., Gielen, B., Calfapietra C., Giardina, C.P., King, J.S., Ledford, J., McCarthy, H.R., Moore, D.J.P., Ceulemans, G.E., De Angelise, P., Finzii, A.C., Kamosky, D.F., Kubiske, M.E., Lukac, M., Pregitzer, K.S., Scarascia-Mugnozzan, G.E., Schlesinger, W.H., Oren, R., 2005. Forest response to elevated CO_2 is conserved across a broad range of productivity. *Proceedings of the National Academy of Science* 102:18052–18056.

Peltzer, D.A.., Allen, R.B., Lovett, G.M., Whitehead, D., Wardle, D.A. (2010). Effects of biological invasions on forest carbon sequestration. *Global Change Biology* 16:732–746.

Pregitzer, K.S. and Euskirchen, E.S. (2004). Carbon cycling and storage in world forests: Biome patterns related to forest age. *Global Change Biology* 10:2052–2077.

Ryan, M.G. (1991). Effects of climate change on plant respiration. *Ecological Application* 1:157–167.

Ryan, M.G., Harmon, M.E., Birdsey, R.A., Giardina, C.P., Heath, L.S., Houghton, R.A., Jackson, R.B., McKinley, D.C., Morrison, J.F., Murray, B.C., Pataki, D.E., Skog, K.E. (2010). A synthesis of the science on forests and carbon for U.S. Forests. *Issues in Ecology* (13):1–16.

Schnitzer, M., H. Kodama, J.A. Ripmeester (1991). Determination of the aromaticity of humic substances by x-ray diffraction analysis. *Soil Science Society of American Journal* 55(3):745–750.

Taylor, G., Tallis, M.J., Giardina, C.P., Percy, K.E., Miglietta, F., Gupta, P.S. Gioli, B., Calfapietra, C., Gielen, B., Kubiske, M.E., Scarascia-Mugnozza, G.E., Kets, K., Long, S.P., Kamosky, D.F. (2008). Future atmospheric CO_2 leads to delayed autumnal senescence. *Global Change Biology* 14:264–275.

Trahan, N.A., Dynes, E.L., Pugh, E. et al. (2015). Changes in soil biogeochemistry following disturbance by girdling and mountain pine beetles in subalpine forests. *Oecologia* 177:981–995.

U.S. Climate Change Science Program (US. CCSP) (2007). The First State of the Carbon Cycle Report (SOCCR): The North American Carbon Budget and Implications for the Global Carbon Cycle. In: King, A.W., Dilling, L., Zimmerman, G., Fairman, D.M., Houghton, R.A., Marland, G., Rose, A.Z., Wilbanks, T.J. (eds). A Report by the U.S. Climate Change Science Program and the Subcommittee on Global Change Research. National Oceanic and Atmospheric Administration, National Climatic Data Center, Asheville N.C., 242 p.

Vitousek, P.M. (1990). Biological invasion and ecosystem processes: Towards an integration of population biology and ecosystem studies. *Oikos* 57:7–13.

Warnock, D.D., Litvak, M.E., Morillas, L. Sinsabaugh, R.L. (2016). Drought-induced pinion mortality alters the seasonal dynamic of microbial activity in pinion-juniper woodland. *Soil Biology* and *Biochemistry* 92:91–101

Zhang, Y., Wolfe, S.A., Morse, P.D. et al. (2015a). Spatiotemporal impacts of wildfire and climate warming on permafrost across a subarctic region, Canada. *Journal of Geophysical Research Earth* 120(11):2338–2356.

Zhang, B., Zhou, X., Zhou, L., Ju, R. (2015b). A global synthesis of belowground carbon responses to biotic disturbance: A meta-analysis. *Global Ecology and Biogeography* 24(2):126–138.

20 Forest Disturbance
Invasive Species

INTRODUCTION

U.S. Presidential Executive Order 1311—Invasive Species[1]

On February 3, 1999, Executive Order 13112 was signed by President Clinton establishing the National Invasive Species Council to oversee the prevention of the introduction of invasive species and provide for their control and to minimize the economic, ecological, and human health impacts that invasive species cause. The Executive Order's key definitions are defined as follows:

- Alien species—means, with respect to a particular ecosystem, any species, including its seeds , eggs, spores, or other biological material of propagating that species, which is not native to that ecosystem.
- Control—means, as appropriate, eradicating, suppressing, reducing, or managing invasive species is populations, preventing spread of invasive species from areas where they are present, and taking steps such as restoration of native species and habitats to reduce the effects of invasive species and to prevent further invasions.
- Ecosystem—means the complex of a community of organisms and its environment.
- Federal agency—means an executive department or agency but does not include independent establishments as defined by 5 U.S.C. 104.
- Introduction—means the intentional or unintentional escape, release, dissemination, or placement of a species into an ecosystem as a result of human activity.
- Invasive species—means an alien species woe introduction does or is likely to cause economic or environmental harm or harm to human health.
- Native species—means, with respect to a particular ecosystem, a species that, other than as a result of an introduction, historically occurred or currently occurs in that ecosystem.
- Species—means group or organisms all of which have a high degree of physical and genetic similarity, generally interbreed only among themselves, and show persistent differences in members of allied groups of organisms.
- Stakeholders—means, but is not limited to, state, tribal, and local government agencies, academic institutions, the scientific community, non-governmental

[1] Executive Order 13112—Invasive Species (1999). United States Department of Agriculture (USDA) Accessed 6/12/23 @ https://ww.usda.gov.

DOI: 10.1201/9781003432838-23

entities including environmental, agricultural, and conservation organizations, trade groups' commercial interests, and private landowners.

- United States—means the 50 states, the District of Columbia, Puerto Rico, Guam, and all possessions, territories, and territorial seal of the United States.

The main thing to remember regarding invasive species is the USDA/USFS (2014) definition derived from Executive Order 13112 issued 1999.

A species is considered invasive if it meets these two criteria:

1. It is non-native to the ecosystem under consideration, and
2. Its introduction causes or is likely to cause economic or environmental harm or harm to human health.

DID YOU KNOW?

In the first five years after a wildfire, our rehabilitation program works to prevent these problems and jump-start the landscape recovery process by:

- Spreading native plant seed or planting native seedlings.
- Applying herbicides to kill invasive plants, removing them by hand, or introducing bacteria to control them.
- Using heavy equipment to disrupt the growth of targeted plant species or contour landscapes to control runoff (USDA, 2023. *Burned Area Rehabilitation.* Washington, DC: United States Department of Agriculture).

INVASIVE SPECIES: THE PROBLEM

In forestlands, invasive species can alter nutrient and C cycling, as well as soil physical properties, all of which can affect SOC stacks and soil properties, all of which can affect SOC stocks. It's all about Q and Q—Quantity and Quality—because exotic invasions impact factors important for OM Q and Q and microbial activity, such as nutrient mineralization, N-fixation by soil bacteria, mycorrhizal inoculation, decomposition and aeration of soils by earthworms (suborder Lubricina), and aggregation of soils by fungi (Wolfe and Klironomos, 2005).

Regular disturbance in many ecosystems is on-going and to a degree not only is this disturbance expected, but it is also needed to foster plant renewal and regeneration. The problem is that disturbance can also be detrimental by promoting invasion of non-native and weedy plants (Hobbs and Huenneke, 1992). Chronic disturbance causes initial invasiveness which disrupts the native nutrient and OM cycling that increases plant nutrient availability (Norton et al., 2008).

In efforts to explain why exotic plants have been so successful in disturbed ecosystems several different views have been put forward: (1) inherent properties of the invading plant—that is, earlier colonization than native vegetation; (2) vegetation factors—that is, species composition, richness, and heterogeneity; (3) soil microbial dynamics; and (4) climate factors such as precipitation amount and timing, aridity, and humidity (Blank and Sforza, 2007). When there is no disturbance, there are other factors such as plant-fungal interactions that may alter soil nutrient dynamics (Brundrett, 2009) and understory plant invasion can occur (Jo et al., 2018).

Plant-Fungal Symbiosis

Plant-fungal symbiosis is common to flowering plants and plays an important role in plant nutrition (Brundrett, 2009). This mutualistic relationship is characterized by host plants receiving mineral nutrients via their root system associated with fungal hyphae, and in return fungi gain a substantial amount of energy (carbon[C]) assimilated from photosynthesis (Smith and Reed, 2008; van der Heijden et al., 2015).

Arbuscular mycorrhizal (AM) tree dominant forests, which are characterized by thin litter layers and a low soil C/N ratio relative to ectomycorrhizal-dominant forests, are invaded by exotic plants to a greater extent (Jo et al., 2018). Note that Nuzzo et al. (2009) point out that other factors that may influence the invasion of exotic plants may be more indirect, for example, external factors such as deer (Odocoileus spp.) browsing or earthworm invasion. Plant-fungal interactions, which differ depending on the dominant mycorrhizal types, can have significant effects on soil nutrient dynamics and community structure in forest ecosystems (Phillips et al., 2013; Bennett et al., 2017; Wurzburger et al., 2017).

THE BOTTOM LINE

As exotic plant species invade ecosystems, ecologists have been attempting to assess the effects of these invasions on native communities and to determine the factors influencing invasion processes (Wolfe and Klironomos, 2005). Plant invasions lead to a shift in plant species composition which can influence N accumulation and cycling, SOC storage, water availability and runoff, and disturbance regime (Berryman et al., 2020).

REFERENCES

Bennett, J.A., Maherali, H., Reinhart, K.O., Lekberg, Y., Hart, M.M., Khronomos, J. (2017). Plant-soil feedback and mycorrhizal type influence temperate forest population dynamics. *Science* 335:181–184.

Berryman, E. et al. (2020). Soil Carbon. Accessed 5/8/23 @ https://www.fs.usda.gov/rm/pubs-journals/2020/rmrs-2020-berrymn-eco/pdf.

Blank, R.R. and Sforza, R. (2007). Plant-soil relationships of the invasive annual grass *Taeniatherum caput-medusae*: A reciprocal transplant experiment. *Plant Soil* 298(1–2):7–19.

Brundrett, M.C. (2009). Mycorrhizal associations and other means of nutrition of vascular plants: Understanding the global diversity of host plants by resolving conflicting information and developing reliable means of diagnosis. *Plant Soil* 320:37–77.

Hobbs, R.J. and Huenneke, L.F. (1992). Disturbance, diversity, and invasion: Implications for conservation. *Conservation Biology* 6(3):324–337.

van der Heijden, M.G.A., Martin, F.M., Selosse, M.A., Sanders, J.R. (2015). Mycorrhizal ecology and evolution: The past, the present, and the future. *New Phytologist* 205:1406–1423.

Jo, I., Domke, G.M., Fei, S. (2018). Dominant forest tree mycorrhizal type mediates understory plant invasions. *Ecology Letters* 21:217–224.

Norton, S.A., Coolidge, K., Amirbahman, A., Bouchard, R., Kopachek, J., Reinhardt, R. (2008). Fractions of Al, Fe, and P in recent sediment form three lakes in Main, USA. *Science of the Total Environment* 404:276–283.

Nuzzo, V.A., Maerz, J.C., Blossey, B. (2009). Earthworm invasion as the driving force behind plant invasion. *Conservation Biology* 23(4):966–974.

Phillips, R.P., Brzotek, E., Midgley, M.G. (2013). The mycorrhizal-associated nutrient economy: A new framework for predicting carbon-nutrient coupling in temperate forests. *New Phytologist* 199:41–51.

Smith, S.E. and Read, D. (2008). *Mycorrhizal Symbiosis*. Cambridge, MA: Academic Press.

USDA/ USFS (2014). *National Strategy and Implementation for Invasive Species Management*. Washington, DC: United States Department of Agriculture and United States Forest Service.

Wolfe, B.E. and Klironomos, J.N. (2005). Breaking new ground: Soil communities and exotic plant invasion. *Bioscience* 44(6):477–487.

Wurzburger, N., Brookshire, E.N.J., McCormick, M.L., Lankau, R.A. (2017). Mycorrhizal fungi as drivers and modulators of terrestrial ecosystems processes. *New Phytologist* 213:996–999.

Index

Note: **Bold** page numbers refer to tables and *italic* page numbers refer to figures.

abiotic processes 214
ablation till 154
absorber 10
absorption 10, 72, 154
  units 11
  vs. adsorption 73
AC (alternating current) 11
acid + bases and salts 196
  aliphatic hydrocarbons 199
  hydrocarbons 199
  organic chemistry 197–198
  organic compounds 198–199
  pH scale 196–197, **197**
acid deposition 11
acid precipitation 11
acid rain 11, 154
acid soil 154
actinomycetes 154
acyclic-chain compounds 44
adaptation 11
adaptive management 11
adhesion 154
adsorption 11, 72
  site density 72
  vs. absorption 73
aeration, soil 154
aerobic 154
agglomerating character 11
aggregates (soil) 155
agriculture 12
agroforestry 249–250
  fixing nitrogen 250–252
agronomy 155
air capacity 155
air pollution abatement equipment 12
air porosity 155
air temperature adjustment 12
albedo 12, 67, **68**
aliphatic hydrocarbons 199
alkali 155
alloys 39–41
alluvium 155
alternating current (AC) 12
alternative fuel 12
ambient 12
amendment, soil 155
ammonification 155
ampere (Amp) 13
anaerobic 155
anaerobic decomposition 13
anion 13
annual removals 13
anode 13
anthropogenic 13
API gravity 13
appropriate use 14
aquatic food chain 90
arbuscular mycorrhizal (AM) tree dominant forests 267
aspect (of slopes) 155
atmosphere
  carbon dioxide in 50
  reservoir 50–51
atmospheric chemicals and productivity 260
atmospheric global greenhouse gas abundances 210
atom structure 189
atterburg limits 155
autotrophic respiration 202
autotrophs 155
available water 155

bare rock succession 14
bedrock 155
BEF *see* biomass expansion factors (BEF)
below-ground net primary production (BNPP) 245
benzene 14
biobased product 14
biochar production 62
biochemical conversion 14
biodiesel 14
bioenergy 14
biofuels 14
biogas 15
biogenic 15
biogenic emissions 15
biogeochemical cycles 41
biological diversity/biodiversity 15
biological function 155
biological integrity 15
biological pump 58, 59
biologic carbon sequestration 78
biology of carbon 55–56

biomass 15, 155, 202
  equations 97–98
  estimation, stand level 96–97
  terrestrial 89
  waste 15–16
biomass expansion factors (BEF) 96
biomass gas (biogas) 15
biomass weight considerations, forestry volume unit to 95–96
biomaterials 16
biome 155
biopower 16
biorefinery 16
biota 16
blow-out 155
blue carbon 68, 70
blue carbon sink 68–69
BNPP *see* below-ground net primary production (BNPP)
bonds 39–41
breccia 155
British thermal unit (Btu) 16
Btu conversion factor 16
Btu per cubic foot 16–17
bulk density 17
bunker fuels 17
burning of fossil fuels 46, *47*
burnup 17
butane (C_4H_{10}) 17
butylene (C_4H_8) 17

calcareous soil 155
calcium sulfate 17
caliche 156
calorie 17
Calvin-Benson cycle 258
CAM *see* Crassulacean acid metabolism (CAM)
Can Carbon Dioxide Influence Climate? (Callendar) 37
Canopy photosynthesis declines 259
capillarity 184
capillary water 156
carbon
  equivalent 202
  intensity 16
  "King of the Elements" 38
  output rate 19
  rainbow 66–67
  sources 55
  stock change 202–203
  stored in harvested wood products 62
  symbol for the element 38, *38*
carbon accounting 18
carbonate pump 59
carbon budget 18
carbon capture and sequestration (CCS) 77–82
carbon compounds 39–41
  inorganic 40
carbon-containing rock 42
carbon cycle 5, *5*, 18, 39. *40*, 41–42, 57, 202
  fast 43–49
  slow 42–43
carbon dioxide (CO_2) 5, 18
  in atmosphere 50
  emissions of 48, *48*
  fertilization 203
  flux/flow of 224–225
  at Mauna Loa 8–10, *9*
  physical properties **6**
  solid 6
  uses 6–10
carbon dioxide equivalent (CDE) 18, 203
carbon fractions (CFs) 257–258
carbon-hydrogen bond (C–H bond) 38
carbon monoxide (CO) 19
carbon sequestration 19, 78, 202, 223
  biologic 78
  geologic 78, 82–83
  terrestrial 78–79
  urban forests and 79–82
carbon sink 19, 78, 103
  blue 68–69
  Oceans 58–63
carbon status 223
  flux/flow of carbon dioxide 224–225
carbon storage 89–90, 223
  in forestland 260
  terrestrial 79
Carnot cycle 19
catena 156
cathode 19
cation 19, 156
Cave of the Dogs 3–7
CCS *see* carbon capture and sequestration (CCS)
CDE *see* carbon dioxide equivalent (CDE)
cellulose 19
cellulosic ethanol 19
cement production 49, *49*
CFC *see* chlorofluorocarbon (CFC)
CFs *see* carbon fractions (CFs)
CH_3-CH_2OH *see* ethanol
chelate 156
chemical bonding 190–191
chlorofluorocarbon (CFC) 20
chronic N addition experiments 253
class, soil 156
clay 156
  soil 169
clear soil horizons 171
climate 20
  effects 20
  forcing 210
climate change 20
  and forest SOC 226–227

specifies shift and 216–217
vs. white knight 70–71
cloud condensation nuclei 20
CNG *see* compressed natural gas (CNG)
coarse-grained soils 180
coarse materials 20
coccolithophores 59
co-firing 20
cogeneration 20, 21
cohesion 156
colloidal 156
comparative yield monitoring method 146
compressed natural gas (CNG) 21
concentrating solar power 21
concentrator 21
conservation 21
program 21
status 21
convection 156
conventional oil production 21
cover board monitoring method 144–145
conversion factor, Btu 16
conversion system, solar energy 21
Crassulacean acid metabolism (CAM) 258
criteria pollutant 22
critical habitat 22
crust 22
cubic foot, Btu per 16–17
cyrospheric melting 67

Daubenmire monitoring method 141–142
DC (direct current) 22
dead trees 256
decomposers 90
decomposition, anaerobic 13
deforestation 22
degasification system 22
degrees of freedom 111
demand indicator 22
denitrification 156
density 183
monitoring method 144–145
dependable capacity 22
depleted resources 23
depletion factor 23
desulfurization 23
detritus 156
DIC *see* dissolved inorganic carbon (DIC)
diesel fuel 23
diffusion 156
dissolved inorganic carbon (DIC) 58
dissolved organic C (DOC) 262
disturbance *vs.* vulnerability 215–216
diversity exchange 23
double sampling 131–133
double-weight sampling 145
drainage 156
dry weight basis calculations 94–95
dry-weight rank method 141
duff 156

E-10 23
E-85 23
E-95 23
ecological system 23
ecosystem 23
respiration 203
terrestrial 79
efficiency 24
energy 24–25
electric energy 24
electric utility 24
elements, periodic classification of 189–190
emissions 24
of carbon dioxide 48, *48*
factor 24
greenhouse gas (GHG) 77
human-caused CO_2 78
emissions coefficient 24
endothermic 24
energy 24–25
efficiency 24–25
loss 25
radiant 35
renewable 36
source 25
enhanced oil recovery (EOR) 78
enthalpy 25
environment 25
environmental chemistry 200
environmental health (abiotic aspects) 25
environmental impact statement 25
environmental restoration 26
environmental restrictions 26
EOR *see* enhanced oil recovery (EOR)
erosion 156
ET *see* evapotranspiration (ET)
ethane (C_2H_6) 26
ethanol 26
ether 26
ethyl alcohol *see* ethanol
ethylene (C_2H_2) 26
eutrophication 156
evaporation of water 185
evapotranspiration (ET) 27, 156
exfoliation 156
exothermic 27

Fahrenheit 27
Farm Security and Rural Investment Act (FSRIA) 14
fast carbon cycle 43–49
Federal Conservation Reserve Program (CRP) 31
feedstocks, petrochemical 35

fens 62–63
fermentation 27
fertility, soil 156
fertilization, forest 249
fire effects on SOC 231–232
Fischer-Tropsch fuels 27
fixation 156
fluvial 156
flux 27, 203
flyway 27
food chain, aquatic 90
food web 90
Forest and Rangeland Renewable Resources Planning Act of 1974 206
forest biomass
  fuels, heat content ranges **94**
  inventory 134
  sampling 104–105
forest carbon dynamics 230
forest disturbance
  invasive species 267–266
  tree mortality 255–262
forest ecosystems, long-term capacity of 257
forest fertilization 249
forest harvesting 238–239
forestland 27
  carbon storage in 260
  disturbance 260
  nitrogen-fixing herbaceous plants in 252
Forest Paradox 240
forestry volume data, estimation of biomass weights from 95–96
forestry volume unit to biomass weight considerations 95–96
forests
  fuel/heat value of 92–95
  mangrove 69
Forest Service Forest Inventory and Analysis (FIA) program 89
forest soil organic carbon, climate change and 226–227
forest stands, thinning 239–240
fossil fuels 28
  burning of 46, *47*
fractionation 28
fraible 156
frequency monitoring methods 140–141
FSRIA *see* Farm Security and Rural Investment Act (FSRIA)
fuel
  cell 28
  cycle 28
  heat value of forests 92–95
  moisture content 93
  moisture on wood heat content 93–95, **94**
  ratio 28
fumarole 28

gallon 28
gas 28
  laws 193–194
  natural 34
  stoves, turn of 47–48
gaseous cycles 41
gasification 29
gasohol 29
generation (electricity) 29
geographic information systems (GIS) 29
geologic carbon sequestration 78, 82–83
geologic hazards 29
geologic sequestration 29
  potential impact of 84
GHG *see* greenhouse gas (GHG)
gigawatt (GW) 30
gigawatt-electric (GWe) 30
gigawatt-hour (GWh) 30
global positioning system (GPS) 30
global warming 30
GPP *see* gross primary production (GPP)
grain alcohol *see* ethanol
graphite forms 38, *39*
gravity, API 13
greenhouse effect 30
greenhouse gas (GHG) 30, 49, 79
  emissions 77
green plants 59
gross calorific value *see* higher heating value (HHV)
gross energy *see* higher heating value (HHV)
gross primary production (GPP) 203, 245, 258–261
ground slope 162
groundwater 163
growing stock 30

habitat 30
  conservation 31
  fragmentation 31
HABs *see* harmful algae blooms (HABs)
half-life of carbon 60
hardwoods 31
harmful algae blooms (HABs) 46
harvested wood products (HWPs) 203, 225
  carbon stored in 62
harvest monitoring method 145–146
heat content ranges, for forest biomass fuels **94**
heating effect 210
heating value 31
heat rate 31
heaving 157
herbaceous plants, nitrogen-fixing 252
herbivores 246
heterotrophic respiration 203
heterotrophs 56, 157
higher heating value (HHV) 92, **93**

higher heating value as-fired (HHV-AF) of a wood sample 93
horizon, soil 157
human-caused CO_2 emissions 78
human influences 162–163
humus 157
HWPs *see* harvested wood products (HWPs)
hydration 157
hydraulic conductivity 157
hydraulic fracturing 31
hydrocarbon 31, 199
  aliphatic 199
  fuels, liquid 27
hydrologic cycle 184–185
hydrolysis 157
hygroscopic coefficient 157

Industrial Revolution 43, 49
industrial wood 32
infiltration 157
inorganic carbon compounds 40
invasive species
  plant-fungal symbiosis 267
  problem 266–267
  U.S. Presidential Executive Order 1311 265–266
inventory 105
ion exchange 32, 72
ions 157
isobutane (C_4H_{10}) 32
isobutylene (C_4H_8) 32
isohexane (C_6H_{14}) 32
isomerization 32

Joule 32

kBtu 32
Keeling Curve 9, *9*
kilowatt (kW) 33
kilowatt-hour (kWh) 33
kinetic energy 33
"King of the Elements," carbon 38

land
  classification 157
  reservoir 51–52
leaching 157
LHV *see* lower heating value (LHV)
Liebig's law 157
line intercept monitoring method 142
liquid hydrocarbon fuels 27
liquid limit (LL) 180
livestock grazing 246–247
loam 157
loess 157
The Lord of the Rings (Tolkien) 224
lower heating value (LHV) 92, **93**
low-speed shaft 33

mangrove forests 69
mantle 33
MAP *see* mean annual precipitation (MAP)
marine reservoir 51
marl 157
MAT *see* mean annual temperature (MAT)
matrix of monitoring techniques 139
matter, physical and chemical properties of 192–193
mature forests 103
MBC *see* microbial biomass C (MBC)
MC *see* moisture content (MC)
mean annual precipitation (MAP) 245
mean annual temperature (MAT) 245
megawatt (WM) 33
melting, cyrospheric 67
mercuric oxide, heating 188
meta-analysis 233–234
methane 33
methanogens 33
methanol 33
methyl alcohol *see* methanol
microbial biomass C (MBC) 262
Milankovitch hypothesis 45
mineral 33–34
mineralization 157
mineral matter 167
mineral soil
  organic carbon 207–209
  soil organic carbon (SOC) in 204
moisture content (MC) 93, 94–95
moisture regimes, diagnostic horizons and temperature and 173–174
mole 34
monitoring methods
  comparative yield method 146
  cover board method 144–145
  Daubenmire method 141–142
  density method 144–145
  double-weight sampling 145
  dry-weight rank method 141
  frequency methods 140–141
  harvest method 145–146
  line intercept method 142
  matrix of 139
  pin frames 143
  point frames 143
  point-intercept method-sighting devices 143
  step-point method 142–143
  visual obstruction method-Robel Pole 146–147
mutualistic relationship 267

naphtha 34
National Forest Management Act 206
nation's soil organic carbon stores 205
natural gas 34
  production 21

natural processes 34
natural sinks 34
net calorific value *see* lower heating value (LHV)
net primary production (NPP) 258–261
niche 34
nitrogen
  cycle 250
  fixation 157, 250
  fixing 250–252
nitrogen-fixing
  herbaceous plants in forest lands 252
  plants 251
  shrubs 251
NOAA
  Global Greenhouse Gas Reference network 9
  Global Monitoring Laboratory 8
nonmetals 187
NPP *see* net primary production (NPP)
number of sampling units
  estimation of 119–120
  optimum allocation 119
  proportional allocation 119

ocean pumps 58
  biological pump 58, 59
  carbonate pump 59
  solubility pump 58–59
open-chain compounds 44
optimum allocation with sampling costs 120–121
organic carbon
  degradable 22
  mineral soil 207–209
organic chemistry 197–198
organic compounds 40, 198–199, *199*
organic matter (OM) quality 207
organic non-fossil material 15
organometallic compounds 40
osmosis 157
oxidation 157

paradigm shift 257
paradox: dead/alive 256
parent material 157
peat 34
peat bogs 62
ped 158
pedogenic/pedagogical process 158
periodic classification of elements 189–190
periodic table *54*, 54–55
permeability 34
petrochemical feedstocks 35
pH 158
  scale 196–197, **197**
photosynthesis 35, 60, 103, 158
physical weathering 171
phytoplankton 58
pin frames 143
planetary albedo 35
plant-fungal symbiosis 267
plant growth medium, soil 160
plants 59–61
plasticity index (PI) 180
plastic limit (PL) 180
point frames 143
point-intercept method-sighting devices 143
pool/reservoir 203
population 105
porosity, soil 158
power system, solar thermal 21
precipitation 185
prescribed fire 35
primary producers 56
profile, soil 158
propane (C_3H_8) 35
propylene (C_3H_6) 35
PyC *see* pyrogenic carbon (PyC)
pyrogenic carbon (PyC) 61, 62, 231–232
pyrolysis 35

quadrillion Btu (Quad) 35

radiant energy 35
radiative forcing 210
rainfall duration 162
rainfall intensity 162
rainwater 183
random sampling, sample size in stratified 121–122
raw materials, recyclers of 163–164
reclamation 36
recycling 36
red carbon 67
reduction 158
regolith 158
regression
  coefficient 124
  equations 97
  estimation 124–128
renewable energy 36
  advantages 77
  resources 36
reservoir
  atmosphere 50–51
  land 51–52
  marine 51
riparian 36
rock 158
  cycle 158
runoff 158

salinization 158
salt marshes 68
sample 105
sampling 105
  protocols and vegetation attributes 134
  unit 105

sampling unit, estimation of number of 119–120
sand 158
 forms 169
Science of Environmental Pollution 46
seagrass 70
 meadows 69
seaweed 70
sequestration 36, 203
 terrestrial 36
SG *see* specific gravity (SG)
shrub characterization 106
silt 158
silvopasture 244, 252
simple random sampling 106–107
simple random sampling methods 107–115
 confidence limits for small samples 110
 effect of plot size 113–114
 sample selection 107–108
 sampling with replacement 109–110
 size of sample 111–113
 standard errors 108–109
sinks 203
site disturbance 260–261
slope 158
slow carbon cycle 42–43
SOC *see* soil organic carbon (SOC)
soil 61–62, 152–154, 158
 air 158
 basics 151–154, 165–167
 C accounting 206
 calcareous 155
 carbon vulnerability, assessing 209–210
 characterization 172–173
 coarse-grained 180
 compaction 181
 composition 162
 compressibility 181
 failure 181
 families and series 176–177
 formation *170,* 170–171
 functions of 159–160
 great groups and subgroups 176
 health 203–204
 horizon 158, 171
 index property of **179**
 mechanics 177
 moisture 162
 moisture regimes **174**
 orders 175, **176**
 particle characteristics **179,** 179–180
 physical properties 167–168
 plant growth medium 160
 productivity, vegetation species composition on 260
 profile 158
 recyclers of raw materials 163–164
 regulator of water supplies 161–163
 resources 83
 separates 168–169, **169**
 stability 247
 stress 180–181
 structure 158–159
 suborders 176
 taxonomy 173, 174–175
 taxonomy classification system **175**
 temperature regimes **175**
 texture 158–159
 water 183, 185–186
soil chemistry 186–195
 chemical bonding 190–191
 chemical formulas and equations 191
 classification of elements 187–188
 elements and compounds 187
 gas laws 193–194
 liquids and solutions 194
 molecular weights, formulas, and the mole 192
 molecules and ions 190
 periodic classification of elements 189–190
 physical and chemical changes 188–189
 physical and chemical properties of matter 192–193
 specific heat 195
 states of matter 193
 structure of the atom 189
 thermal properties 195
soil organic carbon (SOC) 203–205, 209, 247
 fire effects on 231–232
 management and 234–235
 mineralization rates 261
 in mineral soils 204
 stock 247, 249
soil organic content (SOC) pools 214
soil organic matter (SOM) 61, 151, 167, 208
soil organisms **165**
 habitat for 164
soil physics 182
 properties 183–184
 water and soil 183
soil types
 stable rock 181
 Type A soil 181
 Type B soil 181
 Type C soil 182
solar energy conversion system 21
solar power, concentrating 21
solar thermal power system 21
solid carbon dioxide 6
solid fuels, lower and higher heating values of 93
solubility pump 58–59
soluble 158–159
solum 159
SOM *see* soil organic matter (SOM)
specific gravity (SG) 95
stand level biomass estimation 96–97
step-point monitoring method 142–143

straight-chain compounds 44
stratified random sampling 115–116
sustainable carbon fraction management 258
symbiosis, plant-fungal 267
symbiotic relationship 250–251

TBCF *see* total below-ground carbon flux (TBCF)
terrestrial biomass 89
terrestrial carbon
  sequestration 78–79
  storage 79
terrestrial ecosystems 79
terrestrial sequestration 36
  potential impact of 83
tertiary recovery 78
till 159
tilth 159
topsoil 159
total below-ground carbon flux (TBCF) 245–246
tree mortality, forest disturbance 255–262
trend 106
tropes 56

urban forest
  and carbon sequestration 79–82
  defined 80
urban wood 80
USDA Forest Service's Forest Inventory and Analysis (FIA) program 205
U.S. Geological Survey (USGS) 78

vegetation attributes 134
  biomass (production) 138
  cover 136–137
  density 137
  frequency 135–136
  sampling protocols and 134
  structure 138–139
vegetation cover 162
vegetation species composition (genotype effects), on soil productivity 260
viscosity 184
visual obstruction method-Robel Pole 146–147
volatile organic compounds (VOCs) 245
vulnerability *vs.* disturbance 215–216

waste, biomass 15–16
water
  evaporation of 185
  and soil 183
water cycle (hydrologic cycle) 184–185
watt (electric) 36
watt (thermal) 37
weathering 159
weight basis calculations, wet and dry 94–95
weight-volume or space and volume relationships 177–179
wet weight basis calculations 94–95
white knight *vs.* climate change 70–71
wind speed, shading and reduction of 80
wood alcohol *see* methanol
wood heat content
  effect of fuel moisture on 93–95, **94**
  fuel moisture on 93–95
woods (aka forests) 89
woods ecosystems in United States 89
wood soils 90
woody biomass 203

Printed in the United States
by Baker & Taylor Publisher Services